E&M   Endocrinology and Metabolism

Progress in Research and Clinical Practice

Margo Panush Cohen   Piero P. Foà
Series Editors

Endocrinology and Metabolism
Progress in Research and Clinical Practice

Margo Panush Cohen Piero P. Foà
  Series Editors

Margo Panush Cohen   Piero P. Foà
Editors

The Brain as an Endocrine Organ

With 60 Figures

Springer-Verlag
New York Berlin Heidelberg
London Paris Tokyo

Margo Panush Cohen, M.D., Ph.D.
Professor of Medicine
University of Medicine and
  Dentistry of New Jersey
Newark, New Jersey 07103
Director, Institute for
  Metabolic Research
University City Science
  Center
Philadelphia, Pennsylvania
  19104
USA

Piero P. Foà, M.D., Sc.D.
Professor Emeritus of Physiology
Wayne State University
Chairman Emeritus
Department of Research
Sinai Hospital
Detroit, Michigan 48202
Mailing address:
  2104 Rhine Road
  West Bloomfield, Michigan 48033
USA

LIBRARY OF CONGRESS
Library of Congress Cataloging-in-Publication Data
The Brain as an endocrine organ / Margo Panush Cohen, Piero P. Foà,
  editors.
    p. cm. –(Endocrinology and metabolism ; 3)
  Includes bibliographies and index.
  ISBN 0-387-96644-7
    1. Neuroendocrinology. 2. Brain–Physiology. I. Cohen, Margo P.
II. Foà, Piero P. (Piero Pio), 1911- . III. Series.
    [DNLM: 1. Brain–physiology. 2. Neurosecretion. 3. Pituitary
Hormone Releasing Hormones–secretion.   WL 300 B81233]
QP356.4.B68   1988
612'.82—dc19
DNLM/DLC   88-16091
for Library of Congress   CIP

Printed on acid-free paper

© 1989 by Springer-Verlag New York Inc.
Softcover reprint of the hardcover 1st edition 1989

Typeset by TC Systems, Inc., Shippensburg, Pennsylvania

9 8 7 6 5 4 3 2 1

ISBN-13: 978-1-4612-8118-4   e-ISBN-13: 978-1-4612-3480-7
DOI: 10.1007/978-1-4612-3480-7

Preface

In the middle of the 17th century, the great French philosopher Rene Descartes wrote (*L'Homme,* J. Le Gras, Paris, 1669) that a suitable stimulation of the brain results in two types of "movements": exterior movements, designed to seek desirable ends and to avoid undesirable or harmful ones and interior movements or "passions" which through the release of "animal spirits" regulate the heart, the liver, and other organs. When it appears appropriate to meet a threat with force, the passion of rage causes the release of strong spirits, whereas when avoidance appears to be the better choice, the passion of fear causes the brain to release weak spirits. We do not know what influence, if any, Descartes had on the thinking of Walter B. Cannon (*Bodily Changes in Pain, Hunger, Fear and Rage,* Appleton and Co., New York, 1920), of Hans Selye (*The Story of the Adaptation Syndrome,* Acta, Inc., Montreal, 1952), of G.W. Harris or of R. Guillemin (*Hypothalamic-Hypophysial Interrelationships. A Symposium.* C.C. Thomas, Springfield, 1956), but it is interesting to reflect upon the durable value of great ideas which constantly resurface even if modified by other ideas and by new techniques, as if propelled by a preordained intellectual imperative.

This volume was not motivated by such historical musing, but rather by the belief that "The Brain as an Endocrine Organ" would be an interesting and timely topic especially if presented in a manner apt to fulfill the stated goals of this series of monographs: to present new information and new perspectives with a view to their actual or possible clinical applications. Thus, this volume contains two chapters about "typical" hypothalamic hypophysiotropic hormones: one of them discusses the physiopathology of gonadotropin releasing hormone and its numerous analogs, their action as stimulators or suppressors of gonadal function, and their usefulness in the management of such diverse conditions as infertility and hirsutism; the other guides the reader through the intricacies of ACTH secretion and its regulation and discusses the usefulness of corticotropin releasing hormone measurements in the differential diagnosis of the various forms of Cushing's syndrome and hypercortisolemic non-Cushingoid states. A third chapter discusses growth hormone and presents evidence that insulin-like growth factor 1 (IGF-1) participates in the regulation of its secretion and discusses the significance of these factors in starvation, malnutrition, cachexia, and Laron dwarfism. A chapter on the pineal gland, a well-known transducer of optic stimuli into endocrine events, discusses the influence of the light and dark cycle and of other environmental variables (such as low-frequency electric fields) on the gland's secretory activity and the role of pineal malfunction in a number of clinical syndromes. A comprehensive review of hypothalamic and neurotransmitter dysfunction in the pathogenesis of eating disorders leads to the formulation of the challenging new hypothesis that anorexia nervosa is a form of auto-addiction to endogenous opioids, suggesting a new therapeutic approach to this baffling disease. The metabolic functions of neuropeptides synthesized in the brain itself (including the endogenous opioids) and, in particular, their newly discovered role in regulating glucose metabolism and its alterations in severe stress, head injury, brain tumors, or myocardial infarction are discussed in still another chapter. Finally, one chapter presents a comprehensive review of a relatively new topic: the ubiquitous role of adenosine as a central neurotransmitter or neuromodulator and its possible role in such diverse phenomena as locomotor activity, cardiovascular function and respiration, sleep, food intake, and the physiopathology of premenstrual tension and of cortisol-induced emotional disturbances.

New York, New York   Margo P. Cohen
Detroit, Michigan   Piero Foà

Contents

Contributors

GEORGE P. CHROUSOS, M.D.
Senior Investigator, Developmental Endocrinology Branch, National
Institute of Child Health and Human Development, National Institutes
of Health, Bethesda, Maryland, USA

TADASHI KATAGIRI, M.D.
The Third Department of Internal Medicine, Yamagata University
School of Medicine, Yamagata, Japan

D. LYNN LORIAUX, M.D., PH.D.
Clinical Director, National Institute of Child Health and Human
Development, National Institutes of Health, Bethesda,
Maryland, USA

ELLIOT D. LUBY, M.D.
Professor of Psychiatry and Law, School of Medicine and Law, Wayne
State University, Chief, Department of Psychiatry, Harper-Grace
Hospitals, Detroit, Michigan, USA

MARY ANN MARRAZZI, PH.D.
Associate Professor, Department of Pharmacology, Wayne State
University School of Medicine, Associate, Department of Psychiatry,
Harper-Grace Hospitals, Detroit, Michigan, USA

SEIJIRO MARUBASHI, M.D.
The Third Department of Internal Medicine, Yamagata University
School of Medicine, Yamagata, Japan

SHLOMO MELMED, M.D.
Associate Professor of Medicine, UCLA School of Medicine, Director,
Division of Endocrinology and Metabolism, Cedars-Sinai Medical
Center, Los Angeles, California, USA

KAMRAN S. MOGHISSI, M.D.
Professor and Associate Chairman, Department of Obstetrics and
Gynecology, Chief, Division of Reproductive Endocrinology and
Infertility, Wayne State University School of Medicine/Hutzel
Hospital, Detroit, Michigan, USA

LYNNETTE K. NIEMAN, M.D.
Clinical Investigator, Developmental Endocrinology Branch, National
Institute of Child Health and Human Development, National Institutes
of Health, Bethesda, Maryland, USA

JOHN W. PHILLIS, B.V.Sc., PH.D., D.Sc., D.V.Sc.
Professor and Chairman of Physiology, Wayne State University School
of Medicine, Detroit, Michigan, USA

RUSSEL J. REITER, PH.D.
Professor of Neuroendocrinology, Department of Cellular and
Structural Biology, University of Texas Health Science Center, San
Antonio, Texas, USA

HIDEO SASAKI, M.D.
Professor of Medicine and Chairman, Third Department of Internal
Medicine, Yamagata University School of Medicine, Yamagata, Japan

MAKOTO TOMINAGA, M.D.
Third Department of Internal Medicine, Yamagata University School of
Medicine, Yamagata, Japan

KEIICHI YAMATANI, M.D.
Third Department of Internal Medicine, Yamagata University School of
Medicine, Yamagata, Japan

YOSHIKAZU YAWATA, M.D.
Third Department of Internal Medicine, Yamagata University School of
Medicine, Yamagata, Japan

1

Gonadotropin Releasing Hormones: Physiopathology and Clinical Applications

KAMRAN S. MOGHISSI

In 1971, Schally and associates reported the isolation, identification, and synthesis of gonadotropin releasing hormone (GnRH), also called luteinizing hormone releasing hormone (LHRH) or luteinizing releasing factor (LRF). Since then, a large number of publications dealing with the physiology, pathology, and actual or potential clinical applications of GnRH and its analogs have appeared. During this period, over 1700 analogs of GnRH, both agonists and antagonists, have been synthesized, among them numerous highly potent substances which have been used to elucidate the pathophysiology of various endocrine disorders and for the management of several clinical entities. This chapter will review the physiology and neuroendocrine effects of GnRH and will discuss clinical applications of the native substance and its analogs.

Chemical and Biologic Properties of GnRH

GnRH is produced and released in pulsatile fashion from the arcuate nucleus and preoptic anterior hypothalamic area. It reaches the anterior pituitary through the portal system and is believed to bind to specific receptors in the anterior pituitary, where it stimulates the synthesis and secretion of luteinizing hormone (LH) and follicular stimulating hormone (FSH) in both male and female. FSH and LH in turn are essential for testicular and ovarian function (1). Gonadotropin releasing hormone is a decapeptide with an identical structure in all mammals, including humans. Like several other brain peptides, GnRH is synthesized as part of a much larger precursor peptide. When administered to women or men, GnRH stimulates a prompt and large release of LH and a smaller secretion of FSH.

GnRH is rapidly degraded by peptidase and cleared by glomerular filtration. Its half life in peripheral circulation is only 4 to 20 minutes (2). In order to increase the potency and duration of action of GnRH, analogs with agonistic or antagonistic properties have been synthesized. Substitution of an amino acid at the 6 or 10 position results in analogs with agonistic activity, whereas modifications at the 2 or 3 position result in analogs with antagonistic properties. Administration of GnRH agonists produces an initial stimulation of pituitary gonadotropes, resulting in secretion of FSH and LH and the expected gonadal response. However, continuous or repeated administration of an agonist in an inappropriate (nonpulsatile) fashion, or at nonphysiologic doses, ultimately produces an inhibition of the pituitary–gonadal axis. Functional changes resulting from this inhibition include pituitary GnRH receptor "down-regulation," gonadal gonadotropin receptor down-regulation, attenuated gonadotropin secretion, and decreased steroidogenesis and gametogenesis. The inhibitory effects of analogs are fully reversible. Agonists may also act directly on extrapituitary reproductive and nonreproductive target sites such as gonads, uterus, and prostate (3,4).

GnRH antagonists have a direct inhibitory effect on reproductive processess. They compete for and occupy pituitary GnRH receptors, thus blocking the access of endogenous GnRH or exogenously administered agonists to their required recognition site (3).

GnRH agonists are potent therapeutic agents with considerable advantages for clinical use. They cannot be administered orally but may be given parenterally, by nasal spray or in vaginal pessaries. Implants containing GnRH analogs and capable of slow drug release have been developed and are currently undergoing clinical trials. Biologic tolerance of agonists is excellent and their side effects are minimal. GnRH antagonists have been studied in animals and, recently, in humans but clinical indications for their use remain to be clarified.

Physiology of GnRH

Pituitary secretion of gonadotropin is controlled by GnRH. Early studies indicated that LH is secreted in a pulsatile fashion and that the pulses of LH represent episodic secretion of GnRH. Because GnRH is released directly into the pituitary portal system, the levels of GnRH in the peripheral circulation do not correspond to its release pattern. It is therefore assumed that LH pulses reflect pulses of GnRH. In a series of elegant experiments, Knobil and his associates demonstrated the physiologic importance of pulsatile GnRH stimulation of the pituitary for optimal gonadotropin secretion (1). In rhesus monkeys with hypothalamic lesions that destroyed the arcuate nucleus and abolished endogenous gonadotropin release by the pituitary gland, gonadotropin secretion could be restored by chronic intermittent injections, but not by continuous infusion, of GnRH. The initiation of continuous GnRH administration in lesioned animals, in which gonadotropin secretion had been reestablished by intermittent GnRH replacement, resulted in desensitization or down-regulation of pituitary gonadotropin. Resumption of pulsatile administration of GnRH was followed again by normal release of gonadotropin. In other experiments, Knobil and co-workers altered the frequency or dose of the GnRH pulses. and observed a variety of gonadotropin release patterns. Considered together, these experiments clearly established that the arcuate nucleus is the central component of the control system. Its basic unregulated operation consists of generating a signal approximately once per hour (in monkey), which eventuates in the release of a bolus of GnRH into the pituitary portal circulation. Unquestionably, in a physiologic context, the neural component of the control system is profoundly influenced by higher centers as well as by gonadal and other hormones. Thus, the arcuate nucleus may be viewed as a transducer of neuronal signals into endocrine signals, and appears to translate frequency, the language of the nervous system, into the language of the endocrine system, namely, a change in hormone levels in the circulation.

The pattern of gonadotropin release in females varies during different phases of the menstrual cycle (5). Typically, in the early follicular phase, gonadotropins show pulsations of slow frequency (approximately every 90 minutes) and of moderate amplitude. However, most women studied during this phase of the cycle exhibit a striking and near total suspension of pulsatile gonadotropin release during sleep. Even brief periods of awakening are associated with a reversal of this sleep-related absence of pulsatile gonadotropin release. By mid-follicular phase, the pattern of gonadotropin release changes markedly. The frequency of LH pulsation increases to approximately once every 60 minutes. However, the amplitude of each pulse decreases and the sleep-related suspension of pulsatility observed in the early follicular phase disappears. In the late follicular phase, the amplitude of LH pulses increases markedly. Following the formation of the corpus luteum and the secretion of progesterone, the pattern of gonadotropin secretion changes again. In the early luteal phase, there is considerable slowing of the LH pulse frequency which becomes evident as the progesterone level rises.

Another more subtle change in the early luteal phase is the appearance of a bimodal pattern of LH secretion with large, infrequent pulses of an amplitude >15 MIU/ml being interspersed with pulses of <5 MIU/ml amplitude; often these smaller LH pulses appear to closely follow the larger ones. In addition, during the early luteal phase, the secretion of progesterone is constant and at a low level, with no apparent relationship to the appearance of either the large or the small LH pulses. By the mid-luteal phase, all of the changes in the gonadotropin secretion pattern seen in the early luteal phase become exaggerated. The LH pulses slowly become more marked. The bimodal LH pulse pattern becomes more apparent, with the smaller LH pulses now comprising nearly 50% of all pulses. The pattern of progesterone secretion in the mid-luteal phase also changes significantly, with the appearance of pulses that correspond with the LH pulses. In fact, it appears that the majority of the smaller LH pulses occur at the time of the peak progesterone levels, which in turn follow the larger LH pulses (6). At the end of the luteal phase, the LH pulse frequency continues to slow until only one or two large LH pulses occur in a 24-hour period. However, the pulsatile secretion of progesterone persists into the late luteal phase, albeit at a lower level (6). Finally, during the transition from the late luteal to the early follicular phase, there is a change from the low-amplitude LH pulses to a more regular, higher amplitude pattern. In addition, in the early follicular phase a sleep-associated suspension of LH pulsatility reappears. It seems that, during the follicular phase, the rising levels of serum estrogens modulate the hypothalamic–pituitary axis and render the anterior pituitary more responsive to GnRH through negative feedback effect, followed by positive feedback. However, there also appears to be a secondary effect of estrogen that increases the frequency of GnRH-induced gonadotropin pulsations. It is not clear whether the latter effect is primarily due to an increase in GnRH pulse frequency occurring at the level of the hypothalamus or to a facilitated expression of GnRH pulses at the level of the anterior pituitary. Progesterone, on the other hand, appears to mediate its primary endocrine effects by slowing the frequency of GnRH discharge.

In one study of normal men (5), the range of GnRH pulse frequency was found to vary from 7 to 17 pulses per 24 hours and to be associated with a fall of plasma testosterone to levels as low as 91 ng/dl during the periods of lengthened LH interpulse intervals. In another study, an LH pulse frequency of 3 to 18 pulses per 24 hours was observed (7). On the average, normal men exhibit a pulse frequency of approximately 12 pulses per 24 hours with an amplitude of 9 to 11 MIU/ml. Some apparently normal men exhibit striking differences from the expected pattern of GnRH secretion with a predominance of LH pulses clustered during the sleeping hours.

Abnormal GnRH Secretion

Females

The female gonad poorly tolerates deviations from the normal progression of amplitude and frequency in GnRH-induced gonadotropin secretion which occurs during the menstrual cycle, and amenorrhea ensues when this pattern is disturbed (5). In a heterogeneous group of disorders termed hypothalamic amenorrhea (HA), often associated with disorders of nutrition, exercise, stress, and emotional problems, at least four patterns of gonadotropin secretion have been described (5). In group one, the patients exhibited a nonpulsatile pattern of GnRH-induced gonadotropin secretion. The second group showed an onset or augmentation of GnRH secretions during sleep. This pattern was reminiscent of that associated with puberty and was termed "developmental arrest." A third group showed a pattern of gonadotropin secretion characterized by abnormal amplitude of LH secretion. Finally, a subset of patients was identified in which LH pulsations were observed at a frequency that was inappropriately slow during the early follicular phase and was similar to that seen in the luteal phase of normal cycling women. An interesting observation was that the abnormalities of GnRH-induced gonadotropin secretion could change within a given patient over time.

Males

Alteration of GnRH pulse frequency or amplitude may also induce abnormal gonadotropin secretion in the male. In a series of studies, Crowley and associates (5) evaluated the pattern of GnRH-induced LH pulses in men with idiopathic hypogonadotropic hypogonadism (IHH). Following baseline evaluations, these men were placed on a regimen of 25 ng/kg of GnRH administered subcutaneously at 2-hour intervals, a dosage previously shown to yield LH pulses within the range of normal males. Chronic administration of GnRH for 3 months was required to achieve adequate secretion of gonadotropins and sex steroids in men with IHH. Because these men lack endogenous GnRH, they represent an excellent experimental model for studies of abnormalities of GnRH-induced gonadotropin secretion. Intravenous administration of GnRH at a frequency which was progressively increased from every 2 hours to every 15 minutes resulted in a progressive decline and blockage of the pituitary responsiveness to each GnRH pulse. As the frequency of GnRH administration was increased further, pituitary responsiveness became highly erratic, resulting in wide fluctuations of serum LH levels, and reminiscent of the early changes accompanying pituitary down-regulation.

Clinical Applications of GnRH and of Its Analogs

GnRH and its analogs have been used or are being tried in a large number of clinical conditions (Fig. 1.1). These will be briefly reviewed.

Diagnostic Use

Acute injection of native GnRH to men or women elicits an immediate response that may be used to evaluate the status of hypothalamic–pituitary–gonadal function in a variety of neuroendocrine conditions in which amenorrhea and/or infertility are presenting symptoms (4,8). These provocative tests may differentiate hypothalamic disorders from primary pituitary deficiencies. Repeated daily infusions of GnRH rather than

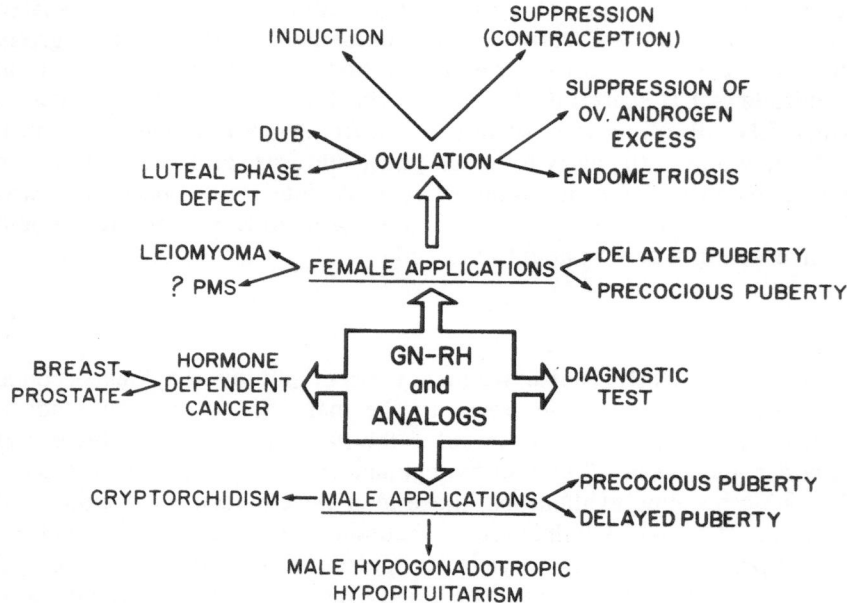

FIGURE 1.1. Clinical indications for the use of GnRH and its analogs.

single bolus administration have been recommended to overcome pituitary resistance and to activate a quiescent pituitary gland.

Gonadal Stimulation

Pulsatile administration of GnRH in an appropriate amount and at a frequency that mimics endogenous release stimulates both male and female gonads. In earlier attempts, GnRH was given by repeated parenteral injection or by continuous infusion for the induction of ovulation, without much success. Nor did the use of GnRH analogs with greater potency and long duration of action yield better results. In a series of 169 anovulatory patients, only 43% ovulated and 15% achieved pregnancy with such regimens (9).

After many years of discouraging results, the landmark discovery of Knobil et al. (1) that, in the Rhesus monkey, a pulsatile secretion of GnRH was needed for optimal stimulation of pituitary gonadotropin secretion and for the avoidance of pituitary desensitization that is paradoxically produced by long-acting superactive analogs of GnRH, provided the rationale for a new therapeutic strategy and stimulated a renewed interest in the clinical use of GnRH for the stimulation of ovulation and fertility. Indeed, the use of long-term pulsatile therapy by means of portable computerized infusion pumps made possible the induction of normal menstrual cycles, ovulation, and pregnancy in amenorrheic women. For example, Leyendecker and Wildt (10) used small infusion pumps to administer 115 treatment courses of 2.5 to 20 ng of GnRH every 90 minutes for prolonged periods of time to women with hypothalamic amenorrhea, and obtained a 100% ovulation rate. Twenty-eight conceptions occurred in 34 patients. Six of them were multiple pregnancies.

Combined therapy with GnRH and human chorionic gonadotropin (hCG) has also been used (10). Pulsatile GnRH is given first to induce follicular growth and maturation, is discontinued after ovulation, and is followed by repeated doses of hCG to support luteal function. Continuing pulsatile GnRH throughout the luteal phase is equally effective in supporting corpus luteum function and has the added advantage that the time of follicular maturation, when hCG would have to be administered, need not be monitored. Some investigators have advocated the subcutaneous route of administration for prolonged pulsatile GnRH therapy of anovulatory infertility (11). Karin and his co-workers (12) infused a dose of 5 to 20 μg of GnRH every 90 minutes via an indwelling catheter inserted in the subcutaneous (s.c.) fatty tissue of the lower abdominal wall. Nineteen treatment courses were given to 14 infertile amenorrheic women. Thirty-six ovulatory cycles were induced in 12 of the women, and eight of them conceived. The s.c. GnRH therapy was given with the same pulse frequency until either menstruation or pregnancy occurred, and hCG was not added. Pulsatile GnRH treatment may be continued without interruption to induce repeated ovulatory cycles until pregnancy occurs. No serious adverse effects have been observed with s.c. therapy. However, during i.v. therapy, occasional severe complications such as bacteremia with high fever and phlebitis have been reported (11).

Crowley and his associates (5) induced ovulation in 17 patients with IHH or hypothalamic amenorrhea with pulsatile GnRH given at a dose of 25 or 100 ng/kg. The results of these studies indicated that 80% of the patients responded to the smaller dose, whereas the 100 ng/kg dose induced ovulation in all cases. However, ultrasound studies at mid-cycle indicated that this larger dose induced the ripening of two or more follicles.

In women with polycystic ovarian disease (PCOD), the pulses of LH secretion are more frequent and have greater amplitude than those of normal women during the early follicular phase, more so than in the mid-follicular and late follicular phases. Induction of ovulation with pulsatile GnRH in patients with PCOD is not as successful as in those with hypothalamic amenorrhea.

GnRH agonists at very low doses have not been tested for pulsatile therapy since they carry the inherent risk of gradually desensitizing the pituitary gland, although the suppression of pituitary function in patients with secondary amenorrhea or PCOD with such agonists has resulted in more effective induction of ovulation by human menopausal gonadotropin (hMG)-hCG (13).

Precocious Puberty

Maturation of the pituitary–gonadal system in both male and female requires pulsatile GnRH stimulation. Idiopathic precocious puberty may be viewed as a disease characterized by premature hypothalamic GnRH activity. Suppression of pituitary gonadal function in this condition has been the aim of various therapeutic modalities. Chronic administration of GnRH agonists has proven to be a remarkably safe and effective therapy for children of both sexes with precocious puberty. Within 6 to 18 months after beginning daily treatment with an agonist in doses of 4 to 8 ng/kg s.c., pubertal levels and patterns of secretion of gonadotropins and sex steroids were found to revert to prepubertal levels and patterns. A more striking aspect of this therapy was the regression of secondary sexual characteristics and the cessation of menstrual bleeding. All effects of therapy are usually reversed when treatment is discontinued. Thus far, over 150 children have been treated successfully in this manner (4,5,14).

Delayed Puberty

Chronic pulsatile administration of GnRH may initiate puberty in subjects with delayed puberty. Thus, in 22 men with IHH, gonadotropin levels increased and development of secondary sexual characteristics and spermatogenesis were observed (5). Similar findings have been reported in females with delayed puberty or with hypothalamic hypogonadotropic hypogonadism.

Cryptorchidism

Treatment of cryptorchidism with hCG is well established. As an alternative therapy, endogenous secretion of gonadotropin can be temporarily activated by GnRH. Several investigators have reported favorable results with intranasal spray of GnRH after 4 weeks of treatment (4,15). Good results have been obtained also with a combination of GnRH and hCG (4).

Contraception

Administration of GnRH analog superagonists to normally ovulating females results in down-regulation of pituitary function and suppression of ovulation, and has been used extensively in both experimental animals and humans to inhibit fertility. In practice, these peptides are given daily from the beginning of the menstrual cycle. The mode of action of these agonists is complex. Initially, they cause the release of supraphysiologic amounts of gonadotropins, resulting in ovarian desensitization. Later, the pituitary gland also becomes desensitized and loses its ability to mount the LH surge needed for ovulation. Although extensive data on the contraceptive effects of GnRH agonists indicate that they are potent inhibitors of fertility in the female, there are potential drawbacks to their use, such as amenorrhea, irregular bleeding, and development of a menopause-like state with associated vasomotor symptoms and bone loss. When the treatment is prolonged, there is also the possibility that continued, unopposed, even if low level, endogenous estrogen may result in the development of endometrial abnormalities (3,4,16).

Post-implantation administration of GnRH agonists brings about depletion of luteal receptors in the rat, rabbit, and monkey. In women, their administration shortens the luteal phase. However, this is reversed by hCG which provides luteal support. Thus, therapeutic abortion cannot be induced with a superagonist given in early pregnancy (3,4,16).

Several recent studies have demonstrated that GnRH antagonists are also effective in reducing fertility in several animal species (17). It is believed that these compounds bind to GnRH receptors and produce an immediate unidirectional antigonadal effect with no initial stimulatory phase. Their evaluation in the human awaits further studies regarding their safety.

In males, daily administration of agonists readily reduces pituitary LH and FSH secretion and testosterone levels with a decrease in sperm density and motility (3,4,18). Unfortunately, only oligospermia but not azoospermia results from such treatment. Moreover, the dramatic androgen deficiency is associated with a loss of libido and potency and the appearance of andropause-type vasomotor symptoms. Regimens involving administration of an agonist plus replacement with testosterone enanthate, while relieving the effects of agonist-induced androgen deficiency, only produced varying degrees of oligospermia insufficient to produce an acceptable contraceptive response (19).

A final but important concern regarding the use of GnRH agonists for contraceptive purposes is their lack of biologic activity when they are administered orally. Rapid development of implant technology, however, may obviate this inconvenience.

Endometriosis

The ability of GnRH to produce amenorrhea and anovulation has provided the basis for the use of GnRH agonists in the management of endometriosis. A relationship between endometriosis and infertility has been documented repeatedly. Approximately 30 to 60% of women with unexplained infertility are found to have pelvic endometriosis. Mild and moderate stages of endometriosis are usually treated with medication, whereas the more severe cases are managed surgically. Currently, medical management of endometriosis includes the use of danazol, a synthetic derivative of testosterone, and progestogens, with or without estrogens. Pharmacologic doses of GnRH agonists have been shown to induce amenorrhea, anovulation, and regression of endometriosis and its associated clinical symptoms. Two compounds have received considerable clinical attention. Intranasal sprays of buserelin, or GnRH ethylamide, in doses of 1200 μg/day effectively suppressed endometrial implants (20). Nafarelin, another GnRH superagonist, has received extensive clinical trials. In a parallel double-blind/double-dummy study design, 204 patients with endometriosis were randomly treated with danazol, or with 400 μg or 800 μg doses of nafarelin for 6 months. A similar degree of relief of symptoms and regression of endometrial implants was observed in all groups. Side effects of nafarelin were milder and fewer compared to those of danazol (21). Very likely, these compounds will be most attractive for the management of endometriosis when they become available.

Uterine Leiomyoma

Uterine myomata are the most common benign tumors of the female reproductive tract. Asymptomatic fibroids are not usually treated unless their large size, rapid growth or degeneration causes symptoms requiring intervention. Similarly, small myomata associated with menorrhagia or other menstrual disorders require prompt attention.

Surgical removal is currently the only effective therapy for fibroids. It has long been recognized that estrogens may stimulate the growth of myomata. Indeed, in hypoestrogenic states such as menopause, myomata regress.

Recent observations indicate that the hypoestrogenic condition induced by GnRH analogs may bring about a marked decrease in the size of myomata (22). Thus, this treatment may become a valuable adjunct for pre-operative management of larger myomata or for small symptomatic fibroids.

Hormone-Dependent Tumors

Suppression of gonadotropin secretion by high doses of GnRH agonists has been found to be beneficial in some hormone-dependent and malignant tumors of the breast and the prostate (3,4). It is believed that the mechanism of action of these drugs involves a decrease in the secretion of pituitary gonadotropin and gonadal steroids, resulting in medical castration. It is also conceivable that agonists may have a direct effect on steroidogenesis.

MAMMARY CARCINOMA

In some estrogen-dependent breast cancers, oophorectomy or antiestrogenic compounds have been found to be beneficial. Inhibition of estrogen secretion by GnRH agonists has, in fact, given encouraging results (4).

PROSTATIC CANCER

Approximately 80% of human prostatic cancers are hormone-dependent when initially diagnosed (23). Standard treatment of prostatic carcinoma has included surgical castration or the use of high-dose estrogen to suppress testosterone secretion. Administration of GnRH superagonists in pharmacologic doses has been found to be highly effective. After an initial transient stimulation of androgen production lasting approximately 1 week, these agents suppress testicular testosterone production to levels observed in castrated individuals for as long as 2 years. However, the adrenal contribution to the androgen pool remains intact, since the agonists exert no effect on adrenal secretion of androgens (23). The GnRH agonists provide an alternative or adjunct modality to existing therapies (3,4,23). A combination of GnRH agonists and antiandrogens has also been proposed for the management of prostatic cancer and may offer additional advantages.

Management of Hirsutism

Hirsutism may result from multiple causes. Among these are excessive androgen production by the ovaries or the adrenal glands and increased sensitivity of the hair follicle to normal circulating androgen levels. Most hyperandrogen states in women are associated with increased ovarian androgen production, frequently associated with polycystic ovarian disease. Thus, Andreyko et al. (24) treated six hirsute women with nafarelin for 6 months with good clinical results, associated with a significant decrease of serum gonadotropin, total testosterone, free testosterone, and androstenedione levels.

Conclusions

Since the identification and synthesis of GnRH 15 years ago, over 1700 GnRH analogs have been synthesized and evaluated for the therapy of a variety of conditions requiring temporary reversible suppression or stimulation of gonadotropin secretion. An effective stimulation of the gonads requires pulsatile administration of a GnRH agonist, preferably the native decapeptide itself. For gonadal suppression, superagonists have been proven to be highly effective. Intensive studies currently underway promise new and more innovative clinical applications of these compounds.

References

1. Knobil E (1980) The neuroendocrine control of the menstrual cycle. Recent Prog Horm Res 36:53.
2. Jeffcoate SL, Greenwood RH, Holland DT (1974) Blood and urine clearance of luteinizing hormone releasing hormone in man measured by radioimmuno-assay. J Endocrinol 60:305.
3. Corbin A, Frederick J, Jones B et al Comparison of LHRH agonists and antagonists antifertility and therapeutic developments. In: Labrie F, Belanger A, Dupont A eds. LHRH and its analogs. New York: Elsevier Science Publishers. 1984 p 95.
4. Sandow J (1983) Clinical applications of LHRH and its analogues. Clin Endocrinol 18:571.
5. Crowley WF, Filicori M, Spratt DI et al (1985) The physiology of gonadotro-pin releasing hormone (Gn-RH): secretion in men and women. Recent Prog Horm Res 41:473.
6. Filicori M, Butler JP, Crowley WF (1984) Neuroendocrine regulation of the corpus luteum in the human: evidence for pulsatile progesterone secretion. J Clin Invest 73:1638.
7. Veldhuis JD, Rogol AD, Evans WS et al (1986) Spectrum of the pulsatile characteristics of LH release in normal men. J Androl 7:83.
8. Wentz AC (1977) Clinical applications of luteinizing hormone releasing hormone. Fertil Steril 28:901.
9. Hammond CB, Wiebe RH, Haney AF et al (1979) Ovulation induction with luteinizing hormone releasing hormone in amenorrheic, infertile women. Am J Obstet Gynecol 135:924.
10. Leyendecker G, Wildt L Pulsatile administration of GnRH in hypothalamic amenorrhea. In: Labrie F, Belanger A, Dupont A eds. LHRH and its analogues. New York: Elsevier Science Publishers. 1984 p 457.
11. Nillius SJ (1983) Clinical use of pulsatile LHRH in infertility. Med Bull 17:1.
12. Skarin G, Nillius SJ, Wide L (1983) Pulsatile subcutaneous low-dose gonado-tropin releasing hormone treatment of anovulatory infertility. Fertil Steril 40:454.
13. Fleming R, Haxton MJ, Hamilton MPR et al (1985) Successful treatment of infertile women with oligomenorrhea using a combination of an LHRH agonist and exogenous gonadotropins. Br J Obstet Gynecol 92:369.
14. Boepple PA, Mansfield MJ, Wierman ME et al (1986) Use of potent long acting agonist of gonadotropin releasing hormone. Endocrine Rev 7:24.
15. Hadziselimovic F, Girard J, Hocht B (1980) Effects of LHRH treatment of hypothalamo-pituitary-gonadal axis and Leydig cell ultrastructure in cryptor-chid boys. Horm Res 13:358.
16. Nillius SJ, Berquist C, Gudmundsson JA et al Superagonists of LHRH for contraception. In: Labrie F, Belanger A, Dupont A, eds. LHRH and its analogues. New York: Elsevier Science Publishers. 1984 p 261.

17. Zarate A, Canales ES, Schally AV et al The use of LHRH agonists and antagonists as antifertility agents in the human female. In: Zatuchni GI, Shelton JD, Sciarra JJ eds. LHRH peptides as female and male contraceptives. Philadelphia: Harper and Row Publishers. 1981 p 277.
18. Nieschlag E, Akhtar FB, Schurmeyer T et al LHRH agonists for male fertility control: experiments in monkeys and men. In: Labrie F, Belanger A, Dupont A eds. LHRH and its analogues. New York: Elsevier Science Publishers. 1984 p 277.
19. Swerdloff RS, Bhasin S Hormonal effect of GnRH agonist in the human male: An approach to male contraception using combined androgen and GnRH agonist treatment. In: Labrie F, Belanger A, Dupont A eds. LHRH and its analogues. New York: Elsevier Science Publishers. 1984 p 287.
20. Lemay A, Maheux R, Faure N et al Efficacy and safety of LHRH agonist treatment in 10 patients with endometriosis. In: Labrie F, Belanger A, Dupont A eds. LHRH and its analogues. New York: Elsevier Science Publishers. 1984 p 383.
21. Henzl M, Corson S, Moghissi K et al (1988) Administration of nasal nafarelin versus oral danazol for endometriosis. A multicenter double-blind comparative clinical trial. N Eng J Med, 318(8): 485–489.
22. Maheux R, Guilloteau C, Lemay A et al (1985) Luteinizing hormone releasing hormone agonist and uterine leiomyoma: a pilot study. Am J Obstet Gynecol 152:1034.
23. Klein LA Prostatic Cancer (1979) N Engl J Med 300:824.
24. Andreyko JL, Monroe SE, Jaffe RB (1986) Treatment of hirsutism with a gonadotropin releasing hormone agonist (Nafarelin). J Clin Endocrinol Metab 63:854.

2

Corticotropin-Releasing Factors

LYNNETTE K. NIEMAN, D. LYNN LORIAUX, AND
GEORGE P. CHROUSOS

Historical Perspectives

Hypothalamic regulation of pituitary adrenocorticotropic hormone
(ACTH) secretion was proposed in 1948 (1) and confirmed in 1955 by the
demonstration that a crude extract of the stalk median eminence induced
ACTH secretion in the rat (2,3). The putative neurotransmitter was
named corticotropin-releasing factor, or CRF. Attempts to isolate a single
CRF from such extracts with purification chromatography revealed as
many as four biologically active substances (3). This early indication of
multiple ACTH secretagogues was confirmed by later work showing that
ACTH release is modulated by multiple humoral factors (for reviews, see
refs. 4–7). We will review here the evidence for these CRFs in man, and
examine their physiologic and potential diagnostic and therapeutic roles.

Vasopressin was one of the first candidates proposed as a hypothalamic
CRF. One of the fractions of crude hypothalamic extract with chromatographic properties similar to vasopressin (8) accounted for about 30% of
stalk median eminence CRF activity (9). The corticotropin-releasing
activity of this low-molecular-weight substance could be blocked by
antivasopressin antisera (9). The presence of vasopressin-containing
pathways in the hypothalamus (10) and the modulation of hypothalamic
vasopressin levels by adrenalectomy (11) and glucocorticoid administration (12) further suggested a role for vasopressin in ACTH secretion.
The in vitro corticotropin-releasing action of hypothalamic extract, however, could not be reproduced by the addition of synthetic vasopressin to
cultured pituicytes (13). The addition of other stalk median eminence
fractions was required for full corticotropin-releasing activity. These
findings stimulated the efforts to isolate and characterize an additional
CRF.

This search culminated in the isolation and sequencing of CRF-41 from
sheep hypothalami by Vale in 1981 (14). Soon after, the structures of
porcine (15) and rat (16) corticotropin-releasing hormone (CRH) were
reported, and the amino acid sequence of human CRH was deduced after
sequencing of the CRH gene (17,18). Human CRH has a 100% homology
with rat CRH and an 83% homology with ovine CRH (oCRH), differing by

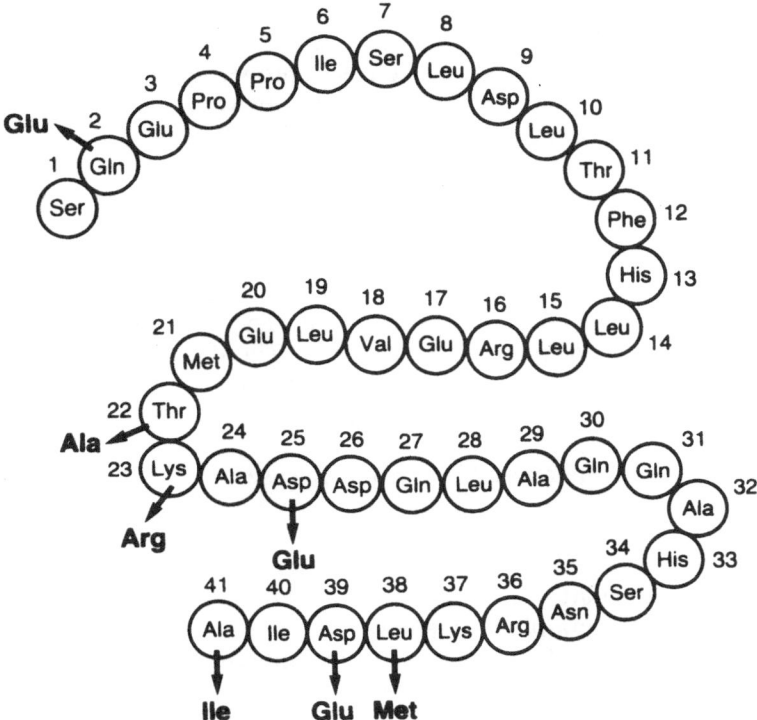

FIGURE 2.1. The chemical structure of ovine and human (*arrows*) corticotropin-releasing hormone. [Reproduced, with permission, from (104)].

only seven amino acid residues (Fig. 2.1). This striking inter-species homology suggests an important function for this evolutionarily conserved peptide.

Corticotropin-Releasing Factors

CRF-41 fulfills many of the criteria for a CRF (19). It is present in the median eminence (11) and can be measured in hypothalamic–pituitary portal venous blood (6). It stimulates the specific increase of ACTH and other POMC-derived compounds from the pituitary gland (14). Its potency, however, is not as great as that of crude hypothalamic extracts, suggesting that full corticotropin-releasing activity involves other factors (13). As suggested above, vasopressin may be one of these factors. Synthetic vasopressin has less corticotropin-releasing activity in a dispersed rat pituicyte system than does synthetic CRH; it enhances the CRH-induced ACTH release by 2- to 4-fold (20,21).

Oxytocin also has corticotropin-releasing properties in some species. In the rat, it amplifies the effect of CRH (but not vasopressin) about 2- to 4-fold (22). Its potency has been estimated as 0.01 to 1 times that of vasopressin (22–24), with a greater apparent potency in the presence of CRF-41. The corticotropin releasing activity of oxytocin in man is not well established. The intravenous morning administration of exogenous oxytocin was followed by decreased ACTH levels consistent with the time of day (25). In a subsequent study oxytocin and saline administration had similar effects on ACTH levels (26). These results suggest that oxytocin does not affect ACTH secretion. The doses of oxytocin employed, however were less than those required for vasopressin enhancement of oCRH action in man and may not have been sufficient to provoke an effect.

The role of adrenergic agents in the release of ACTH remains controversial (27,28). Some studies using in vivo rat models have demonstrated ACTH release after administration of β-adrenergic substances, while others showed no effect (for review, see 29). In man, noradrenergic agents may stimulate ACTH release through an indirect α_1-adrenergic mechanism at a supra-pituitary site (30,31).

Lactate stimulates ACTH release from cultured rat anterior pituicytes at levels similar to those observed during strenuous exercise in man (personal communication, Anton Luger). Other factors known to stimulate ACTH release from in vitro rat pituitary systems include angiotensin II (32,33), the carboxy-terminal octapeptide of cholecystokinin (CCK-8) (34), vasoactive intestinal peptide (VIP) (35), gastrin releasing peptide (36), atrial natriuretic factor (37), and serotonin (38). Analogues of γ-aminobutyric acid both stimulate and inhibit ACTH release (39). Prostaglandin E_2 may inhibit ACTH release (40). The physiologic relevance of these peptides in both rat and man remains to be elucidated.

Tissue Distribution of CRH

CRH has been demonstrated in a variety of tissues by immunohistochemical techniques and by measurement of the biologic activity or CRH immunoreactivity of tissue extracts. The anatomic correlation provided by such work has been instrumental in understanding the role of CRH and vasopressin in the intact animal. Both are present in hypothalamic neuronal cell bodies whose axons project to the median eminence (ME). Presumably, the hormones can be secreted from the axon terminals of these neurons into the hypophysial portal capillaries.

The parvocellular neurons in the paraventricular nucleus (PVN) are the major source of CRH in the hypothalamus of rat (41), man (42), and bird

(43). The importance of this nucleus has been confirmed in the rat by radioimmunoassay and immunohistochemical localization of CRH, and by surgical ablation (44,45). The magnocellular neurons of the PVN, the supraoptic nucleus (SON), and the accessory magnocellular nuclei account for the remaining 10% of CRH in hypothalamic-median eminence pathways (46).

The parvocellular neurons of the rat PVN also contain vasopressin (47). In normal rats, as many as 40% of them contain both vasopressin and CRH (48–50). The remainder contain only CRH. The proportion of cells containing vasopressin and its mRNA increases after adrenalectomy (51,52).

Vasopressin is also transported to the ME by the magnocellular neurons, the cell bodies of which are located in the PVN, SON, and accessory nuclei. The concentration of vasopressin in these cells is not affected by adrenalectomy. It is unclear whether the high levels of vasopressin in the pituitary portal blood derive from the magnocellular or from the parvocellular axons.

CRH immunoreactivity is present in other areas of the rat brain. The limbic system, parts of the neocortex, and some areas in the hindbrain corresponding to several noradrenergic centers such as the locus coeruleus and the nucleus tractus solitarius contain CRH (53). In man, CRH is present outside of the central nervous system in the normal pancreas, stomach, duodenum, and adrenal medulla (54). It is produced by the placenta of a variety of species (55,56), and has been localized by immunohistochemical stains to several "ectopic" sites including tumors of the bronchus, ovary, lung, colon, and prostate (57–60).

CRH in Blood and CSF

Radioimmunoassay has been used to determine the circulating levels of human CRH. These assays, however, are usually relatively insensitive, with detection limits of about 20 ng/L. Immunoreactive hCRH generally cannot be detected in healthy men at rest (55,56,61,62). Plasma cortisol and ACTH levels increase but hCRH remains low during minor surgery or strenuous exercise (63,64). Plasma corticotropin-releasing activity as measured by in vitro ACTH release, however, increases in rats with hypothalamic lesions undergoing major surgery (65). This suggested that circulating factors not recognized by the CRH radioimmunoassay might mediate, in part, the hypercortisolism associated with trauma and burns. Brodish has suggested that these substances are released by tissue injury and has coined the term "tissue CRH" to distinguish them from hypothalamic CRH. Lactate (64) may be one such CRH-releasing substance.

Immunoreactive plasma CRH levels increase during pregnancy from <10 pg/ml to levels of up to 2700 pg/ml at term (66–69). CRH levels increase further during labor and delivery, and decline rapidly after delivery. The placenta is the most likely source of this material. It contains and secretes CRH in vitro (70). Pregnancy-associated CRH has been reported to be biologically active in in vitro systems (66,67,69). Others have concluded that maternal CRH is probably not biologically active because of the lack of correlation between maternal CRH and cortisol levels (68). Thus, the physiologic significance of CRH in pregnancy remains unknown.

CRH has not been measured in the hypophysial portal circulation of man; lower animals have high portal venous levels (up to 500–1500 pg/ml) with peripheral plasma levels below 10 pg/ml (71). Immunoassayable CRH is present in the cerebrospinal fluid (CSF) of man (72) but the source of the CSF peptide is unknown.

CRH Receptors

CRH receptors are present in the anterior pituitary gland, the neocortex, the limbic system, and the medulla oblongata of the rat (73,74). They are coupled to adenylate cyclase (75). Following adrenalectomy the number of pituitary CRH receptors decreases. The administration of glucocorticoids restores a normal concentration of receptors. The mechanism regulating the decrease in receptor number after adrenalectomy is unknown.

Receptors for CRH are also present in the adrenal medulla and sympathetic ganglia. CRH binding to these tissues results in a stimulation of adenylate cyclase that is similar to that seen in the central nervous system (76). CRH-induced stimulation of catecholamine secretion from bovine chromaffin cells in culture suggests a physiologic role for the peptide in the regulation of the autonomic nervous system (76).

Administration of Exogenous CRH

Healthy Men and Women

The intravenous bolus administration of oCRH to healthy men and women at doses of 0.1 to 30 mcg/kg results in a dose-dependent increase in plasma ACTH and cortisol that lasts longer than 3 hours (Fig. 2.2) (77,78). oCRH is also active when given subcutaneously, but with decreased potency (79). Although plasma levels of ACTH, β-endorphin and other pro-opiomelanocortin (POMC)-derived molecules increase after intravenous oCRH administration, (80–82) other pituitary products are unaffected (81,83). Transient side effects such as flushing, a metallic taste, and tachypnea occur in about 20% of individuals given a dose of 1 mcg/kg. These side effects increase in frequency with higher doses.

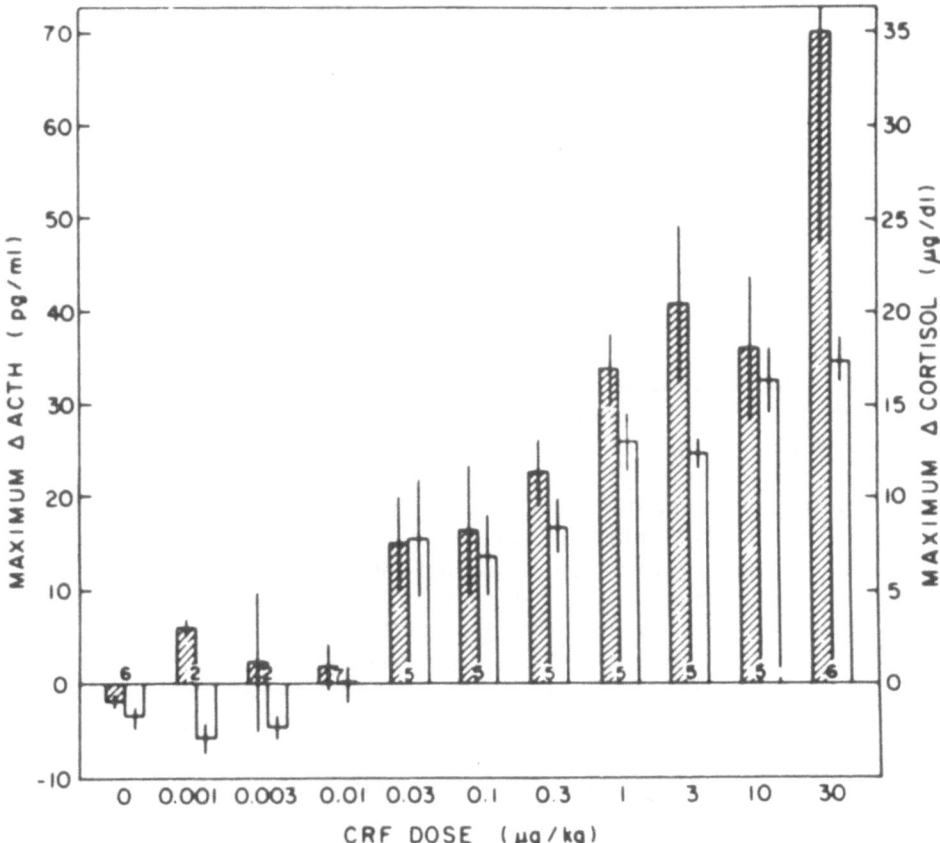

FIGURE 2.2. Maximum changes in plasma ACTH and cortisol concentrations after intravenous administration of synthetic ovine CRH to healthy men. The maximum change in concentration (either decrement or increment, whichever was greater) during the 120 minutes after CRH injection from the level immediately before CRH injection was calculated for each individual. The height of the bar represents the mean of individual values, the bracket represents the SEM. The number of subjects in each group is indicated at the base of each pair of bars. Shaded bars represent ACTH concentrations. Open bars represent cortisol values. [Reproduced from *The Journal of Clinical Investigation,* 1983; 71:587–95, by copyright permission of the American Society for Clinical Investigation.]

Doses of 30 mcg/kg have provoked hypotension (81,83). Studies in dogs (84) and nonhuman primates (85) suggest that this phenomenon is due to dilatation of the mesenteric vessels and sequestration of blood in the abdominal viscera. Since maximal stimulation of plasma cortisol levels occurs at a dose of 1 mcg/kg, most investigators have chosen doses between 1 and 2 mcg/kg for provocative testing.

The pattern of the cortisol response to oCRH in normal volunteers is dependent on the time of day. The incremental hormone response is greatest in the evening when basal levels are low. Peak stimulated values are similar in the morning and in the evening (81,83). ACTH responses are similar in the morning and evening. Individual ACTH responses tend to vary inversely with the basal cortisol level, suggesting that cortisol exerts a negative feedback effect on the ability of the pituitary gland to respond to oCRH (81,86).

Human CRH, 1 mcg/kg, elicits a response that is similar in magnitude but shorter in duration than that seen after an equimolar dose of oCRH; peak hormone levels are slightly lower and the duration of the response is only 1 hour (87,88). The pattern of the ACTH and cortisol responses to hCRH is similar to that of the endogenous hormone pulses (Fig. 2.3). This similarity supports the argument that CRH is the major hypothalamic secretagogue responsible for basal ACTH secretion.

The discrepancy in the duration of action of the two peptides appears related to their different pharmacokinetic properties. oCRH has an initial half-life of 6 to 12 minutes (probably representing distribution into the plasma volume) followed by a second half-life of 45 to 70 minutes (probably reflecting metabolic clearance). Human CRH has a more rapid clearance, the second half-life being about 25 minutes.

The hormonal response to CRH can be modulated. Exogenous vasopressin (10 pressor units) augments the ACTH response to oCRH to a level greater than that seen with either agent alone (89–91). Hypertonic saline potentiates the hormonal response to hCRH, presumably by increasing endogenous vasopressin secretion (92). Naloxone (20 mg i.v.) enhances the response to CRH (93) while pre-treatment with morphine sulphate decreases it (94). The met-enkephalin analog FK 33-824 also blocks the response to hCRH in normal men (95). These data suggest that endogenous opiate secretion may contribute to an ultra-short loop negative feedback effect on the secretion of POMC-derived substances.

Both chronic and short-term hypercortisolism suppress the pituitary response to exogenous CRH. This is demonstrated by the lack of response to CRH in hypercortisolemic patients with the ectopic secretion of ACTH (96,97) and the blunted responses of patients with Cushing's

FIGURE 2.3. *A.* Plasma immunoreactive ACTH (IR) and cortisol elevations after intravenous administration of 1 mcg of ovine (oCRH, triangles) or human corticotropin-releasing hormone (hCRH, circles) to normal men and women (mean ± SE). *B.* Comparison of natural, spontaneously occurring plasma ACTH, and cortisol secretion episodes (open circles) and responses to human CRH (solid circles). Samples were drawn every 10 minutes. All spontaneous or human CRH-induced episodes were centered according to their highest (peak) plasma ACTH value at 0 minutes. [Reproduced, with permission, from (104)].

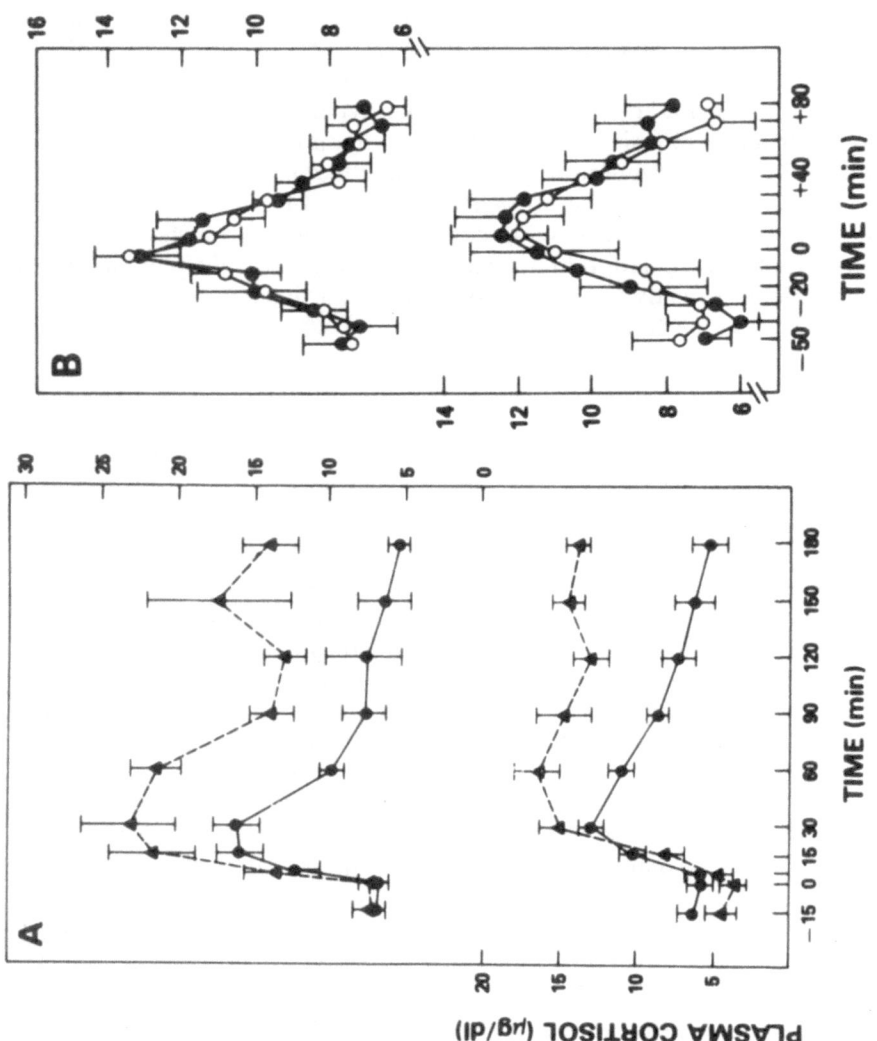

syndrome following the surgical cure of hypercortisolism (98,99). Exogenous glucocorticoids also blunt the hormonal response to CRH. The ACTH response of patients with Cushing's disease can be blocked by pre-treatment with 8 mg of dexamethasone per day for 2 days (100). Similarly, non-Cushingoid patients treated with supraphysiologic doses of glucocorticoids fail to respond to oCRH (101). If a physiologic dose of glucocorticoids is given on alternate days, the ACTH response to oCRH is decreased on the day of glucocorticoid administration but is normal on the day off of therapy (102). Thus, while hypercortisolism can suppress corticotroph responsiveness to CRH, the minimal dose and duration of hypercortisolism required to produce this effect is unknown.

The continuous infusion of oCRH at a dose of 1 mcg/kg/hour for 24 hours to normal men resulted in cortisol levels 50 to 70% above baseline with a muted circadian rhythm (103). This suggests that the pituitary gland has variable sensitivity to CRH or to glucocorticoid negative feedback depending on the time of day.

Cushing's Syndrome

THE CRH STIMULATION TEST

An important diagnostic application for CRH may be in the differential diagnosis of Cushing's syndrome. Patients with Cushing's disease show a normal or exaggerated ACTH and cortisol response to CRH (96,97,100,104–108), while patients with ectopic ACTH secretion and non-ACTH dependent causes of hypercortisolism fail to respond to CRH (96,97,100). While most patients respond in these ways, three patients with the ectopic secretion of ACTH have been reported to have a false positive response, and some patients with Cushing's disease have a false negative response.

More recently, a report of CRH responsiveness in ectopic CRH-secreting tumors has led to renewed concern over the use of CRH testing in the differential diagnosis of Cushing's syndrome (109,110). We suggest several reasons for this controversy (apart from inherent biologic differences). The lack of test standardization and the lack of criteria for judging

--▷

FIGURE 2.4. *A*. Adrenocorticotrophic hormone (ACTH) and cortisol responses to ovine corticotropin-releasing hormone (oCRH) in patients with Cushing's syndrome. Responses exceeding 4 coefficients of variation (CV) units above baseline values (dotted line) are considered positive. *B*. Twenty-four hour urinary 17-hydroxysteroid responses to high dose dexamethasone in patients with Cushing's syndrome. Responses below 50% of the mean baseline level (*dotted line*) are considered positive. [Reproduced, with permission, from (96)].

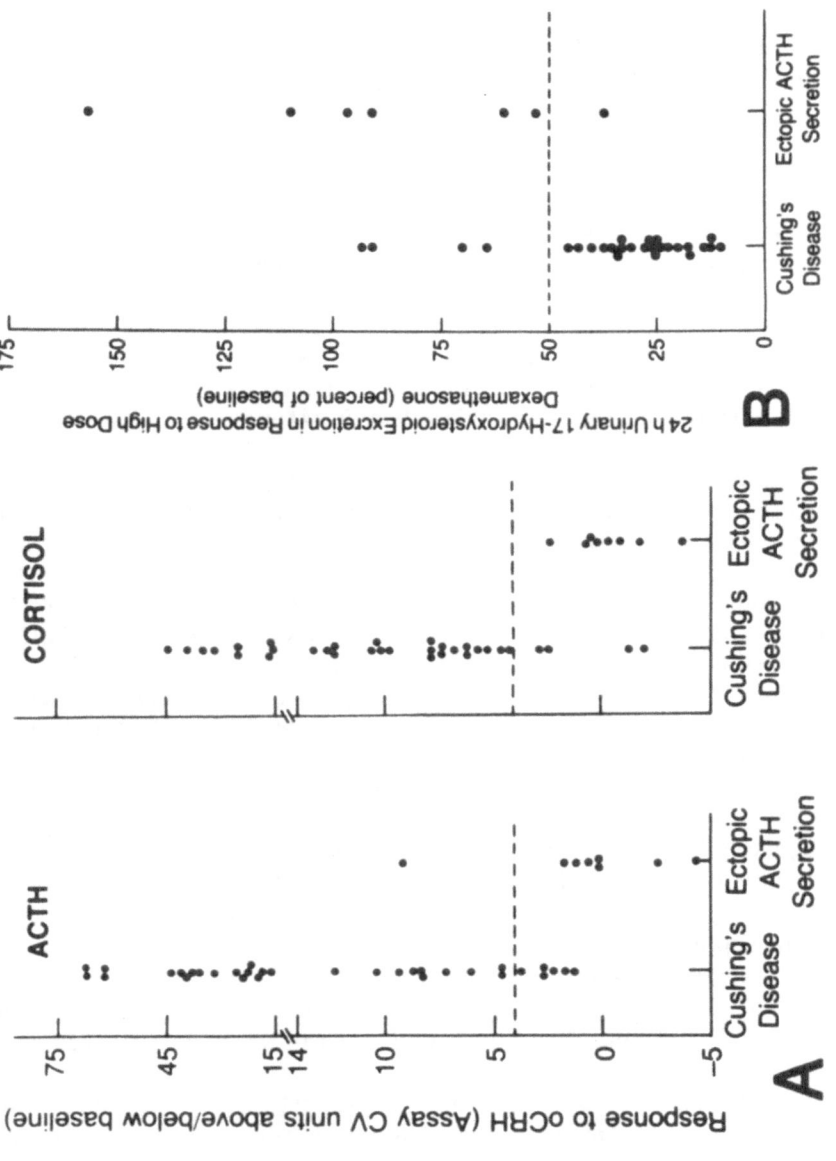

a response have made it difficult to compare results from different testing centers. We have used a weight-adjusted dose (1 mcg/kg) of CRH to avoid differences in the delivered dose between individuals. Since most normal men and women (and patients with Cushing's disease) have peak ACTH and cortisol responses within 60 minutes of CRH administration, we measure plasma ACTH and cortisol levels 15, 10, 5, and 0 minutes before and 5, 15, 30, 45, and 60 minutes after injection of oCRH.

Our early experience in the interpretation of CRH test results showed that a patient's responses might be judged as "flat" by one observer and "present" by another. To eliminate this subjectivity we have set criteria for a response to CRH based on the performance parameters of our ACTH and cortisol assays as shown by the coefficient of variation (96). A response is considered to be present if the mean post-CRH hormone level exceeds the mean pre-CRH level by 4 or more assay coefficient of variation (CV) units. (4 assay CV units was taken as an arbitrary and conservative cut-off). Using this approach we compared the diagnostic accuracy of the oCRH stimulation test to that of a standard low and high dose dexamethasone suppression test in 41 patients with Cushing's syndrome.

Twenty-seven of the 33 patients with Cushing's disease had an ACTH or cortisol response to CRH, and 29 responded to high dose dexamethasone with a decrease in urinary 17-hydroxysteroid excretion of at least 50% (Fig. 2.4). The four patients who failed to suppress with dexamethasone had a positive ACTH (n = 4) or cortisol (n = 3) response to oCRH. The two patients with neither an ACTH nor a cortisol response to oCRH each had a positive response to 8 mg of dexamethasone.

Only one of eight patients with the ectopic ACTH syndrome responded to dexamethasone. This patient, who had a bronchial carcinoid, also had an ACTH, but not a cortisol, response to CRH.

The sensitivity, specificity and diagnostic accuracy of the two tests were similar (Table 2.1), although a negative response to either did not adequately exclude Cushing's disease. Because of this, the diagnostic accuracy of the combination of tests was also evaluated. For these purposes a positive result was defined as a cortisol response to CRH and/or suppression with 8 mg of dexamethasone; a negative response to both tests were judged as negative. This criteria achieved a 100% sensitivity and 88% specificity for Cushing's disease. The predictive value of a negative response was 100% compared to the 85 to 90% diagnostic accuracy of either test alone.

The diagnostic sensitivity of the oCRH test and the dexamethasone suppression test is similar. The oCRH test, however, requires only 2 hours to complete and can be safely performed in an outpatient setting. The convenience and cost-effectiveness of the test is enhanced further, without loss of diagnostic usefulness, by assessing only cortisol responses. The combination of the oCRH test and the dexamethasone

TABLE 2.1. Performance parameters of the dexamethasone suppression and oCRH tests in the differential diagnosis of Cushing's syndrome. [Reproduced, with permission, from (96).]

	Sensitivity (%)	Specificity (%)	Predictive Value (%)		Diagnostic accuracy (%)
			+ Response	− Response	
Dexamethasone Suppression test	88	86	97	67	88
Ovine CRH					
Cortisol or ACTH response	94	86	97	75	93
Cortisol response	88	100	100	64	90
Dexamethasone and ovine CRH[a]	100	86	97	100	98

[a] Results are here defined as positive if a response is seen to dexamethasone, oCRH, or both; results are defined as negative if the response to both tests is negative.

suppression test may yield enhanced diagnostic accuracy in the setting of ACTH-dependent Cushing's syndrome. The test is also useful in the evaluation of non-ACTH-dependent Cushing's syndrome; we and others have reported that hypercortisolemic patients with an adrenal adenoma or carcinoma fail to respond to CRH.

One important variable in the interpretation of the CRH test is the degree of hypercortisolism at the time of testing. The distinction between the appropriately suppressed pituitary gland (i.e., nonpituitary etiology) and the nonsuppressed pituitary gland (Cushing's disease) is dependent on glucocorticoid negative feedback at the level of the pituitary. We have observed conversion of hormonal responses to CRH from absent to present within 1 week of medical correction of hypercortisolism in patients with the ectopic secretion of ACTH. Patients with mild or intermittent hypercortisolism due to ectopic ACTH secretion might theoretically retain responsiveness to CRH if the corticotrophs were not fully suppressed.

PETROSAL SINUS SAMPLING

Simultaneous bilateral petrosal sinus sampling with adjunctive administration of CRH is an important tool in the differential diagnosis of Cushing's syndrome (111). This procedure takes advantage of the anatomic symmetry of the pituitary gland, where each half of the gland is drained by a venous plexus into the ipsilateral sinus (Fig. 2.5). Following bilateral femoral vein catheterization, the catheters are advanced to the inferior petrosal sinuses and blood is sampled from each side and from a peripheral site. Baseline samples are drawn from all sites simultaneously for measurement of ACTH. Ovine CRH is then infused at a 1 mcg/kg dose over one minute and blood is drawn from all sites 3, 5, and 10 minutes later.

Patients with an ACTH-secreting pituitary tumor show a gradient in ACTH concentration between the peripheral and petrosal samples so that there is a "step-up" of ACTH levels in the central samples (Fig. 2.5). A gradient of 1.7 seems to best distinguish central from ectopic ACTH secretion. Patients with a central increase in ACTH levels also show differential ACTH secretion in the petrosal sinuses, resulting in a right or left-sided "lateralization." Both the central-to-peripheral and right-to-left gradients are accentuated after CRH administration. This lateralization of

FIGURE 2.5. Left, diagramatic representation of bilateral catheterization of the inferior petrosal sinuses. Right, results of bilateral petrosal sinus sampling in patients with Cushing's disease with surgically verified microadenomas (*closed circles*). Open circles show results from a patient with corticotroph hyperplasia. IR = immunoreactive [Reproduced, with permission, from (104)].

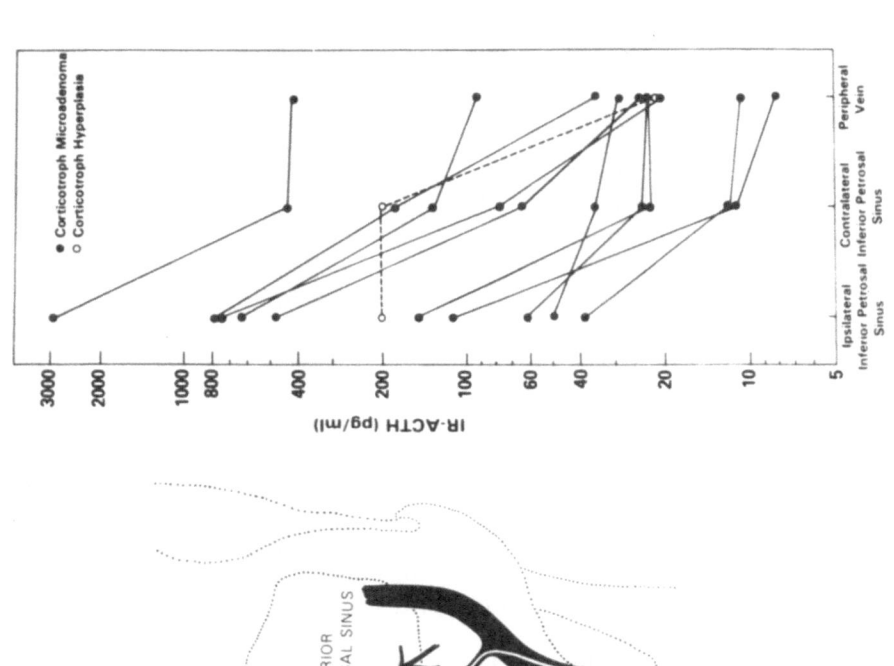

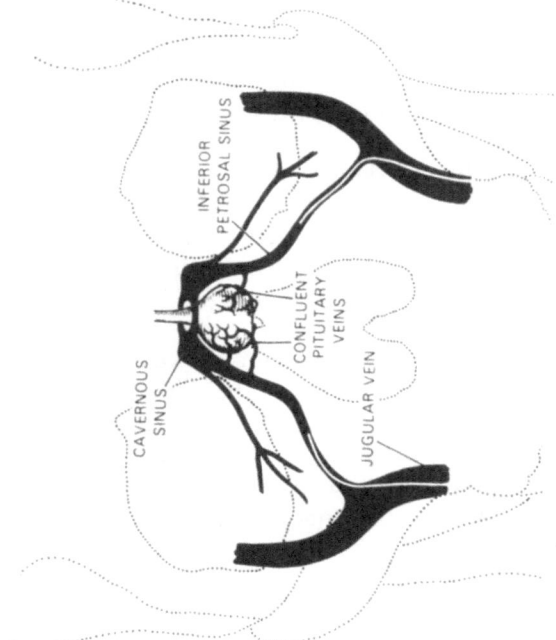

ACTH secretion allows for localization of the tumor. If a tumor is not seen at surgery, cure can be achieved by hemihypophysectomy of the side of the gland having the greater ACTH secretion. This approach to the problem of inapparent tumors at surgery has led to an increased surgical cure rate without the potential adverse effects of total hypophysectomy.

POSTOPERATIVE EVALUATION

Hypocortisolism follows the successful surgical cure of Cushing's syndrome. This transient secondary adrenal insufficiency is presumably caused by suppression of the CRH neuron and/or the pituitary gland by longstanding hypercortisolism. During the first year after surgery most patients gradually recover normal adrenal function; others may experience recurrent Cushing's disease (99,112,113). oCRH has been used in the postoperative setting to assess the responsiveness of the pituitary, the recovery of the axis and the risk of recurrence of Cushing's disease.

We have followed 29 patients cured of Cushing's disease for 11 months (range 3 to 36) (99). Twenty-three had subnormal ACTH and cortisol responses while six had normal responses to oCRH 1 to 2 weeks after selective microadenomectomy. Three of the patients with normal responses had recurrences, while all of the patients with subnormal responses remain cured. A normal response to oCRH in the early postoperative period may thus identify a subgroup of patients at risk for recurrence of Cushing's disease.

The recovery of the hypothalamic–pituitary–adrenal axis can be assessed by the 1-hour ACTH test. We have compared the hormonal responses to exogenous ACTH and oCRH in seven patients studied for one year after surgical cure of Cushing's syndrome (99). The cortisol response to each stimulus showed a similar gradual improvement with time. The correlation between the peak cortisol response to exogenous ACTH and oCRH during the first postoperative year ($r = 0.98$) suggests that the pituitary does not recover before the adrenal glands, but rather that the two recover in parallel. A 1-hour morning ACTH stimulation test is more convenient and less costly than a 3-hour evening oCRH test. After the immediate postoperative period the tests yield similar information. For these reasons, the oCRH test cannot be recommended as a substitute for the 1-hour ACTH test in assessing the recovery of the adrenal axis.

Adrenal Insufficiency

Adrenal insufficiency results when the adrenal glands cannot respond to ACTH stimulation or when ACTH stimulation is deficient. The inability to respond to ACTH reflects an intrinsic glandular abnormality and is referred to as primary adrenal insufficiency. The lack of adequate ACTH stimulation reflects an abnormality in the release of ACTH, either from abnormal pituitary or abnormal hypothalamic function, or both, and is

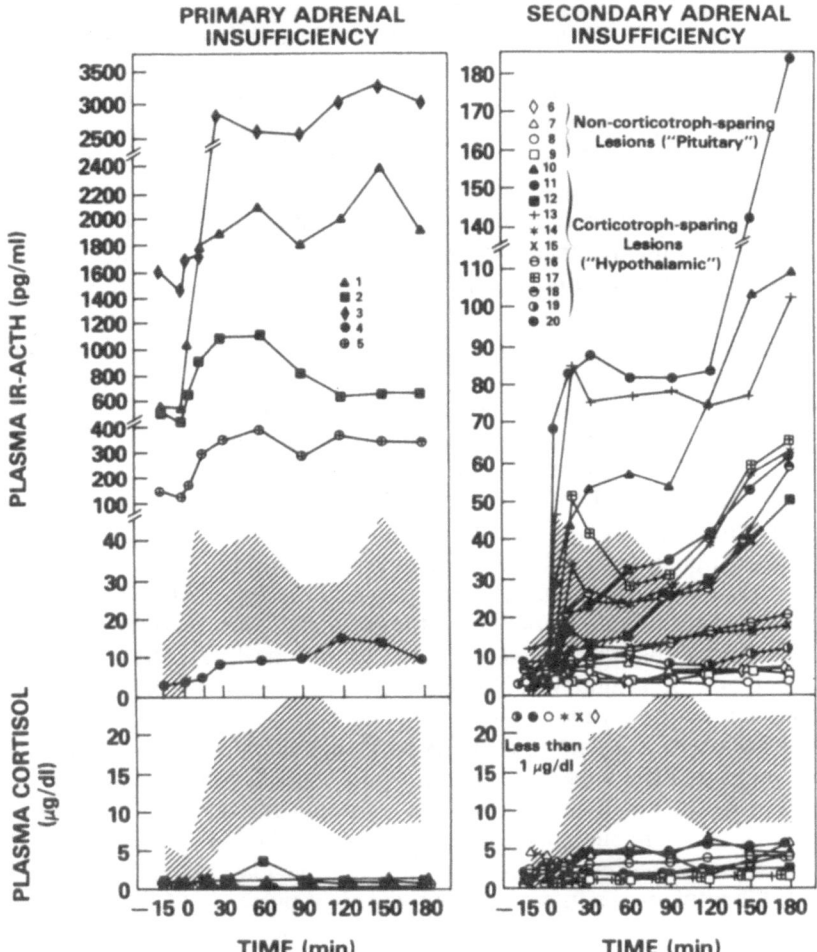

FIGURE 2.6. Plasma immunoreactive (IR) ACTH and cortisol responses to ovine corticotropin-releasing factor in patients with primary and secondary adrenal insufficiency. The hatched area represents the normal range from 27 normal men and women. Solid symbols represent patients with corticotroph-sparing lesions ("hypothalamic") and open symbols represent those with non-corticotroph-sparing lesions ("pituitary"). [Reprinted with permission from ref. 114, © The Endocrine Society, 1984.]

termed secondary adrenal insufficiency. The differentiation of these forms of adrenal insufficiency requires the administration of ACTH for two days to assess the responsiveness of the adrenal glands. Patients who respond with increased urinary glucocorticoid excretion have secondary adrenal insufficiency, and those without a response have primary adrenal insufficiency. Because this test is cumbersome, we evaluated the usefulness of CRH stimulation in distinguishing between these two groups of patients (114). Patients with adrenal insufficiency received CRH (1 mcg/kg CRH i.v.) at 2000 hours. All had received a prolonged ACTH infusion to assess the responsiveness of the adrenal gland. Glucocorticoid replacement was withheld for 24 to 36 hours before CRH was given.

All the patients had a subnormal cortisol response to CRH, regardless of their ACTH response or the etiology of adrenal insufficiency (Fig. 2.6). Four of five patients with primary adrenal insufficiency had elevated basal ACTH levels and an exaggerated ACTH response to CRH. The fifth had a near-normal ACTH pattern. The patients with secondary adrenal insufficiency had low to normal basal plasma ACTH levels and a more heterogeneous ACTH response. Peak and time integrated ACTH values ranged from undetectable to supranormal; one patient had a response approaching that seen in individuals with primary adrenal insufficiency. Of the patients with secondary adrenal insufficiency, those who did not show an ACTH response to CRH had known pituitary disease, while those with an ACTH response tended to have a structural lesion in the hypothalamus. This apparent anatomic correlation suggested that the ACTH response to CRH might localize an anatomic defect in patients with secondary adrenal insufficiency. While the basal ACTH value gave the best discrimination between the primary and secondary adrenal insufficiency, the CRH test may be useful in differentiating adrenal insufficiency of pituitary origin from adrenal insufficiency arising from lesions above the pituitary gland.

The ability of a subset of patients with secondary adrenal insufficiency to respond to CRH suggested that intermittent pulsatile administration of the compound might restore pituitary-adrenal function. We tested this hypothesis in eight such individuals who had an ACTH response to oCRH (115). After the secretory capacity of the adrenal glands was restored with an ACTH infusion, each patient received eight pulses of hCRH (1 mcg/kg) a day for 1 day. The pattern of injections was chosen to mimic the normal circadian pulse pattern of ACTH. This treatment resulted in a normalization of the circadian pattern and of mean daily levels of cortisol. ACTH levels were mildly elevated (Fig. 2.7). While pulsatile CRH therapy is less practical than glucocorticoid replacement, such an approach would potentially avoid the side effects of overtreatment with steroids. This

---
▷

FIGURE 2.7. Mean (± SE) plasma ACTH (*top*) and cortisol (*bottom*) responses to pulses of hCRH at the times indicated by the arrows. The shaded areas represent the range in eight normal subjects studied in the basal state. *Upper panel,* the range of the detection limit for the plasma ACTH determinations. Insets, individual areas under the curve of plasma ACTH (*top*) and cortisol (*bottom*) during the 24 hours of pulsatile hCRH administration. [Reprinted with permission from ref. 115, © The Endocrine Society, 1985.]

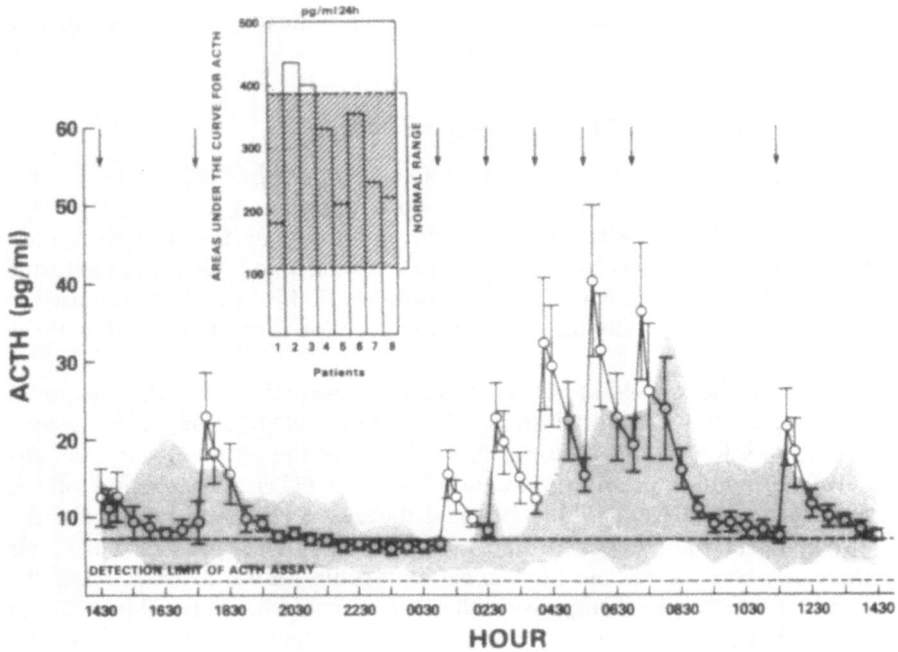

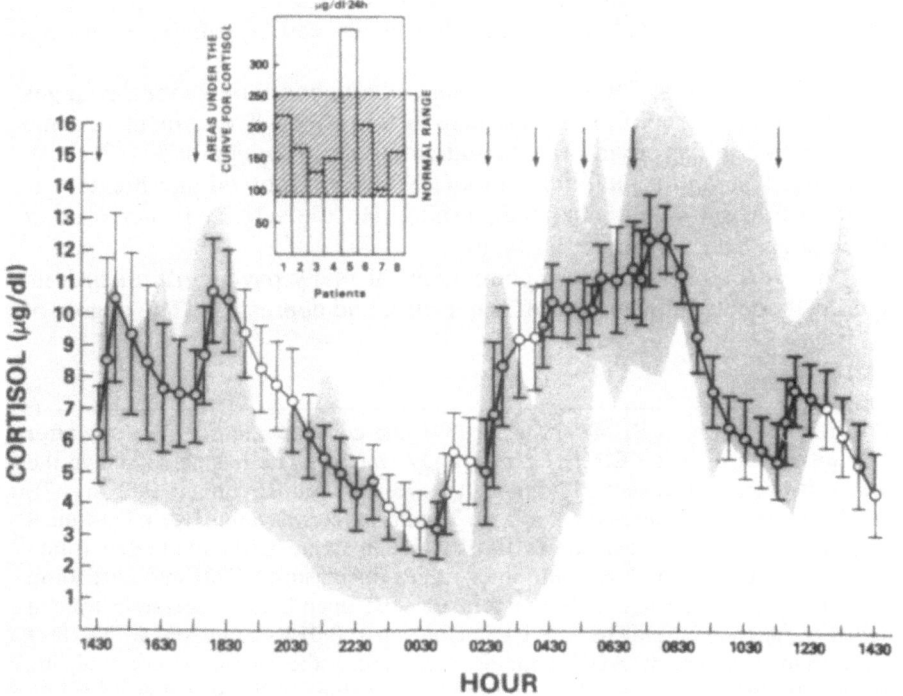

model may also provide insight into the role of glucocorticoid feedback in the regulation of pituitary function.

Hypercortisolemic Non-Cushingoid States

Transient hypercortisolism without Cushingoid stigmata commonly accompanies "stressful" conditions such as test-taking and surgery. This hypercortisolism probably represents activation of the hypothalamic–pituitary–adrenal axis by the organism's response to internal and external events. Little is known of the physiology of this transient hypercortisolism except that daily urinary steroid excretion is rarely more than three times normal.

CRH has been used in the rat to study its potential role in the response to physical and emotional stress. Intracerebroventricular (icv) administration of CRH to rats results in stimulation of the pituitary–adrenal axis (116), activation of the sympathetic system (117), suppression of the pituitary–gonadal axis (118,119), and increased electrical activity of the amygdala (120). Heart rate and blood pressure rise and cortisol levels increase. Both naloxone and anti-β-endorphin antisera can prevent the gonadal suppression, suggesting that β-endorphin mediates this effect of icv CRH by decreasing luteinizing hormone releasing hormone (LHRH) secretion. The increase in the electrical activity of the amygdala is a "kindled" phenomenon. Repetitive administration of decreasing doses of icv CRH cause amygdala seizures that finally can become spontaneous (120).

Administration of CRH icv also causes a number of behavioral changes that are reminiscent of the human response to stress. The animals become restless in familiar environments and catatonic in foreign ones (121,122). Feeding (121) and sexual behaviors (119) are suppressed and memory is enhanced, resulting in enhanced learning. Hostile activity is increased at the highest doses of icv CRH (120).

Hypercortisolism is also a component of many psychiatric disorders, among them alcoholism, anorexia nervosa, and depression. The degree of

---

FIGURE 2.8. Mean ($\pm$ SEM) plasma ACTH and cortisol values before and after administration of ovine CRH (1 μg/kg body weight). The left panel shows the results in untrained subjects (UT, n = 5), moderately trained runners (MT, n = 7), and highly trained runners (HT, n = 5). The hormone was administered at time 0 (6 p.m.). The right panel shows the basal and ovine CRH-stimulated (time-integrated stimulation above base line) values for plasma ACTH and cortisol in the three groups of subjects (untrained subjects, open bars; moderately trained subjects, hatched bars). The asterisk indicates p < 0.01 as compared with values in untrained subjects; AUC denotes area under the curve. [Reprinted, by permission of The New England Journal of Medicine (1987); 316:1309–15.]

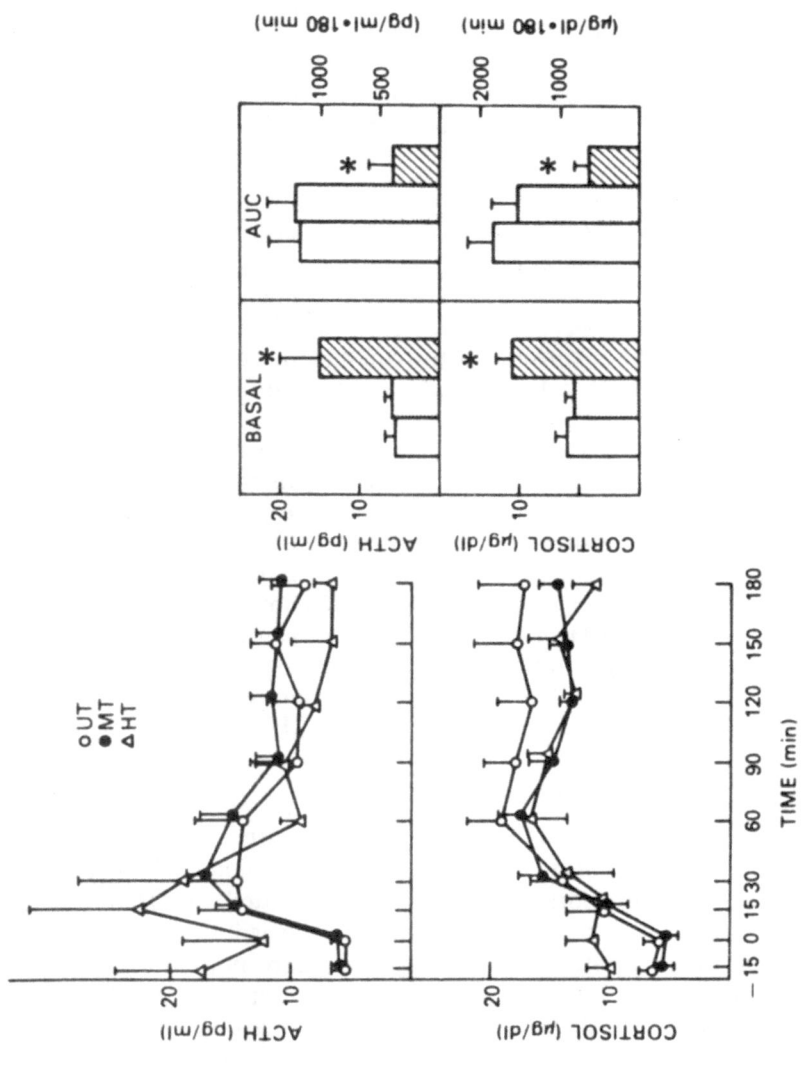

hypercortisolism may be indistinguishable from that of Cushing's syndrome. Thus, depression (123,124) and alcoholism (125) are often referred to as pseudo-Cushing's states. An additional difficulty in the distinction between true and pseudo-Cushing's states is the overlap in clinical findings. Obesity, depression, and hypertension are common in otherwise healthy adults as well as in patients with Cushing's syndrome and may provoke an evaluation for hypercortisolism. Exogenous CRH and measurement of immunoreactive CRH in biologic fluids have been used to investigate the physiology of chronic hypercortisolemic states such as strenuous exercise, depression, and anorexia nervosa.

EXERCISE

Strenuous running (>45 miles/week) increases basal plasma cortisol levels in both men and women without inducing the physical changes of Cushing's syndrome (126,127). The adrenal physiology of male runners was investigated by administering oCRH (64). The subjects were divided into three groups according to the amount of running performed each week. Sedentary individuals did not engage in organized exercise. Running more than 15 but less than 25 miles/week was classed as moderate and running more than 45 miles/week was considered strenuous exercise. The groups had a similar age but differed in maximal oxygen consumption (VO_2 max), body weight and the percent of body fat, with maximal oxygen consumption increasing and body weight and percent fat decreasing with increasing exercise. ACTH and cortisol responses after oCRH (1 mcg/kg, i.v.) administration at 1800 hours are shown in Fig. 2.8. The response to CRH in men running less than 25 miles per week was similar to that of sedentary men. Men running more than 45 miles per week had elevated baseline cortisol levels when compared to controls and a blunted ACTH and cortisol response to CRH.

PRIMARY AFFECTIVE DISORDER

The hormonal responses of drug-free depressed patients to evening administration of oCRH (1 mcg/kg) has been studied (128). Basal plasma ACTH levels were normal but the time-integrated ACTH responses to oCRH were suppressed. Despite this, basal cortisol levels were elevated, and the time-integrated cortisol responses were normal (Fig. 2.9). One explanation of the discrepancy between an attenuated ACTH response and a normal cortisol response is that in depression, endogenous secretion of CRH is excessive and the corticotroph response to CRH is blunted by chronic hypercortisolism. This is supported by a negative correlation between basal cortisol levels and ACTH responses to CRH. CSF concentrations of CRH are elevated in depression (129,130).

In general, the ACTH response to CRH differs between normal individuals, hypercortisolemic depressed patients and patients with Cush-

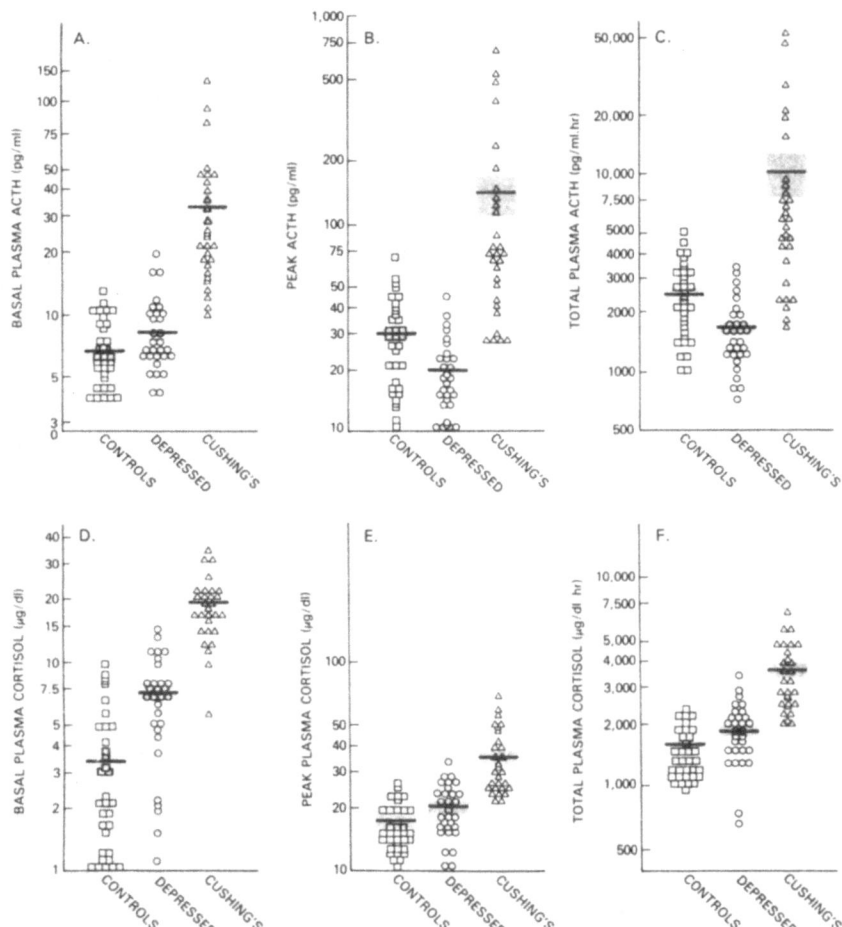

FIGURE 2.9. Individual basal levels of plasma ACTH and cortisol and their responses to ovine CRH in controls, depressed patients, and patients with Cushing's disease. Panel *A* shows basal plasma ACTH; *B* shows peak plasma ACTH response; *C*, total integrated plasma ACTH response to oCRH. Panel *D* shows basal plasma cortisol; *E* peak plasma cortisol response to oCRH; and *F*, the total integrated plasma cortisol response to oCRH. All data are plotted on a logarithmic scale; horizontal dark lines and shading indicate means ± SEM. [Reprinted, by permission of The New England Journal of Medicine (1986); 314:1329–35.]

ing's disease. There is, however, considerable overlap between the lower range of the normal response and the responses of depressed individuals. The upper range of the normal response has significant overlap with the responses seen in Cushing's disease. Thus, while the oCRH test may be a useful adjunct in the differential diagnosis of these disorders, it does not have a high diagnostic accuracy.

ANOREXIA NERVOSA

Underweight patients with anorexia nervosa are hypercortisolemic but not Cushingoid (131,132). The pathophysiology of the hypercortisolism is poorly understood. Gold et al. have recently studied the hormonal responses to oCRH in these patients (133). When chronically under-weight, the women did not show normal suppression of plasma cortisol after midnight administration of dexamethasone. Evening plasma cortisol levels were elevated but the mean plasma ACTH level was normal. Both the ACTH and cortisol responses to oCRH (1 mcg/kg, 2000 hours) were blunted. In contrast, 6 months to 2 years after normal weight was achieved, the hormonal responses to CRH were normal. CSF CRH levels are elevated during the active phase of the disease and decline with refeeding and appropriate weight gain (134,135).

Conclusions

ACTH secretion from the pituitary gland appears to be regulated primarily by the 41-amino-acid peptide corticotropin releasing hormone. Other factors such as vasopressin and glucocorticoids may modulate the effects of endogenous CRH on the pituitary gland.

Synthetic ovine and human CRH have been used to study the physiology of the hypothalamic–pituitary–adrenal axis in health and disease. The intravenous administration of these compounds at doses of up to 1 mcg/kg is without serious side effects. The greatest clinical utility of CRH may be in the differential diagnosis of Cushing's syndrome where it is most useful in the distinction between its ACTH-dependent forms. In this group of patients, the oCRH test will identify those with pituitary microadenomas, and inferior petrosal sinus sampling will improve the outcome of surgery. The oCRH test can also confirm the diagnosis of ACTH-independent forms of Cushing's syndrome. A CT scan of the adrenal glands will locate the lesion in these patients.

CRH promises to be a useful research tool to study the stress response and disturbances of the hypothalamic–pituitary–adrenal axis in other hypercortisolemic, non-Cushingoid states. The role of CRH in the pathophysiology of these states is unresolved, but remains an exciting area of inquiry.

References

1. Harris GW (1948) Neural control of the pituitary gland. Physiol Rev 28:134–79.
2. Saffran M, Schally AV (1955) Stimulation of the release of corticotrophin from the adenohypophysis by a neurohypophyseal factor. Endocrinology 57:439–44.
3. Guillemin R, Rosenberg B (1955) Humoral hypothalamic control of anterior pituitary: study with combined tissue cultures. Endocrinology 57:599–607.
4. Antoni FA (1986) Hypothalamic control of adrenocorticotropin secretion: advances since the discovery of 41-residue corticotropin-releasing factor. Endocrinol Rev 7:351–78.
5. Rivier CL, Plotsky PM (1986) Mediation by corticotropin-releasing factor (CRF) of adenohypophysial hormone secretion. Ann Rev Physiol 48:475–94.
6. Vale W, Rivier C, Brown MR et al (1983) Chemical and biochemical characterization of corticotropin-releasing factor. Rec Prog Horm Res 39:245–70.
7. Gillies G, Grossman A (1985) The CRFs and their control: chemistry, physiology and clinical implication. Clin Endo Metab 14:821–43.
8. Gillies G, Lowry P (1979) Corticotropin-releasing factor may be modulated by vasopressin. Nature 278:463–4.
9. Gillies G, VanWimersma Greidanus TB, Lowry PJ (1978) Characterization of rat median stalk eminence vasopressin and its involvement in adreno-corticotropin release. Endocrinology 103:528–34.
10. Zimmerman, EA, Stillman MA, Recht LD et al (1977) Vasopressin and corticotropin-releasing factor. An axonal pathway to portal capillaries in the zona externa of the median eminence containing vasopressin and its interaction with corticoids. Ann NY Acad Sci 297:405–19.
11. Holmes MC, Antoni FA, Catt KJ et al (1986) Predominant release of vasopressin vs. corticotropin-releasing factor from the isolated median eminence after adrenalectomy. Neuroendocrinology 43:245–51.
12. Kovacs K, Kiss JZ, Makara GB (1986) Glucocorticoid implants around the hypothalamic paraventricular nucleus prevent the increase of corticotropin-releasing factor and arginine vasopressin immunostaining induced by adrenalectomy. Neuroendocrinology 44:229–34.
13. Gillies GE, Linton EA, Lowry PJ (1982) Corticotropin-releasing activity of the new CRF is potentiated several times by vasopressin. Nature 299:355–7.
14. Vale W, Speiss J, Rivier C et al (1981) Characterization of a 41-residue ovine hypothalamic peptide that stimulates secretion of corticotropin and β-endorphin. Science 213:1394–7.
15. Schally AV, Chang RC, Arimura A et al (1981) High molecular weight peptide with corticotropin-releasing factor activity from porcine hypothalami. Proc Natl Acad Sci USA 78:5197–201.
16. Rivier J, Spiess J, Vale W (1983) Characterization of rat hypothalamic corticotropin-releasing factor. Proc Natl Acad Sci USA 80:4851–5.
17. Furutani Y, Morimoto Y, Shibahara S et al (1983) Cloning and sequence analysis of cDNA for ovine corticotropin-releasing factor precursor. Nature 301:537–40.
18. Shibahara S, Morimoto Y, Furutani Y et al (1983) Isolation and sequence

analysis of the human corticotrophin-releasing factor precursor gene. EMBO J 2:775–9.

19. Guillemin R (1964) Hypothalmic factors releasing pituitary hormones. Rec Prog Horm Res 20:89–130.
20. Buckingham JC (1985) Two distinct corticotrophin-releasing activities of vasopressin. Br J Pharmacol 84:213–9.
21. Vale W, Vaughan J, Smith M et al (1983) Effects of synthetic ovine corticotropin-releasing factor, glucocorticoids, catecholamines, neurohypophyseal peptides and other substances on cultured corticotropic cells. Endocrinology 113:1121–31.
22. Gibbs DM, Vale W, Rivier J et al (1984) Oxytocin potentiates the ACTH-releasing effect of CRF (41) but not vasopressin. Life Sci 34:2245–9.
23. Knepel W, Homolka L, Vlaskovska M et al (1984) Stimulation of adrenocorticotropin/β-endorphin release by synthetic ovine corticotropin-releasing factor in vitro: enhancement by various vasopressin analogs. Neuroendocrinology 38:344–50.
24. Pearlmutter AF, Rapino E, Saffran M (1974) A semi-automated in vitro assay for CRF: activities of peptides related to oxytocin and vasopressin. Neuroendocrinology 15:106–19.
25. Legros JJ, Chiodera P, Demey-Ponsart E (1982) Inhibitory influence of exogenous oxytocin on adrenocorticotropin secretion in normal human subjects. J Clin Endocrinol Metab 55:1035–9.
26. Lewis DA, Sherman BM (1985) Oxytocin does not influence adrenocorticotropin secretion in man. J Clin Endocrinol Metab 60:53–6.
27. Mezey E, Reisine TD, Brownstein MJ et al (1984) Beta-adrenergic mechanism of insulin-induced adrenocorticotropin release from the anterior pituitary. Science 226:1085–7.
28. Cryer PE, Gerich JE, Dallman MF et al (1986) The sympathochromaffin system and the pituitary-adrenocortical response to hypoglycemia. Science 231:501–2.
29. Tilders FJH, Berkenbosch F (1986) CRF and catecholamines: their place in the central and peripheral regulation of the stress response. Acta Endocrinol 276(Supp):63–75.
30. Al-Damjuli S, Tomlin S, Perry L et al (1985) Alpha-adrenergic stimulation of ACTH secretion by a specific mechanism in man. J Endocrinol 104 (Supp):40.
31. Al-Damjuli S, Cunnah D, Grossman A et al (1986) Circulating adrenaline does not enhance basal or CRF-stimulated ACTH secretion in man. J Endocrinol 108 (Supp): 139.
32. Aguilera G, Harwood JP, Wilson JX et al (1983) Mechanisms of action of corticotropin-releasing factor and other regulators of corticotropin release in rat pituitary cells. J Biol Chem 258:8039–45.
33. Capponi AM, Favrod-Coune CA, Gaillard RC et al (1982) Binding and activation properties of angiotensin II in dispersed rat anterior pituitary cells. Endocrinology 110:1043–5.
34. Reisine T, Jensen R (1986) Cholecystokinin-8 stimulates adrenocorticotropin release from anterior pituitary cells. J Pharmacol Exp Ther 236:621–6.
35. Westendorf JM, Phillips MA, Schonbrunn A (1983) Vasoactive intestinal

peptide stimulates hormone release from corticotropic cells in culture. Endocrinology 112:550–7.

36. Hale AC, Price J, Ackland JF et al (1984) Corticotrophin-releasing factor-mediated adrenocorticotropin release from rat anterior pituitary cells is potentiated by C-terminal gastrin-releasing peptide. J Endocrinol 102:R1.

37. Horvath J, Ertl T, Schally AV (1986) Effect of atrial natriuretic peptide on gonadotropin release in superfused pituitary cells. Proc Natl Acad Sci USA 83:3444–6.

38. Spinedi E, Negro-Vilar A (1983) Serotonin and adrenocorticotropin (ACTH) release: direct effects at the anterior pituitary level and potentiation of arginine vasopressin-induced ACTH release. Endocrinology 112:1217–23.

39. Anderson RA, Mitchell R (1986) Effects of gamma aminobutyric acid receptor agonists on the secretion of growth hormone, luteinizing hormone, adrenocorticotrophic hormone and thyroid-stimulating hormone from the rat pituitary gland. J Endocrinol 108:1–8.

40. Vlaskovska M, Hertting G, Knepel W (1984) Adrenocorticotropin and κ-endorphin release from rat adenohypophysis in vitro: inhibition by prostaglandin formed locally in response to vasopressin and corticotropin-releasing factor. Endocrinology 115:895–903.

41. Merchenthaler I, Vigh S, Petrusz P et al (1983) The paraventricular-infundibular corticotropin-releasing factor (CRF) pathway as revealed by immunocytochemistry in long-term hypophysectomized or adrenalectomized rats. Regul Peptides 5:295–306.

42. Brisson J-L, Clavequin M-C, Fellmann D et al (1985) Anatomical and ontogenetic studies of the human paraventriculo-infundibular corticoliberin system. Neuroscience 14:1077–90.

43. Bugnon C, Fellmann D, Gouget A et al (1984) Corticoliberin neurons: cytophysiology, phylogeny and ontogeny. J Steroid Biochem 20:183–95.

44. Swanson LW, Sawchenko PE, Rivier J et al (1983) Organization of ovine corticotropin-releasing factor immunoreactive cells and fibers in the rat brain: an immunohistochemical study. Neuroendocrinology 36:165–86.

45. Bruhn TO, Plotsky PM, Vale WW (1984) Effect of paraventricular lesions on corticotropin-releasing factor-like immunoreactivity in the stalk median eminence: studies on the adrenocorticotropin response to ether stress and exogenous CRF. Endocrinology 114:57–62.

46. Sawchenko PE, Swanson LW, Vale WW (1984) Corticotropin-releasing factor: co-expression within distinct subsets of oxytocin-, vasopressin-, and neurotensin-immunoreactive neurons in the hypothalamus of the male rat. J Neurosci 4:1118–29.

47. Zimmerman EA, Stillman MA, Recht LD (1977) Vasopressin and corticoliberin-releasing factor (CRF): an axonal pathway to portal capillaries in the zona externa of the median eminence containing vasopressin and its interaction with adrenal steroids. Ann NY Acad Sci 297:405–19.

48. Kiss JZ, Mezey E, Skirboll L (1984) Corticotropin-releasing factor immunoreactive neurons of the paraventricular necleus become vasopressin positive after adrenalectomy. Proc Natl Acad Sci USA 81:1854–8.

49. Sawchenko PE, Swanson LW, Vale WW (1984) Co-expression of corticotropin-releasing factor and vasopressin immunoreactivity in parvocellular

neurosecretory neurons of the adrenalectomized rat. Proc Natl Acad Sci USA 81:1883–7.

50. Whitnall MH, Mezey, E, Gainer H (1985) Co-localization of corticotropin-releasing factor and vasopressin in median eminence neurosecretory vesicles. Nature 317:248–50.

51. Young III WS, Mezey E, Siegel RE (1986) Quantitative in situ hybridization histochemistry reveals increased levels of corticotropin-releasing factor mRNA after adrenalectomy in rats. Neurosci Lett 70:198–203.

52. Wolfson B, Manning RW, Davis LG et al (1985) Co-localization of corticotropin-releasing factor and vasopressin mRNA in neurons after adrenalectomy. Nature 315:59–61.

53. Olschowka JA, O'Donohue TL, Mueller GP et al (1982) The distribution of corticotropin-releasing factor-like immunoreactive neurons in rat brain. Peptides (Fayetteville) 3:995–1015.

54. Suda T, Tomori N, Tozawa F et al (1984) Distribution and characterization of immunoreactive corticotropin-releasing factor in human tissues. J Clin Endocrinol Metab 55:861–6.

55. Linton EA, Lowry PJ (1985) Development of an immunoradiometric assay for CRF-41: measurement of CRF-41 in hypothalamic extracts and plasma. J Endocrinol Invest 8 (S3) Abstr OC2, p 7.

56. Suda T, Tomori N, Yajima F et al (1985) Immunoreactive corticotropin-releasing factor in human plasma. J Clin Invest 76:2026–9.

57. Suda T, Tomori N, Tozawa F et al (1984) Immunoreactive corticotropin and corticotropin-releasing factor in human hypothalamus, adrenal, lung cancer, and phaeochromocytoma, J Clin Endocrinol Metab 58:919–24.

58. Nieuwenhuijzen Kruseman AC, Linton EA, Lowry PJ et al (1982) Corticotropin-releasing factor immunoreactivity in human gastrointestinal tract. Lancet 2:1245–6.

59. Carey RM, Varma SK, Drake CR et al (1984) Ectopic secretion of corticotropin-releasing factor as a cause of Cushing's syndrome: a clinical, morphological and biochemical study. N Engl J Med 311:13–20.

60. Jessop DS, Cunnah D, Millar JGB et al (1987) A phaeochromocytoma presenting with Cushing's syndrome associated with increased concentrations of circulating corticotrophin-releasing factor. J Endocrinol 113:133–38.

61. Charlton BG, Leake A, Ferrier IN et al (1986) Corticotropin-releasing factor in plasma of depressed patients and controls. Lancet 1:161–2.

62. Cunnah D, Jessop DS, Besser GM et al (1986) Measurement of circulating CRH in man. J Endocrinol 113:123–31.

63. Udelsman R, Norton JA, Jelenich SE et al (1987) Responses of the hypothalamic–pituitary–adrenal and renin-angiotensin axes and the sympathetic system during controlled surgical and anesthetic stress. J Clin Endocrinol Metab 64:986–94.

64. Luger A, Deuster PA, Kyle SB et al (1987) Acute hypothalamic-pituitary-adrenal responses to the stress of treadmill exercise; physiologic adaptations to physical training. N Engl J Med 316:1309–15.

65. Brodish A (1977) Tissue corticotropin-releasing factors. Fed Proc 36:2088–93.

66. Sasaki A, Shinkawa O, Margioris AN et al (1987) Immunoreactive cortico-

tropin-releasing hormone in human plasma during pregnancy, labor and delivery. J Clin Endocrinol Metab 64:224–9.

67. Linton EA, McLean KC, Nieuwenhuijzen-Kruseman AC et al (1987) Direct measurement of human plasma corticotropin-releasing hormone by a "two-site" immunoradiometric assay. J Clin Endocrinol Metab 64:1047–53.

68. Campbell EA, Linton EA, Wolfe CDA et al (1987) Plasma corticotropin-releasing hormone concentrations during pregnancy and parturition. J Clin Endocrinol Metab 64:1054–9.

69. Goland RS, Wardlow SL, Stark RI et al (1987) Biological activity of corticotropin-releasing hormone in human maternal and fetal plasma during pregnancy. Endocrinol Suppl 120:271.

70. Margioris AN, Grino M, Chrousos GP (1987) Perifused human placenta fragments release immunoreactive corticotropin-releasing factor. Endocrinol Suppl 120:246.

71. Gibbs DM (1985) Measurement of hypothalamic corticotropin-releasing factors in hypophysial portal blood. Fed Proc 44:203–6.

72. Suda T, Tozawa F, Mouri T et al (1983) Presence of immunoreactive corticotropin-releasing factor in human cerebrospinal fluid. J Clin Endocrinol Metab 57:225–6.

73. DeSouza EB, Kuhar MJ (1986) Corticotropin-releasing factor receptors in the pituitary gland and central nervous system: methods and overview. Methods Enzymol 124:560–90.

74. Millan M, Aguilera G, Wynn PC et al (1986) Autoradiographic localization of brain receptors for peptide hormones: angiotensin II, CRF and gonadotropin-releasing hormone. Methods Enzymol 124:590–615.

75. Wynn PC, Hauger RL, Homes MC et al (1984) Brain and pituitary receptors for corticotropin-releasing factor: localization and differential regulation after adrenalectomy. Peptides (Fayetteville) 5:1077–84.

76. Udelsman R, Harwood JP, Millan MA et al (1986) Functional corticotropin-releasing factor receptors in the primate peripheral sympathetic nervous system. Nature 319:147–50.

77. Orth DN, Jackson RV, DeCherney GS et al (1983) Effect of synthetic ovine corticotropin-releasing factor. J Clin Invest 71:587–95.

78. Grossman A, Nieuwenhuyzen-Kruseman AC, Perry L et al (1982) New hypothalamic hormone, corticotropin-releasing factor, specifically stimulates the release of adrenocorticotropic hormone and cortisol in man. Lancet 1:921–2.

79. DeBold CR, Sheldon Jr WR, DeCherney GS et al (1985) Effect of subcutaneous and intranasal administration of ovine corticotropin-releasing hormone in man: comparison with intravenous administration. J Clin Endocrinol Metab 60:836–40.

80. Jackson RV, DeCherney GS, DeBold CR et al (1984) Synthetic ovine corticotropin-releasing hormone: simultaneous release of proopiomelanocortin peptides in man. J Clin Endocrinol Metab 58:740–3.

81. McLoughlin L, Tomlin S, Grossman A et al (1984) CRF-41 stimulates the release of β-lipotrophin and β-endorphin in normal human subjects. Neuroendocrinology 38:282–4.

82. Motomatsu T, Takahashi H, Ibayashi H et al (1984) Human plasma

42   Lynnette K. Nieman, D. Lynn Loriaux, and George P. Chrousos

proopiomelanocortin N-terminal peptide and adrenocorticotropin: circadian rhythm, dexamethasone suppression and corticotropin-releasing hormone stimulation. J Clin Endocrinol Metab 59:495–8.

83. Schulte HM, Chrousos GP, Oldfield EH et al (1985) Ovine corticotropin-releasing factor administration in normal men. Hormone Res 21:69–74.

84. MacCannell KL, Lederis K, Hamilton PL et al (1982) Amunine (ovine CRF), urotensin I and sauvagine, three structurally-related peptides, produce selective dilation of the mesenteric circulation. Pharmacology 25:116–20.

85. Udelsman R, Gallucci WT, Bacher J et al (1986) Hemodynamic effects of corticotropin-releasing hormone in the anesthetized cynomolgus monkey. Peptides 7:465–71.

86. Hermus ARMM, Pieters GFFM, Smals AGH et al (1984) Plasma adrenocorticotropin, cortisol and aldosterone responses to corticotropin-releasing factor: modulatory effect of basal cortisol levels. J Clin Endocrinol Metab 58:187–91.

87. Schurmeyer TH, Avgerinos PC, Gold PW et al (1984) Human corticotropin-releasing factor (hCRF) in man: pharmacokinetic properties and dose-response of plasma ACTH and cortisol secretion. J Clin Endocrinol Metab 59:1103–8.

88. Hermus ARMM, Pieters GFFM, Pesman GJ et al (1984) Differential effects of corticotropin-releasing factor in human subjects. Clin Endocrinol 21:589–95.

89. DeBold CR, Sheldon WR, DeCherney GS et al (1984) Arginine vasopressin potentiates adrenocorticotropin release induced by ovine corticotropin-releasing factor. J Clin Invest 73:533–8.

90. Liu JH, Muse K, Contreras P et al (1983) Augmentation of ACTH-releasing activity of synthetic corticotropin-releasing factor (CRF) by vasopressin in women. J Clin Endocrinol Metab 57:1087–9.

91. Lamberts SWJ, Verleun T, Oosterom R et al (1984) Corticotropin-releasing factor (ovine) and vasopressin exert a synergistic effect on adrenocorticotropin release in man. J Clin Endocrinol Metab 58:298–303.

92. Rittmaster RS, Cutler GB Jr, Gold PW (1987) The relationship of saline-induced changes in vasopressin secretion to basal and corticotropin-releasing hormone-stimulated adrenocorticotropin and cortisol secretion in man. J Clin Endocrinol Metab 64:371–6.

93. Conaglen JV, Donald RA, Espinir EA et al (1985) Effect of naloxone on the hormone response to CRF in normal man. Endocrinol Res 11:39–44.

94. Rittmaster RS, Cutler GB Jr, Sobel DO et al (1985) Morphine inhibits the pituitary-adrenal response to ovine corticotropin-releasing hormone in normal subjects. J Clin Endocrinol Metab 60:891–5.

95. Allolio B, Deuss U, Kavlen D et al (1986) FK 33-824, a met-enkephalin analog, blocks corticotropin-releasing hormone-induced adrenocorticotropin secretion in normal subjects but not in patients with Cushing's disease. J Clin Endocrinol Metab 63:1427–31.

96. Nieman LK, Chrousos GP, Oldfield EH et al (1986) The ovine corticotropin-releasing hormone test and the dexamethasone suppression test in the differential diagnosis of Cushing's syndrome. Ann Intern Med 105:862–7.

97. Orth D, DeCherney G, DeBold C et al (1984) Responses to corticotropin-

releasing hormone (CRH) in Cushing's and Nelson's syndromes [Abstract] Excepta Med Int Congr Ser 162:1158.

98. Hotta MN, Shibasaki T, Suda T et al (1985) The use of the corticotropin-releasing hormone test to monitor recovery of patients with Cushing's disease or Cushing's syndrome due to adrenal adenoma after adenomectomy. Endocrinol Jpn 32:113–25.

99. Avgerinos PC, Nieman LK, Chrousos GP et al (1987) The ovine corticotropin-releasing hormone test in the postoperative evaluation of patients with Cushing's syndrome. J Clin Endocrinol Metab 65:906–13.

100. Muller OA, Hartwimmer J, Hauer A et al (1986) Corticotropin-releasing factor (CRF): stimulation in normal controls and in patients with Cushing's syndrome. Psychoneuroendocrinology 11:49–60.

101. Schuermeyer TH, Tsokos GC, Avgerinos PC et al (1985) Pituitary-adrenal responsiveness to corticotropin-releasing hormone in patients receiving chronic, alternate day glucocorticoid therapy. J Clin Endocrinol Metab 61:22–7.

102. Avgerinos PC, Cutler GB Jr, Tsokos GC et al (1987) Dissocation between cortisol and adrenal androgen secretion in patients receiving alternate day prednisone therapy. J Clin Endocrinol Metab 65:24–9.

103. Schulte HM, Chrousos GP, Gold PW et al (1985) Continuous administration of synthetic ovine corticotropin-releasing factor in man. J Clin Invest 75:1781–5.

104. Chrousos GP, Schuermeyer TH, Doppman J et al (1985) Clinical applications of corticotropin-releasing factor. Ann Intern Med 102:344–58.

105. Pieters GFF, Hermus ARM, Smals AGH et al (1983) Responsiveness of the hypophyseal-adrenocortical axis to corticotropin-releasing factor in pituitary-dependent Cushing's disease. J Clin Endocrinol Metab 57:513–6.

106. Nakahara M, Shibasaki T, Shizume K et al (1983) Corticotropin-releasing factor test in normal subjects and patients with hypothalamic-pituitary-adrenal disorders. J Clin Endocrinol Metab 57:963–8.

107. Lytras N, Grossman A, Perry L et al (1984) Corticotrophin-releasing factor: responses in normal subjects and patients with disorders of the hypothalamus and pituitary. Clin Endocrinol (Oxf) 20:71–84.

108. Muller OA, Stalla GK, Hartwimmer J et al (1985) Corticotropin releasing factor (CRF): diagnostic implications. Acta Neurochir (Wien) 75:49–59.

109. Orth D (1984) The old and the new in Cushing's syndrome. (editorial) N Engl J Med 310:649–51.

110. Suda T, Kondo M, Totani R et al (1986) Ectopic adrenocorticotropin syndrome caused by lung cancer that responded to corticotropin-releasing hormone. J Clin Endocrinol Metab 63:1047–51.

111. Oldfield EH, Chrousos GP, Schulte HM et al (1985) Preoperative localization of ACTH-secreting microadenomas by bilateral and simultaneous inferior petrosal sinus sampling. N Engl J Med 312:100–3.

112. Kuwayama A, Kageyama N, Nakane T et al (1981) Anterior pituitary function after transsphenoidal selective adenomectomy in patients with Cushing's disease. J Clin Endocrinol Metab 53:165–72.

113. Lamberts SWJ, Klijn JGM, deJong FH et al (1981) The recovery of the hypothalamo-pituitary adrenal axis after transsphenoidal operation in three patients with Cushing's disease. Acta Endocrinol 98:580–5.

114. Schulte HM, Chrousos GP, Avgerinos PC et al (1984) The corticotropin-releasing factor stimulation test: a possible aid in the evaluation of patients with adrenal insufficiency. J Clin Endocrinol Metab 58:1064–7.

115. Avgerinos PC, Schurmeyer TS, Gold PW et al (1985) Pulsatile administration of human corticotropin-releasing hormone in patients with secondary adrenal insufficiency: restoration of the normal cortisol secretory pattern. J Clin Endocrinol Metab 62:816–21.

116. Rock JP, Oldfield EH, Schulte HM et al (1984) Corticotropin-releasing factor administered into ventricular CSF stimulates the pituitary-adrenal axis. Brain Res 323:365–8.

117. Brown MR, Fisher LA, Spiess J et al (1982) Corticotropin-releasing factor: actions on the sympathetic nervous system and metabolism. Endocrinology 111:928–31.

118. Rivier C, Vale W (1984) Influence of corticotropin-releasing factor on reproductive functions in the rat. Endocrinology 114:914–21.

119. Sirinathsinghji DJS, Rees LH, Rivier J et al (1983) Corticotropin-releasing factor is a potent inhibitor of sexual receptivity in the female rat. Nature 305:232–5.

120. Weiss SRB, Post RM, Gold PW et al (1986) Corticotropin-releasing factor-induced seizures and behavior: interaction with amygdala kindling. Brain Res 372:345–51.

121. Britton DR, Koob GF, Rivier J et al (1982) Intraventricular corticotropin-releasing factor enhances behavioral effects of novelty. Life Sci 31:363–7.

122. Sutton RE, Koob GF, LeMoul M et al (1982) Corticotropin-releasing factor produces behavioral activation in rats. Nature 293:331–3.

123. Sachar EJ (1975) Twenty-four hour cortisol secretory patterns in depressed and manic patients. Prog Brain Res 42:81–91.

124. Besser GM, Edwards CRW (1972) Cushing's syndrome. Clin Endocrinol Metab 1:451–90.

125. Smals AGH, Njo KT, Knoben JM et al (1977) Alcohol-induced Cushingoid syndrome. J Roy Coll Phycns 12:36–41.

126. MacConnie SE, Barkan A, Lampman RM et al (1986) Decreased hypothalamic gonadotropin-releasing hormone secretion in male marathon runners. N Engl J Med 315:411–7.

127. Warren MP (1980) The effects of exercise on pubertal progression and reproductive functions: role of endogenous corticotropin-releasing factor. Science 231:607–9.

128. Gold PW, Loriaux DL, Roy A et al (1986) Responses to corticotropin-releasing hormone in the hypercortisolism of depression and Cushing's disease. N Engl J Med 314:1329–35.

129. Roy A, Picker D, Paul SM et al (1986) Cerebrospinal fluid corticotropin-releasing hormone correlates positively with post-dexamethasone plasma cortisol levels in depressed patients. Am J Psychiatry 114:641–5.

130. Nemeroff CB, Widerlov E, Bisette G et al (1984) Elevated concentrations of CSF corticotropin-releasing factor-like immunoreactivity in depressed patients. Science 226:1342–4.

131. Boyar RM, Hellman LD, Roffwarg H et al (1977) Cortisol secretion and metabolism in anorexia nervosa. N Engl J Med 296:190–3.

132. Vigersky RA, Loriaux DL, Andersen AE et al (1976) Anorexia nervosa: behavioral and hypothalamic aspects. Clin Endocrinol Metab 5:517–35.
133. Gold PW, Gwirtsman H, Avgerinos PC et al (1986) Abnormal hypothalamic–pituitary–adrenal function in anorexia nervosa. N Engl J Med 314:1335–42.
134. Hotta M, Shibasaki T, Masuda A et al (1986) The response of plasma adrenocorticotropin and cortisol to corticotropin-releasing hormone (CRH) and cerebrospinal fluid immunoreactive CRH in anorexia nervosa patients. J Clin Endocrinol Metab 62:319–24.
135. Kaye WH, Gwirtsman HE, George DT et al (1987) Elevated cerebrospinal fluid levels of immunoreactive corticotropin-releasing hormone in anorexia nervosa: relation to state of nutrition, adrenal function and intensity of depression. J Clin Endocrinol Metab 64:203–8.

3

The Neurobiology of Anorexia Nervosa: An Auto-Addiction?

MARY ANN MARRAZZI AND ELLIOT D. LUBY

Overview of Anorexia Nervosa

Description and Diagnostic Criteria of the Disorder

Anorexia nervosa (AN) is a disorder in which patients refuse to eat, as a result of a morbid fear of obesity. They believe themselves to be "fat and ugly" despite severe emaciation and an appearance shocking to others. They lose weight through caloric restriction, ritualistic exercising, and abuse of laxatives and diuretics. Some patients become involved in bulimic binging and purging cycles in which they transiently lose control of disciplined dieting but avoid weight gain through vomiting. A related eating disorder is bulimia in which the binge-purge cycle occurs without caloric restriction and normal body weight is maintained. The incidence of eating disorders is highest between the ages of 15 and 30 years and is 10-20 times greater in females than males (1). In women, amenorrhea also occurs. The prevalence is highest in middle class Caucasian families. The patients are typically perfectionist and model children until the onset of the disorder which may be triggered by a spectrum of psychosocial stressors. Prevalence of the disorder has been reported as high as 1 in 200 adolescent girls in British schools (2) and the rate is rising. It can be a lethal disorder. A mortality rate of close to 20% was found in two populations of AN patients over approximately a 20-year period (3,4). Suicide is a prominent factor in mortality.

The diagnostic criteria for AN (5) are:

1. Weight loss of at least 25% of original body weight or to at least 15% below ideal body weight over a 6 month period
2. Distorted body image such as fat, morbid fear of obesity, and distorted attitudes towards food
3. Amenorrhea of at least 3 months duration
4. Physical signs such as lanugo, bradycardia, hypothermia
5. Excessive exercising, vomiting, laxative abuse and/or binging episodes
6. Absence of known physical or psychiatric illness to account for weight loss

The psychiatric differential diagnosis includes bulimia, anorexia as a symptom of depression, bizarre eating patterns as part of a schizophrenic delusion or paranoia, or as part of the behavior of a borderline personality disorder.

7. Usual onset at 10–30 years of age.

Some diagnostic schemes differentiate between patients who just restrict caloric intake (restricting AN) and patients who also have binge–purge cycles (bulimia).

The diagnostic criteria for bulimia are

1. Recurrent episodes of binge eating
2. At least three of the following:
    a. Consumption of high-calorie, easily ingested food during a binge
    b. Termination of a binge by abdominal pain, sleep, or self-induced vomiting
    c. Surreptitious eating during a binge
    d. Repeated attempts to lose weight
    e. Frequent weight fluctuations greater than 10 lbs
3. Awareness of abnormal eating patterns and fear of not being able to stop eating voluntarily (in contrast to the marked denial seen in anorexia nervosa)
4. Depressed mood after binges
5. Absence of physical disorders or anorexia nervosa.

In contrast to AN, body weight is near normal although there are frequent significant weight fluctuations due to alternating periods of binging and fasting. The anorexic patient often denies the problem whereas the bulimic patient recognizes the abnormal eating pattern. The anorectic patients grossly overestimate their body size, whereas the bulimic patients give a fairly accurate assessment.

The behavioral changes in AN are different from those that occur in voluntary starvation. In AN, initiative is high, the mood is elated but often labile, and patients are strong-willed and take pride in their personal appearance. In starvation, there is a lack of initiative, a labile quarrelsome mood, indecisiveness, and deterioration of personal appearance. In AN, preoccupation with thoughts of food continues even after weight gain. In contrast to the fatigue and avoidance of physical activity in starvation, anorectic patients are extremely hyperactive, exercise relentlessly and seem to have inexhaustible energy.

Current treatment involves management of the medical consequences, nutritional support, individual psychotherapy, family and group therapy with supportive, insight oriented, and behavior modification approaches. The nutritional support may include supervised oral refeeding, drugs that treat a presumed underlying depression, drugs that increase food intake even indirectly, and parenteral hyperalimentation.

In this chapter, we will describe the physical characteristics of anorexia nervosa, summarize some of the currently held pathogenetic hypotheses and discuss a novel opioid auto-addiction mechanism which we have proposed.

Physical Manifestations

AN patients exhibit all the general effects of starvation including the wasting of subcutaneous adipose and muscle tissue, hypotension, bradycardia, hypothermia and increased tolerance to heat, acrocyanosis, loss of hair, growth of lanugo, constipation, edema, nocturia and polyuria, and sleep disturbances with early morning awakening. In severe cases, the face develops a triangular appearance with forehead bossing and the sunken cheeks seen in starving children during times of famine. Both starved and anorectic patients may suffer anemia, leukopenia, lymphocytopenia and thrombocytopenia, and low blood urea nitrogen (BUN) and albumin. Plasma glucose remains in the normal fasting range, but potassium, chloride, bicarbonate, and phosphate levels are often low as a result of malnutrition, self-induced vomiting, and the abuse of laxatives and diuretics. These electrolyte disturbances can produce metabolic psychosis, cardiac arrhythmias, and cardic arrest. Aspiration of vomitus can also be lethal. Atrophy of the cerebral cortex and ventricular enlargement have been observed and may be associated with cognitive impairments similar to those found in the dementias (6,7). However, cortical glucose metabolism, measured by positron emission tomography, is normal (8).

Endocrine and Metabolic Function

A number of the endocrine and metabolic abnormalities of AN are indistinguishable from those of starvation and return to normal when body weight is restored. Thyroid function is low. Serum triiodothyronine (T3) is low and thyroxine (T4) is preferentially converted to inactive reverse T3 rather than to active T3. The level of thyroid stimulating hormone (TSH), the response to thyroid releasing hormone (TRH) and the absolute levels of T4, however, remain normal (9–13). The tolerance for glucose tends to decrease, as in diabetes (13,14). AN patients are slightly resistant to insulin but recover from the resulting hypoglycemia more slowly than normal. Insulin binding to erythrocytes is increased when the patient is underweight but normalizes with the restoration of body weight (15). The basal level of growth hormone may be elevated, but its response to some secretagogues, such as dopaminergic agonists and insulin hypoglycemia, is reduced (13,16). Diabetes insipidus may also occur (13,16,17). Vasopressin in the cerebrospinal fluid (CSF) and its

CSF/plasma ratio are increased (18,19). The vasopressin response to hypertonic saline challenge is reduced and erratic (18,19).

Changes in the hypothalamic-pituitary-adrenal axis are of particular interest. Cortisol levels in plasma are markedly elevated and escape dexamethasone suppression (13,20). The diurnal rhythm of plasma corticoids is reduced or absent (14). Cushingoid signs of hypercortisolism are not present, perhaps because anorexia patients lack the necessary substrate for lipogenesis and fat deposition, even if stimulated by cortisol (21). Urinary metabolites of cortisol are reduced despite the elevated plasma levels, due to decreases and alterations in cortisol metabolism. The latter is due to the low T3 resulting from starvation. Measurements of corticotropic-releasing factor (CRF) and studies of the adrenocorticotropic hormone (ACTH) response to CRF help localize the defect. Indeed while the level of CRF in the CSF of anorectic patients was found elevated (21,22), the ACTH response was found decreased in underweight anorectic patients (21–23). In two studies, a parallel decrease in the cortisol response to CRF was found, although the degree varied greatly (22,23). In a third study (21), the plasma cortisol response to CRF was high. In the absence of elevated ACTH, the authors inferred an increased response of cortisol to ACTH, perhaps due to a variable degree of adrenal hyperplasia, although the cortisol response to direct ACTH challenge was not measured. These effects were shown in underweight anorectic patients (21–23). Thus, it has been suggested that the hypercortisolism of anorectic patients may be the result of a primary hypothalamic hypersecretion of CRF and of an increased sensitivity of the adrenal cortex to ACTH. The excessive secretion of cortisol, in turn, would inhibit the ACTH response to CRF through an intact hypothalamic feedback mechanism. The negative feedback by cortisol is intact and operating excessively, because of the elevated cortisol, and hence the ACTH response to CRF is blunted. With short-term weight restoration, the basal hypercortisolism is corrected but the abnormal responses to CRF remain (21), whereas with long-term recovery, all these abnormalities are corrected (21,22). Bulimic patients have normal responses to CRF (21). Administration of CRF induces a pattern of 24-hour cortisol secretion and behavioral effects similar to that in AN. Since all these changes in the hypothalamic-pituitary-adrenal axis occur in starvation and return to normal with weight restoration, the authors consider them secondary to the starvation and weight loss of AN and not necessarily part of its etiology. High plasma cortisol in the presence of normal or low plasma ACTH is also found in food-deprived rats (24). Hypercortisolism and escape from dexamethasone suppression also occur in depression and Cushing's syndrome. In AN and depression, the pattern of responses to CRF, the elevation of CRF in CSF and hence the mechanisms of hypercortisolism may be the same. This pattern is in contrast to that seen in patients with Cushing's syndrome, in whom the response of ACTH to

CRF is increased, the levels of ACTH are high, and the elevated cortisol is due to increased ACTH secretion rather than to hyperresponsiveness of the adrenals to ACTH. The relationship of these changes to the endogenous opioids and the auto-addiction model of AN will be discussed in a later section.

Primary or secondary amenorrhea is a cardinal and the initial presenting sign of AN in a significant number of patients. The secretion of reproductive hormones returns to a prepubertal pattern. Luteinizing hormone (LH) and follicle stimulating hormone (FSH) are low resulting in low, noncyclic secretion of the ovarian hormones. The LH response to luteinizing hormone releasing hormone (LHRH) is low and further complicated by low levels and abnormal secretion patterns of LHRH itself (13,25). The FSH response to LHRH is normal or even enhanced, so that the FSH/LH ratio increases (13,25). The response to the hypothalamic hormones is normal and LHRH administration can induce an ovarian cycle, indicating that the changes are at the hypothalamic level. Clomiphene, which increases LH secretion by blocking the estrogen feedback in the hypothalamus, is not effective in underweight anorectic patients, further indicating a hypothalamic defect (13,25). In many cases, amenorrhea is secondary to weight loss below a critical level. In other cases, amenorrhea occurs before this critical level is reached, and/or even after restoration of body weight, i.e. the return of the menstrual cycle and of the normal hormone profile are delayed (1,13,20,25–33). This suggests that in addition to starvation and weight loss, there may be additional causes of the amenorrhea.

In contrast to the decrease in cholesterol levels commonly observed in malnutrition or dieting, the total cholesterol level is elevated in patients with AN (34–38). This is due to an increase of the low density lipoprotein (LDL) fraction, whereas high-density and very-low-density lipoprotein cholesterol levels remain normal (38). This appears to be the only metabolic difference between AN and starvation.

As stated above, a number of these neuroendocrine changes occur during starvation per se. Indeed, in five normal women, three weeks of starvation produced elevated plasma cortisol, escape of dexamethasone suppression, altered thyroid function, and elevated basal and stimulated plasma levels of growth hormone (39,40). The effects of starvation on neurotransmitter and endocrine reproductive function in rats have been reviewed (41).

The number of calories required to gain weight is disproportionately high in patients with AN, especially if nonbulimic (42–47). This increase persists after short-term weight restoration but disappears gradually with long-term recovery (42). The restoration of weight is not linearly related with caloric intake (48), varies greatly among individuals (45) and, even when effective in restoring body weight, the caloric requirements are influenced by the extent of the initial emaciation. Paradoxically, the closer to normal body weight the patient is at the start of refeeding, the greater the caloric requirement (43). These differences in caloric efficiency may be due to the type of tissue formed at the various stages of recovery, to the composition of the diet, and possibly to changes in temperature regulation. Patients who were previously obese seem to gain weight faster than those who were not (44). The significance of these findings for the pathophysiology of starvation and especially for the unique pathophysiology of AN remains to be determined. Although controversial, animal studies do suggest an altered metabolic set point as a result of starvation (49–54).

Hypothalamic Dysfunction

The hypothesis that a dysfunction of the hypothalamus is an etiologic factor in AN is based on the hypothalamic endocrine changes discussed above and on evidence of inadequate temperature regulation.

Indeed, the classic lateral hypothalamic syndrome caused by lesions of the hypothalamus in rats, is characterized by hypophagia and even aphagia requiring forced feeding to prevent death (55). Nevertheless, this syndrome and AN differ in several significant aspects, indicating that hypothalamic dysfunction cannot fully explain AN (55). The lateral hypothalamic syndrome is characterized by four stages: a stage of complete aphagia and adipsia in which animals must be force-fed to survive; a stage of anorexia and adipsia in which animals will ingest a tasty liquid diet but must still be force-fed to survive; a stage in which the animals will consume sufficient food if adequately hydrated and if the diet is highly palatable, and a final stage of recovery in which animals will regulate food intake and maintain normal body weight. In the last stage, they still require a palatable diet and do not respond normally to some regulatory challenges. For example, while caloric dilution, deprivation, and low environmental temperature will cause a normal increase in food intake, insulin- or 2-deoxyglucose-induced hypoglycemia and changes in the hydrational state will not. In contrast, adipsia does not occur in patients with AN, who may actually drink more than a normal amount of water as a result of partial diabetes insipidus or just to reduce the temptation to eat. In addition, patients with AN are extremely hyperactive rather than akinetic and cataleptic and psychomotor retarded as in the lateral hypothalamic rats. Anorexia nervosa patients select low-calorie foods rather than the highly palatable, high-calorie foods selected by rats with the lateral hypothalamic syndrome. Neglect of sensory cues seen in the latter is not seen in AN. Finally, recovery in AN patients does not occur in stages and is not associated with residual regulatory deficits. Thus, the symptoms of hypothalamic dysfunction in AN seem to be the consequence of starvation and severe weight loss rather than their cause, and can be corrected by weight restoration.

Neurotransmitter Involvement

Neurotransmitters, of course, are involved in the central nervous system control of feeding behavior (56–65) and the role of many of them in the etiology of AN has been explored. The most extensively studied neurotransmitter is norepinephrine. Tricyclic antidepressants, drugs which act through catecholaminergic mechanism, produce small increases in food intake and weight gain. Although not specific for AN patients, this side effect is therapeutically useful in such patients. Low levels of endogenous norepinephrine secretion are indicated by low plasma concentration and decreased urinary secretion of norepinephrine metabolites (vanallic acid, normetanephrine, and 3-methoxy-4-hydroxyphenylglycol [MHPG]). These changes are secondary to weight loss and/or the related depressive disorder rather than the cause of AN. Treatment of the depressive symptoms and return to normal body weight normalize the norepinephrine system (66,67). However, CSF norepinephrine levels, as opposed to plasma levels, appear to be normal in underweight and recent weight-restored anorectic patients (19,68). After long-term recovery, CSF and plasma norepinephrine levels, and the urinary excretion of MHPG, the metabolite of norepinephrine that reflects brain metabolism, may be low (19,68,69). The mechanism of these long-term changes in the absence of short-term changes needs to be explored. It is possible that they may be part of a compensation made in long-term recovery or that the short-term changes may be masked by the malnutrition. As expected from the low norepinephrine levels, α-adrenergic receptor binding is increased in underweight anorectics but returns to normal with weight restoration (70). Decreases in CSF levels of 5-hydroxy-indoleacetic acid (HIAA), the serotonin metabolite, and of homovanillic acid (HVA), the dopamine metabolite, were found in underweight anorectic patients and returned to normal with weight recovery (19,68). A further abnormality of serotonin metabolism was an increased accumulation of HIAA in the CSF after a probenecid block of transport. The accumulation was greater in the weight-recovered non-bulimic anorectics than in weight-recovered bulimic patients or in control individuals (when corrected for basal differences in the CSF levels). Since these were weight-recovered patients, differences in the degree of emaciation and heme-concentration cannot account for this effect (71). The probenecid test did not reveal abnormalities in dopamine metabolism (71), despite the decreased absolute CSF level of HVA (68). Decreased urinary metabolites of dopamine (66,72) and serotonin (72) in underweight anorectic patients have also been found but have not been analyzed with respect to nutritional status or the depressive symptoms. The dopaminergic system has also been implicated, although pharmacologic studies have yielded conflicting results (73–77). Additional uncertainty derives from the fact that norepinephrine, dopamine, gamma-aminobutyric acid (GABA), and drugs that modulate their action can either increase or decrease food intake depending on the site of action, i.e., the brain region and the receptor subtype involved in a particular response.

The Auto-Addiction Opioid Hypothesis

Statement of Hypothesis

We have proposed that anorexia nervosa is an addiction to dieting mediated by endogenous opioids (78). According to this hypothesis, endogenous opioids, released during an initial period of dieting, produce a positively reinforcing sense of elation eventually leading to an addiction to dieting. Evidence in support of the hypothesis consists of:

1. the addictive characteristics of the clinical behavior
2. changes in endogenous opioid levels in anorexia nervosa
3. changes in endogenous opioid levels and systems induced by food deprivation
4. relevant opioid actions
5. theoretical considerations regarding the relationship of anorexia nervosa and abnormal opioid function
6. therapeutic benefits derived from the interruption of the addictive cycle with narcotic antagonists.

We will discuss each of these points in the following sections.

Addictive Characteristics of Clinical Behavior

The World Health Organization defines drug dependence as "a state, psychic and sometimes physical, resulting from the interaction between a living organism and a drug characterized by behavioral and other responses that always includes a compulsion to take the drug on a continuous or periodic basis in order to experience its psychic effects and sometimes to avoid discomfort" (79). One needs only substitute "dieting" and "weight loss" for "drug" to realize how the anorexic behavior closely fits the WHO definition of drug dependence.

Behavior in AN is similar in many respects to that observed in the addictive disorders. The similarities include strong denial, compulsiveness, i.e., the organization of one's life around dieting and excessive exercise, and persistence in spite of the severe social and life-threatening medical contraindications of dieting. Many patients describe feelings of "demonic" possession, powerlessness over the weight loss, and a family history of addiction. Patients become experts in nutrition and know the caloric content of every food on their trays. The devotion to dieting and weight loss continues to the exclusion of virtually all other interests and pursuits much like the alcohol- or heroin-dependent person is preoccupied with the acquisition and use of the addictive substance.

A 15-year-old girl, who had reduced her weight from 140 pounds to 56 pounds, entered the hospital in a state of life-threatening emaciation. Soon after her admission, her heart rate became irregular and an EKG revealed a junctional arrhythmia. Although she was told about her dangerous cardiac disorder, she still refused to eat. Looking at the window, she said, "I want to be as thin as the wire on that screen. Even that isn't thin enough. I want to disappear altogether." After a catheter for parenteral feeding was inserted, she was found exercising in bed, stating that she felt compelled to lose even more weight. Another patient left the hospital, with a catheter in place, to buy laxatives in a pharmacy across the street.

Like the addicted, the anorectic patients use denial as a major defense and frequently refuse hospitalization and treatment until overpowered by physical disease or pressure from the family or an employer. They perceive it in a different logical framework as a necessary, disciplined, or even healthful achievement. Thus, two conflicting ego states may coexist: the patient may candidly state the she knows that she cannot wear clothing which reveals her arms and legs because people appear shocked by her wasted appearance, while a moment later, she will insist that she is not anorexic and that she still has not achieved her ideal body weight. In therapy, anorectic patients will frequently identify others as having the disorder but will deny its existence in themselves. A 60-pound girl described a hundred-pound patient as anorectic while firmly denying the disorder in herself. When her logic was questioned, she became angry and refused to discuss the matter further.

Like addicted patients, anorectic patients stubbornly resist treatment and their symptoms are accepted as an egosyntonic way of life rather than manifestations of a pathologic process. When confronted with the pathologic nature of their dieting, they will frequently sign out of the hospital against medical advice. One young woman, pleasant on admission, became explosively hostile as her denial was challenged. Finally, after one week of treatment, she insisted that she had too many obligations outside the hospital and left against medical advice.

We have often observed heroin addicts speak lovingly of the drug and lapse into heroin reveries while heroin-deprived in a hospital setting. In a similar manner, anorectic women treasure their dieting and weight loss, feeling that nothing else in life provides them with this level of satisfaction or sense of achievement. This sense of elation, although not precisely comparable to an opiate or alcohol "high," does represent a powerful reinforcing affect.

Periodically, uncontrolled binging will occur in AN. Some patients will forestall weight gain by induced vomiting. Others will gain massively in a short period of time. The most dramatic example occurred in a patient who became ravenously hungry and ate everything in sight, even devouring paper plates and plastic knives in a kind of psychic blindness. After a rapid fifteen-pound weight gain, she developed the delusion that she was pregnant and that her female therapist was the father of the child. Such episodes of bulimia suggest a ravenous hunger which many anorectics must suppress almost continuously through their obsessive rituals.

Alcoholics and opiate addicts often describe their powerlessness over the use of these substances. Similarly, anorectic patients sometimes feel possessed by some process or demon over which they have no control. In describing her anorectic state of mind, a 67-pound, 5'6" patient wrote, "I am scared because I don't know what is going to happen. I am angry with anyone who tries to interfere with anything that I have accomplished with my eating. I feel superior to others because I do not have to eat. I feel helpless because the anorectic part of me is so strong (demonic). I am alone because it is my decision whether I want to overcome this or not." Thus, like the addicted, the anorectic patients lose control over their habit and fall under its control.

The parents of anorectic patients are frequently alcoholics: 12–19% of the fathers and 7% of the mothers (80). The similarity of the psychologic aspects of AN and alcoholism has been discussed recently by Szmukler and Tantum (81) who described the AN phenomenon as starvation dependence. On the other hand, phenomena like tolerance and withdrawal have not been described in AN. Indeed, Margules (82) believes that tolerance to endogenous opioids does not occur in chronic starvation, although the intense anxiety noted after weight gain and other psychophysiologic changes in patients during nutritional repletion may reveal subtle evidence of a withdrawal state.

Changes in Endogenous Opioid Levels in Anorexia Nervosa

The opioid activity in the CSF was found increased in patients with AN (83), although only in those who were markedly underweight. Since in these studies total opioid activity was measured by means of a radioreceptor assay, based on the competitive binding of enkephalamide by crude brain membranes, it is impossible to say which opioids (β-endorphin, dynorphin, the enkephalins, the neoendorphins or the neodynorphins) were responsible for this increase. In one study, the β-endorphin level in the CSF of AN patients was found to be normal using a specific radioimmunoassay, but the weight of the patients was not reported (84). Thus, in light of the previous study, the observation is difficult to interpret.

CSF opioids are believed to originate in the hypothalamus, or at least in the brain, rather than in the pituitary (85–90). Pituitary origin would require an unlikely reverse circulation from the pituitary to get into the CSF. Cerebrospinal fluid endorphin levels are not influenced by hypopituitarism. Conversely, brain lesions that lower endorphin levels in brain do not affect the pituitary levels. Brain and pituitary endorphin levels are regulated by different factors and appear to belong to independent pools and to have different roles.

Changes in Endogenous Opioids and Opioid Systems Induced by Food Deprivation

The release of endogenous opioid activity induced by dieting is critical to our auto-addiction hypothesis. In animal studies, food deprivation alters the opioid content of various tissues, as measured by specific radioimmunoassay (Table 3.1), although the changes vary greatly depending on the tissue, the opioid, and the conditions of starvation. Decreases in a particular brain region may of course reflect a release into a body fluid or a decrease in synthesis. As noted above, it is generally believed that the level of opioids in the CSF reflects their production by the brain, whereas the plasma level reflects the production by the pituitary.

The study of Knuth and Friesen (91) is of particular interest because food intake was restricted only to 50% of normal and thus may better resemble the initial dieting that, according to our hypothesis, triggers AN. Furthermore, female animals were used and most of the changes in β-endorphin were observed during the starvation-induced anestrus, a possible model of AN-induced amenorrhea. Two of the studies (93,95) indicate that the starvation-induced changes in opioid levels can be reversed by refeeding. The results obtained in sheep (96) may not be applicable since the animal is a ruminant.

Increases in opioid action, i.e., increases in naloxone-sensitive analgesia (97–100) and hypotension (101), provide additional even if indirect evidence of opioid release during food deprivation lasting as long as four days. Since the effect of food deprivation on analgesia may be biphasic (97,102), it is easy to conceptualize pathophysiologies occurring in opposite directions.

In addition to food restriction, the metabolic glucopenia induced either by blocking glycolysis with 2-deoxyglucose (2DG) or by insulin hypoglycemia (to a degree which would induce feeding) results in an increase in plasma β-endorphin (103,104) and in opioid analgesia (105–111). Moreover, there is evidence that opioids are involved in insulin-induced feeding in animals (112), and that naloxone blocks the increase in food intake induced by 2DG in humans, albeit without reducing the subjective feeling of hunger (113).

The behavioral effects of food restriction are in harmony with the possibility that food restriction induces fluctuations in endogenous opioid activity, as we have previously discussed (78). Indeed, food restriction capable of maintaining rats at 80–90% of their normal body weight alters their behavioral response to addictive drugs and to the opiate antagonists naloxone and naltrexone (114–123).

TABLE 3.1 Changes in Radioimmunoassayable opioids induced
by food deprivation

Food Deprivation Conditions, Animal Species and Sex	Opioid Peptide	Brain Region or Body Fluid	Change*	Ref. #
50% Restriction up to 10 days, Rat, female	β-Endorphin	Various hypothalamic nuclei	+, − or no depending on nucleus	91
Complete, 2–3 days, Rat, male	β-Endorphin	Hypothalamus	−	92
Complete, 2–4 days, Rat, male	β-Endorphin	Pituitary	+	93
		Hypothalamus	nc	
		Striatum	+	
		Plasma	nc	
Complete, 4 days, Rat, male	Dynorphin	Pituitary	nc	
		Hypothalamus	−	
		Striatum	−	
Complete, 4 days, Rat, male	Met-enkephalin	Pituitary	nc	
		Hypothalamus	nc	
		Striatum	−	
		Hippocampus	−	
		Cortex	−	
Complete, 1–3 days, Rat, male	β-Endorphin	Hypothalamus	−	94
		Pituitary	+	
		Gut (duodenum)	nc	
	Dynorphin	Hypothalamus	nc	
		Pituitary	+	
		Gut (duodenum)	+	
	α-Neoendorphin	Hypothalamus	nc	
		Pituitary	+	
		Gut (duodenum)	+	
Complete, 3 days, Rat, male	β-Endorphin	Hypothalamus	nc	95
		Pituitary	nc	
		Plasma	+	
	Met-enkephalin	Hypothalamus	nc	
		Pituitary	nc	
Complete, 4–24 h, Sheep, male	Met-enkephalin	Hypothalamus (basomedial)	+	96
		Amygdala	+	
		Olfactory bulb	−	

* +, increase; −, decrease; nc, no change

Food deprivation also alters the response of the animal to opioids. Thus, morphine increases food intake when rats are fed ad libitum, but decreases it after 24 hours of starvation (124). Similarly, the inhibitory effect of naloxone on food intake decreases as the period of food deprivation increases (125), and naloxone suppression of feeding does not occur if rats are adapted to a restricted feeding schedule (126).

Obviously the evidence herein reviewed suggests, but does not prove, that dieting induces opioid release. Experiments designed to provide additional documentation are in progress in our laboratory and include an attempt to correlate changes in discrete brain regions with changes in CSF and plasma and to correlate animal and patient studies.

Relevant Opioid Actions

Two categories of opioid actions appear useful to the starving animal: stimulation of food intake, to correct the starvation, and, when food is not available, adaptation to starvation and improved survival.

The first response has been documented extensively. Narcotics and the endogenous opioid peptides increase food intake when injected systemically or directly into the CSF or the hypothalamus (65,124,127–142). This action appears to be mediated, at least in rats, by multiple subtypes of opiate receptors including the μ, κ, and σ receptors as demonstrated using a variety of preferential agonists (65,127–129,132–139,141,142). The endogenous opioid peptides dynorphin and β-endorphin have the same effect (132,133,135,136).

The suggestion that the endogenous endorphins mediate homeostatic adaptations to starvation (82) is based on the similarities between the actions of the opiates and the physiologic changes which take place in starvation and in AN. They result in the conservation of energy and increased survival of the individual. These changes are as follows:

1. Constipation; this slows the transit through the GI tract and hence increases the extraction of nutrients and water
2. Water retention and famine edema; this is promoted by release of vasopressin and reduction of the digestive secretions. Interestingly, vasopressin in the CSF is increased in AN (18).
3. Decreased body temperature
4. Decreased release of thyroid hormone and decreased calorigenesis
5. Decreased blood pressure and vascular sympathetic tone
6. Depressed respiration and decreased sensitivity of the respiratory center to carbon dioxide and hypoxia
7. Decreased intensity of emotional reactions (fear and rage)
8. Lethargy, drowsiness and passivity
9. Inhibition of reproductive activities: secretion of FSH and LH are reduced by opioids.

Responses 1 and 2 conserve bodily resources. Responses 3 to 8 decrease the metabolic rate and hence the metabolic need. Response 9 reduces species survival function to those necessary for preservation.

Margules argues that after the initial responses of the sympathetic nervous system and glucagon, the endorphinergic system assumes responsibility for the prolonged adjustment to fuel shortage. Indeed, it is known that starvation-related enduring metabolic changes are not dependent on the sympathetic nervous system.

If the opioids can down-regulate metabolism as well as activate food intake, it is not surprising that opiate agonists can cause anorexia as well as hyperphagia (124,142), and that both responses can be blocked by development of tolerance or by naloxone. In general, body weight changes as expected. However, under certain conditions of food deprivation, naloxone prevents the resulting weight loss, preventing the increase in metabolic efficiency, while morphine inhibits the activation of ingestive behavior.

Some of the known glucoregulatory effects of narcotic drugs and endogenous opioid peptides may participate in the proposed role of the opioid system in the adaptation to starvation. Thus, morphine causes hyperglycemia (82,143,144). Morphine and the opioid peptides modulate the release of glucoregulatory hormones from the pancreas, such as insulin, glucagon, somatostatin, and pancreatic polypeptide (82,144–165), and of other glucoregulatory hormones, such as ACTH and hence cortisol, growth hormone, and thyroid hormone (1,82,144,145,151,160,161,164,167,168,169), although the direction of the changes may vary among these studies, they all indicate endorphin effects on these glucoregulatory hormones. The differences may depend on the experimental conditions and the prevailing metabolic state. For example, opiate blockade produces earlier and/or greater increases in glucose-mobilizing hormones and hence in glucose itself in the "starved" insulin hypoglycemic state than under basal conditions (161). The predominant subtype of opioid receptor (μ, δ, κ, etc.) and the levels of various nutrients (see below) may also be significant variables. Opioid peptides also modulate insulin action on glucose fluxes and may thus serve a function in the redistribution of glucose during stress (166). Opiates increase epinephrine-induced glycogenolysis and gluconeogenesis by stimulating glucagon and ACTH release (82). Lipolysis may also be activated (82,170,171), providing an additional source of energy during starvation.

In addition to these rapid adjustments, food deprivation may also cause chronic changes in the metabolic set points (49–52,54). In particular, there is evidence that the efficiency of food utilization is enhanced after food deprivation, so that body weight can be restored with a lower than expected caloric intake (50,51,52,54). This increased efficiency after food deprivation can also be seen when superimposed on the hyperphagic response to increased palatability, resulting in a greater than normal weight gain for the amount of food intake (51). Under some experimental conditions, body weight is not fully restored despite a return to ad libitum feeding after a prolonged period of food deprivation (49,53). This is particularly true under circumstances of "activity-induced self-starvation" (53). Rats restricted to eating for a limited time of the day initially lose weight, which they recover after adjustment to the new feeding schedule. However, when allowed to run on a wheel, they ignore the food and run themselves to death (53). It seems to us that this phenomenon has striking similarities with AN and that a possible role of the opioids should be considered. Another study suggests that endorphins enhance metabolic efficiency (172). Hyperphagia, obesity, and greater oxygen utilization, induced by improving the palatability of food, are blocked by naloxone, indicating an involvement of endogenous opioids (see section "Endocrine and Metabolic Function").

The opiates inhibit reproductive function by suppressing the release of LH and FSH, and hence the gonadal hormones (105,144,151,167,173–178). Thus, the release of endogenous opioids resulting from starvation or other stimuli may account for the amenorrhea seen in patients with AN that occurs both as a consequence of the weight loss below a critical body mass and independently of it (see section "Endocrine and Metabolic Function"). This hypothesis finds additional indirect support from the observations that the hypothalamic β-endorphin level changes during different stages of the estrous cycle (179) and that there are sex-related differences in the opioid system (180–183). Furthermore, estrogen and β-endorphin receptors have been co-localized in some hypothalamic areas (184) suggesting that opioids may be a factor in the marked dominance of AN among females.

Anorexia nervosa generally begins shortly after puberty, and it is interesting to speculate that it may coincide with a defect in the post-translational processing of pro-opiomelanocortin, leading to altered production of β-endorphin and hence altered regulation of gonadotropin secretion (185). In accord with this hypothesis, plasma β-endorphin reaches adult levels at puberty (186), and pubertal maturation is associated with qualitative and quantitative changes in the opiate regulation of gonadotropins (187,188).

Theoretical Considerations Concerning the Relationship of Anorexia Nervosa and Abnormal Opioid Function

Our auto-addiction model proposes that an initial period of dieting releases endogenous opioids. In the normal individual (Fig. 3.1A), the endogenous opioid release would produce elation, increased food intake, and an adaptive metabolic down-regulation. In turn, the increased food intake would interrupt the cycle. In the anorectic and bulimic patients, these responses would become uncoupled, albeit in different ways. In AN

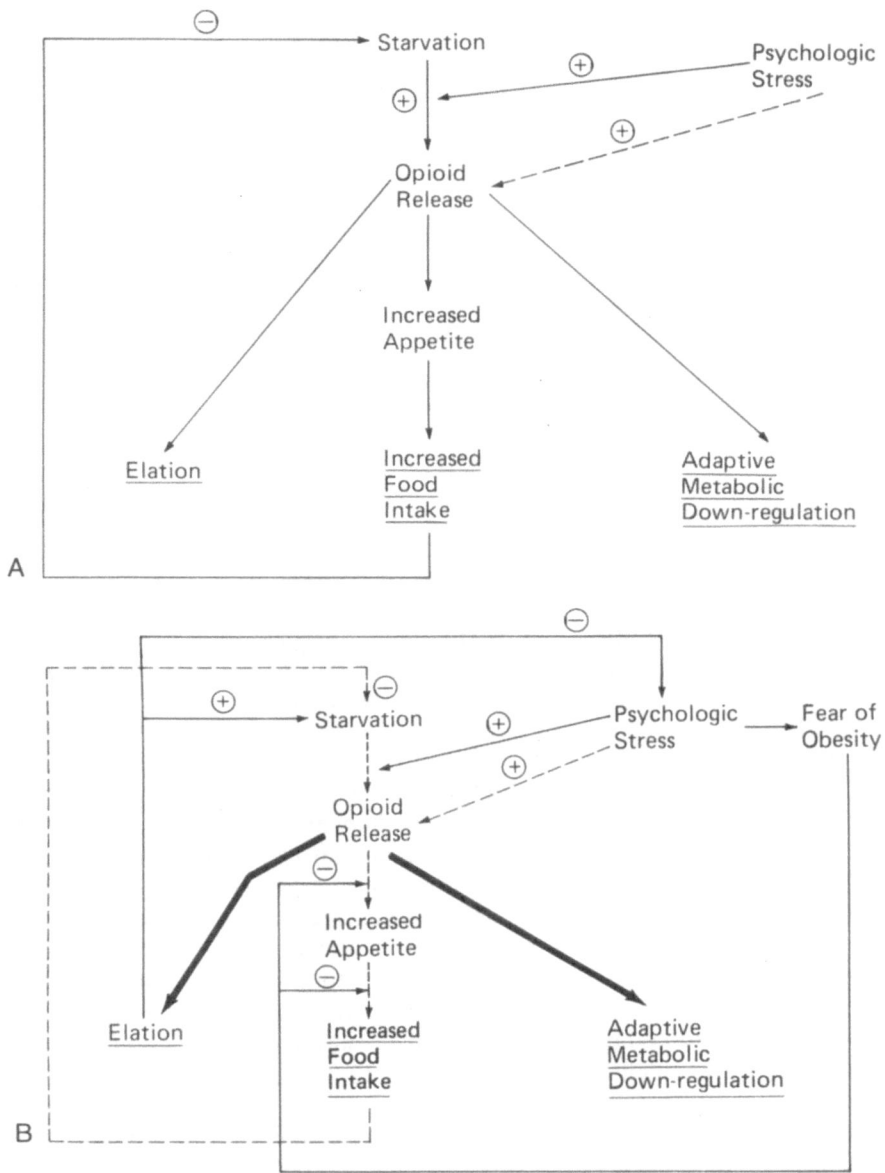

FIGURE 3.1. Effects of opioid release in relation to dieting. *A:* Normal pattern, *B:* Anorexia nervosa. *C:* Bulimia. The thickness of the line suggests the intensity of the response. A broken line indicates the absence or marked reduction in eating disorders of a pathway which is normally present. Inhibitory and stimulatory influences are designated by minus and plus, respectively.

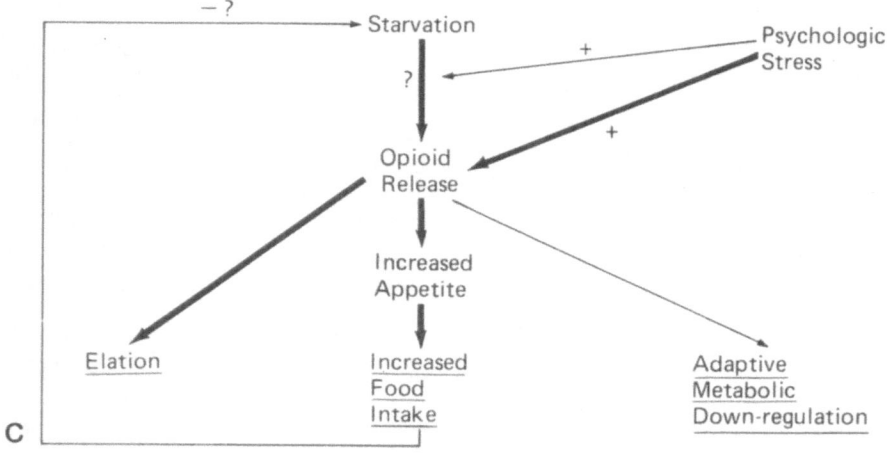

FIGURE 3.1. *Continued*

(Fig. 3.1B), the opioid-induced drive to eat would be decreased relative to the opioid-induced elation and/or the metabolic down-regulation. Starvation and opioid release would continue, and elation would be reinforced leading to an addictive cycle. Either a decrease in the absolute level of the opioid drive to eat or an increase in the other responses could cause the imbalance. The episodes of binge eating that occur in anorectic patients may be due to a periodic breakthrough of the opioid drive to eat. Anorectic patients have suppressed their hunger, which at times overwhelms them. They have the universal fear that any lapse of discipline will result in a ravenous gorging and horrid obesity. The fear of loss of control may relate to excessively high opioid drives rather than unrealistic fears of normal drives. The decrease in the opioid drive to eat could be in the feeding behavior response to appetite with appetite remaining normal.

Bulimia, on the other hand, could represent a markedly elevated opioid drive to eat, reinforced by an opioid-induced elation (Fig. 3.1C) and could represent a "permanent" unsuppressed anorectic binge. Thus, an initial period of starvation or a strong psychologic stress would stimulate the release of endogenous opioids, creating a powerful drive to eat, which is controlled by restriction in the anorectic and by the binge-purge cycle in the bulimic. These diagrams imply that restricting anorectics, binging anorectics and bulimics represent a continuous spectrum of eating disorders. Behaviorally this appears to be the case. Indeed, our preliminary results suggest that narcotic antagonists may be useful in treating both anorectic and bulimic patients (see section "Narcotic Antagonists and Anorexia Nervosa").

The proposed disorder of opioid function may occur as a result of the eating disorder or it may pre-exist and manifests itself when activated by dieting. In either case, the eating disorder would be reinforced, as when the release of opioids is associated with dieting. Alternatively, the defect of opioid function may be associated with, and hence reinforce another form of behavior leading to a different form of psychopathology, as discussed below in section "Atypical Opioid Systems in Relation to Anorexia Nervosa".

The auto-addiction model of AN does not preclude other psychodynamic determinants, which may be responsible for the initial dieting which triggers opioid release and/or may predispose the individual to the need for the positive reinforcement of the opioid-induced elation. Moreover, psychologic stressors may well represent an additional stimulus to opioid release and thus cause it to reach a magnitude that creates the addiction. Psychologic stressors associated with triggering AN include separation from parents, an initial sexual experience, a failure in academic or athletic competition, rape, and the observation of someone dying from a wasting disease.

In our view, AN consists of three-phases: an initial phase of dieting possibly due to (a) one of the above-mentioned psychodynamic factors, (b) a valid need to diet, which goes out of control, or (c) anorexia secondary to another illness; a second phase, during which the anorectic patient develops ritualistic, obsessional behavior organized around dieting and exercising in order to restrict or burn calories; and a final phase, occurring in about fifty percent of AN cases, in which the defect becomes chronic, can no longer be understood solely in psychologic terms, and requires a corollary biologic explanation. Indeed, clinicians have long observed that the longer anorectic patterns exist, the more likely it is that the disorder will become irreversible (189,190). Anorexia nervosa may then be understood as a psychologic process in which dieting will initiate a series of central nervous system changes which perpetuate the disorder and in some instances make it irreversible. It is our hypothesis that the opioid system is of fundamental importance in the biology of the disorder, in its chronicity and intractability, and in the development of its addictive characteristics.

In some respects, this addiction could be thought of as a behavioral and physiologic extension of a conditioned food aversion similar to that obtained experimentally in rats by Bernstein and Borson (191). In some cases, the learned food aversions could be conditioned to disease states. For example, when tumors were implanted in rats, the diet fed while hosting the tumor became aversive. When given a choice of diets which were equally acceptable prior to the tumor implant, the rats avoided the diet previously associated with the presence of the tumor, even after its removal. Such data underline the importance of positive or negative feeling of well-being comparable to the opioid-induced elation which, in our hypothesis can be sufficiently strong to override the metabolic cues and to distort the patient's body image. Patients feel so much better when on a diet-induced high that they believe that this is the way to go.

Physical hyperactivity is a characteristic sign of AN and may represent a conscious effort to lose weight. Exercise is a known stimulus to endorphin secretion. Alternatively, it may represent a physiologic effect that narcotics have under some unusual circumstances (see section II8). Other interactions of hyperactivity and dieting are suggested by the "activity-induced self-starvation" discussed above (53) and by the observation that food deprivation may increase the level of activity (49).

Narcotic Antagonists and Anorexia Nervosa

According to the auto-addiction hypothesis, interruption of the addictive cycle by means of opiate-blocking agents should be beneficial. Indeed, preliminary clinical studies suggest this to be the case (192) and we are currently conducting double-blind controlled clinical trials. The clinical histories of eight patients treated with naltrexone, a long-acting orally active narcotic antagonist, are summarized below and in Fig. 3.2.

Case 1 (Fig. 3.2A) was a 33-year-old white female with a 10-year history of AN during which her weight fluctuated between 70 and 90 lbs. At the time of her visit she had been continuously amenorrheic and had used 10–30 Correctol® laxative pills daily. She had been hospitalized on two prior occasions for treatment of the anorexia. On the fourth day of her hospitalization under our care, she suffered an episode of metabolic encephalopathy with confusion, disorientation, short-term memory impairment and auditory halluciations caused by severe hypophosphatemia. Total parenteral nutrition (TPN) was started at this time. On day 10, after recovery from this episode, naltrexone (25 mg 3 times/day therapy was started, TPN was stopped one week later and naltrexone was continued. During naltrexone treatment, she began to eat on her own, no longer feared the weight gain, and stopped the vomiting and the abuse of laxatives. She became determined to reverse her anorectic emaciation and actively participated in all phases of the treatment program. She gained 20 lbs during five and a half weeks of hospitalization and another 15 lbs as an out-patient on naltrexone, thus reaching her ideal body weight. Two months after discharge, her electrolytes were normal, suggesting that she had not vomited or used laxatives. Shortly thereafter, her menses resumed for the first time in 10 years. The naltrexone was stopped 3 months after discharge. Now, two years later, she is still maintaining her ideal body weight.

Case 2 (Fig. 3.2B) was a 17-year-old white female with a 4-year history of AN who had been previously hospitalized for two 9-month periods. Both times she was discharged when her weight reached 100 lbs but immediately resumed her anorectic behavior and her weight fell back to less than 70 lbs. She had had one menstrual period in her entire life.

Attributing the weight gain to naltrexone, she refused to continue the drug and resumed her pattern of food restriction, weight loss, and excessive exercising. As an out-patient, she is not taking the drug and is not maintaining her weight.

While in these 2 cases TPN may have contributed to the weight gain, it cannot account for the change in anorectic behavior and for the continuing weight gain as in case 1. Discontinuance of TPN normally results in resumption of weight loss.

Case 3 (Fig. 3.2C) was a 21-year-old white female with a 5-year history of AN and three prior hospitalizations. The maximum weight that she ever attained was 80 lbs and her lowest adult weight was 54 lbs. She was

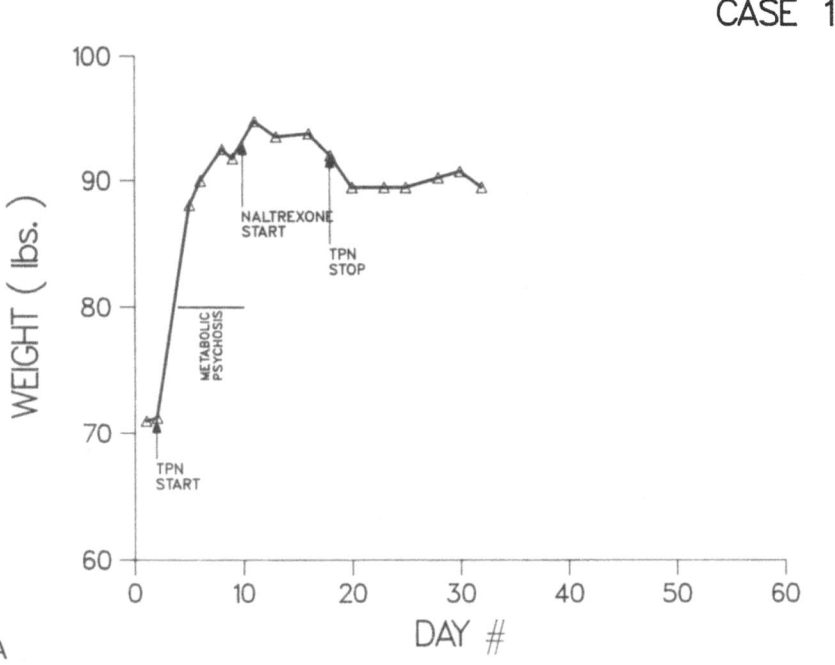

FIGURE 3.2. Effect of naltrexone in patients with anorexia nervosa. Solid lines indicate hospitalization on our unit. Broken lines indicate the out-patient course of naltrexone following the discharge. The extended out-patient course for case 6 is shown separately in Figure 3.2G. For cases 4, 5, and 6, multiple admissions to our unit are shown as multiple curves with different symbols designating the different admissions. It should be noted that in one admission, only naltrexone, without TPN, was used (admission 2 of cases 4 and 5, admission 3 of case 6). In another admission, only TPN without naltrexone was used (admission 1 of cases 4 and 5, admission 2 of case 6). In the third admission of case 5, a combination of naltrexone and TPN were used. In the first admission of case 6, a combination of lithium and Elavil were used. Comparison of the multiple admissions for each patient are very interesting.

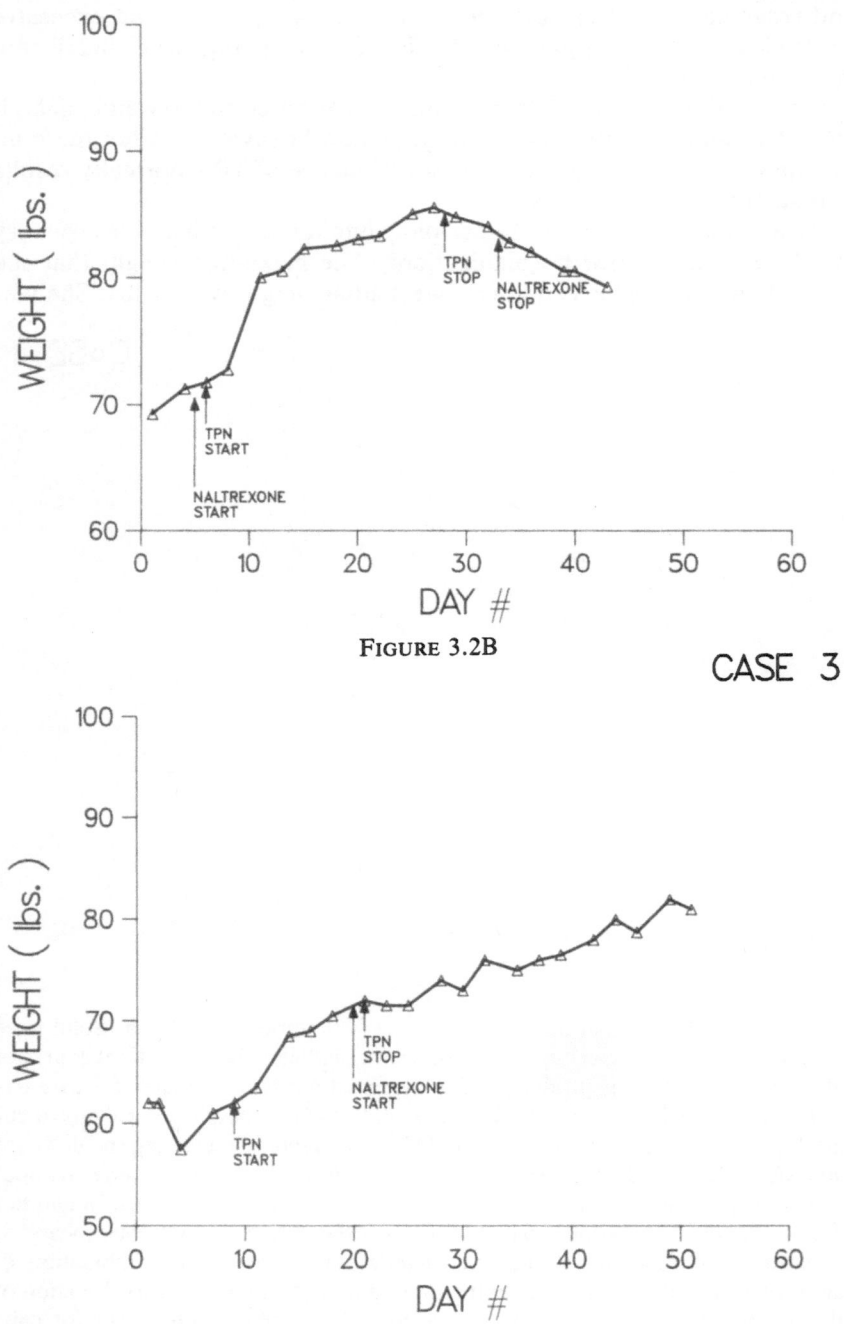

CASE 2

FIGURE 3.2B

CASE 3

FIGURE 3.2C

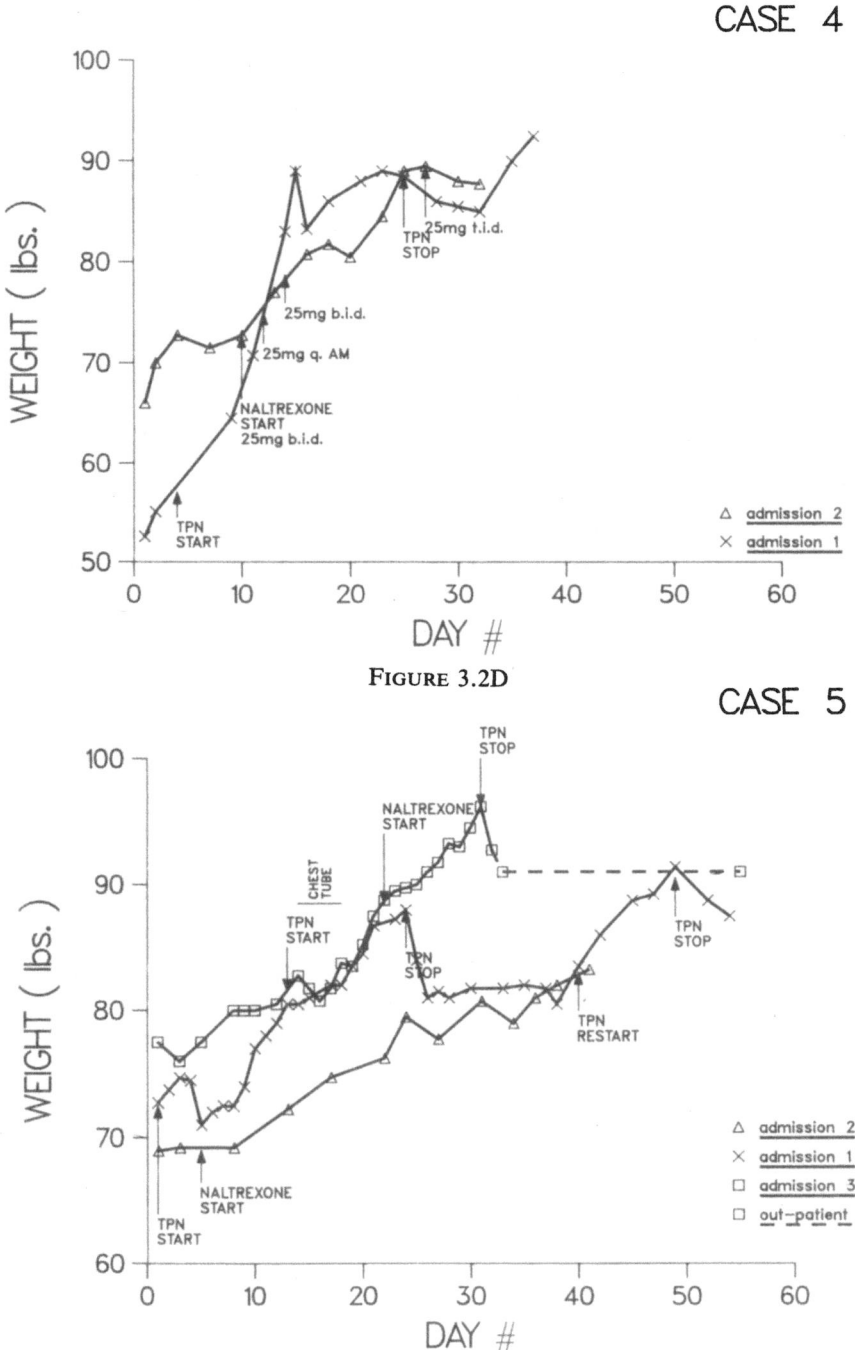

CASE 4

FIGURE 3.2D

CASE 5

FIGURE 3.2E

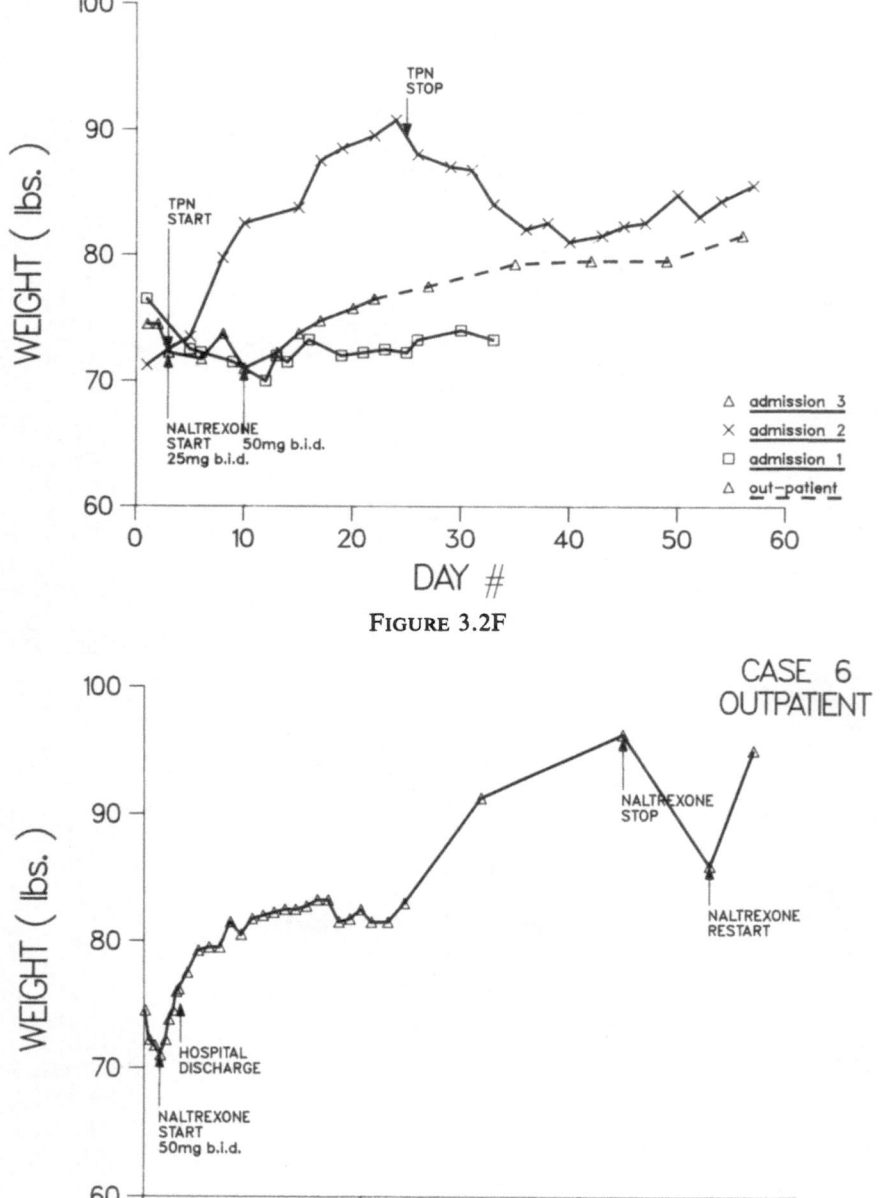

FIGURE 3.2F

FIGURE 3.2G

CASE 7

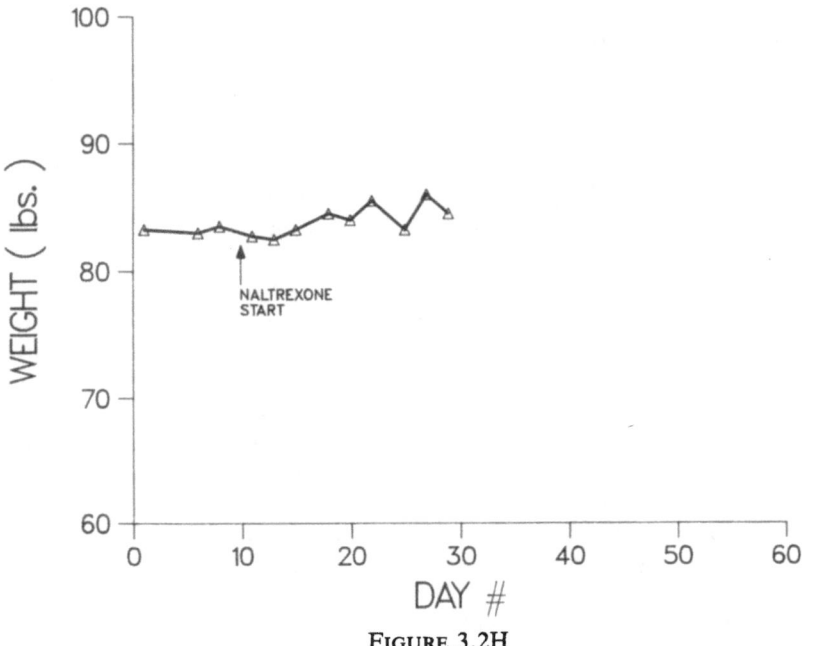

FIGURE 3.2H

CASE 8

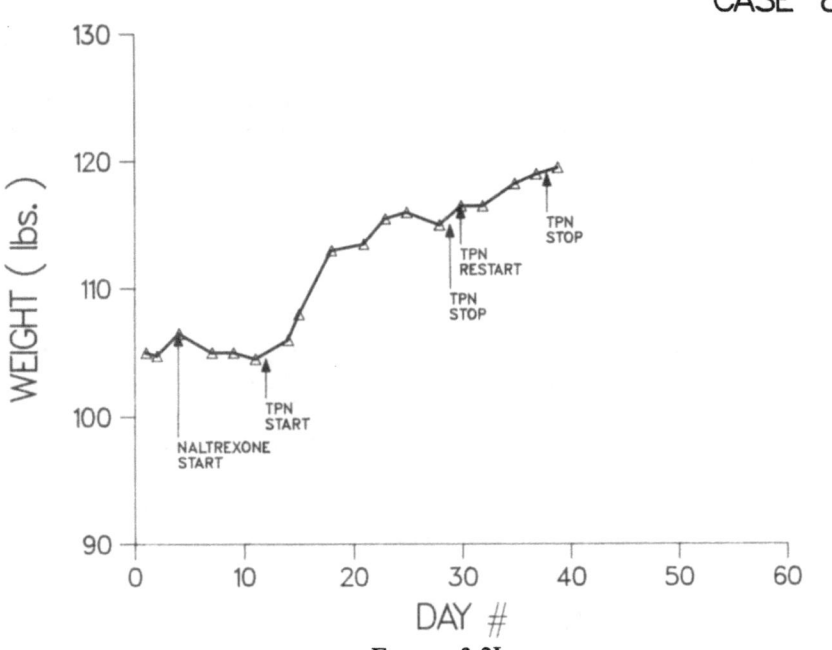

FIGURE 3.2I

transferred to our ward from the forensic center, where she had been remanded for repeated petty thefts of food. While in the hospital she stole food from other patients only to hoard it. She also had a history of depression, including a suicide attempt, and had a borderline personality disorder. She was placed on TPN, but she attempted to injure herself with the TPN apparatus so that it had to be discontinued. She was placed on naltrexone and maintained the same rate of weight gain reaching a weight of 82 lbs, her highest weight ever. Upon return to the forensic center, naltrexone was discontinued, antipsychotic drugs were initiated and she rapidly began to lose weight again.

Cases 4–6 represent multiple admissions to our unit, allowing comparison of treatment with and without naltrexone in the same patient. On one admission, each of the three received TPN without naltrexone. Body weight rose substantially but fell when TPN was stopped; TPN did not alter anorectic behavior. On another admission for each patient, naltrexone without TPN was used and a significant steady weight gain was obtained. This was associated with a significant decrease in anxiety over weight gain and in the obsessive quality of dieting.

Case 4 (Fig. 3.2D) was an 18-year-old white female with a 4-year history of AN and of five prior hospitalizations. On hospital discharge she refused naltrexone and her weight dropped from 88 to 70 lbs.

Case 5 (Fig. 3.2E) was a 22-year-old white female with a 5-year history of AN and of three hospitalizations in our unit over a period of two years. At 69 lbs she still jogged 10 miles per day. She fiercely resisted treatment but the resistance appeared to diminish with naltrexone. Her active participation in psychotherapy was minimal and cannot account for any improvement. As an out-patient, after the first admissions with naltrexone, she did not take the drug as prescribed and did not gain weight above 80 lbs. During the third hospitalization, she reached 96 lbs, her highest weight in five years. She is continuing to take the drug and is now actively participating in out-patient psychotherapy and showing significant improvement.

Case 6 (Fig. 3.2F) was a 36-year-old white female admitted to our unit three times over the past two years. She could not control her anorectic behavior despite the threatened loss of her marriage and of the custody of her three children. Her maximum non-pregnant body weight was 128 lbs. During the first hospitalization, lithium and amitriptyline but no TPN or naltrexone were used and her weight fluctuated without significant improvement. On the second admission, she received naltrexone and achieved a small but steady weight gain. She was discharged at 76 lbs because she seemed unwilling to accept the idea of additional weight gain. However, she continued taking naltrexone (50 mg twice daily) as an out-patient, continued to gain weight despite stressful life events (Fig. 3.2G) and eventually reached 96.5 lbs, the highest in four years, and had premenstrual symptoms. She became frightened of the weight gain, secretly discontinued the naltrexone, immediately lost 10 lbs, and her anorectic fears intensified greatly. On confrontation, she admitted that she had discontinued the naltrexone, accepted resumption of the drug, and her weight returned to 95 lbs. At the time of this writing, eight months later, she is maintaining that weight.

In two cases (7 and 8), no effects of naltrexone were observed. One (Fig. 3.2H) was an 18-year-old white female with a history of anorexia for less than one year. She was our only acutely ill patient, an indication that perhaps, the opioids may be involved only in the chronic phase of AN. The other was our only male patient (Fig. 3.2I). He was 36 years old, weighed 105 lbs on admission, and was known at work as "Dr. Death" because of his severe emaciation. The use of TPN and naltrexone were so intermingled that we could not determine if naltrexone had any effect.

Naltrexone is a competitive antagonist and hence its dose must be titrated against the individual level of endogenous opioids. This may vary greatly among patients. In case 6, improvement was not seen until the dose was increased from 25 to 50 mg bid. Thus, therapeutic failure may be a matter of dosage.

In all, eight patients were treated with naltrexone in addition to TPN, individual psychotherapy, and family therapy. In seven of these eight patients previous treatment had been ineffective and in six the disorder had lasted four to ten years. In six patients who were chronically ill, the drug appeared to have some beneficial effect on weight gain, hyperactivity, and attitude. The obsessive preoccupation with dieting, anxiety about weight gain, resistance to treatment, and positively reinforcing elation decreased. In responsive patients, naltrexone relieved the constipation and abdominal distension, probably through blockade of opioid-mediated mechanisms. Two subjects, cases 1 and 6, continued taking naltrexone as out-patients and gained additional weight. One has maintained her ideal body weight for over two years.

The weight gain was due to increased caloric intake (Fig. 3.3), but whether, in addition, naltrexone increased the efficiency of food utilization remains to be determined. The problem is being investigated in our controlled clinical trial. Comparison of the weight gain and caloric intake in one patient treated sequentially with TPN and naltrexone (Fig. 3.3D) leads one to speculate that TPN and naltrexone may have different effects on metabolic efficiency. However, since this patient tampered extensively with her TPN line, these data must be interpreted with caution. As discussed in a previous section of this review, alterations of metabolic set point have been reported in AN during both the malnourished and the short-term weight-recovery phases. Accordingly, any change observed after opiate blockade may reflect either a direct effect of the drug or a change in the pathophysiology of AN.

A few years ago another group of investigators (193,194) reported that naloxone, given by IV infusion, produced a 10-fold increase in weight gain

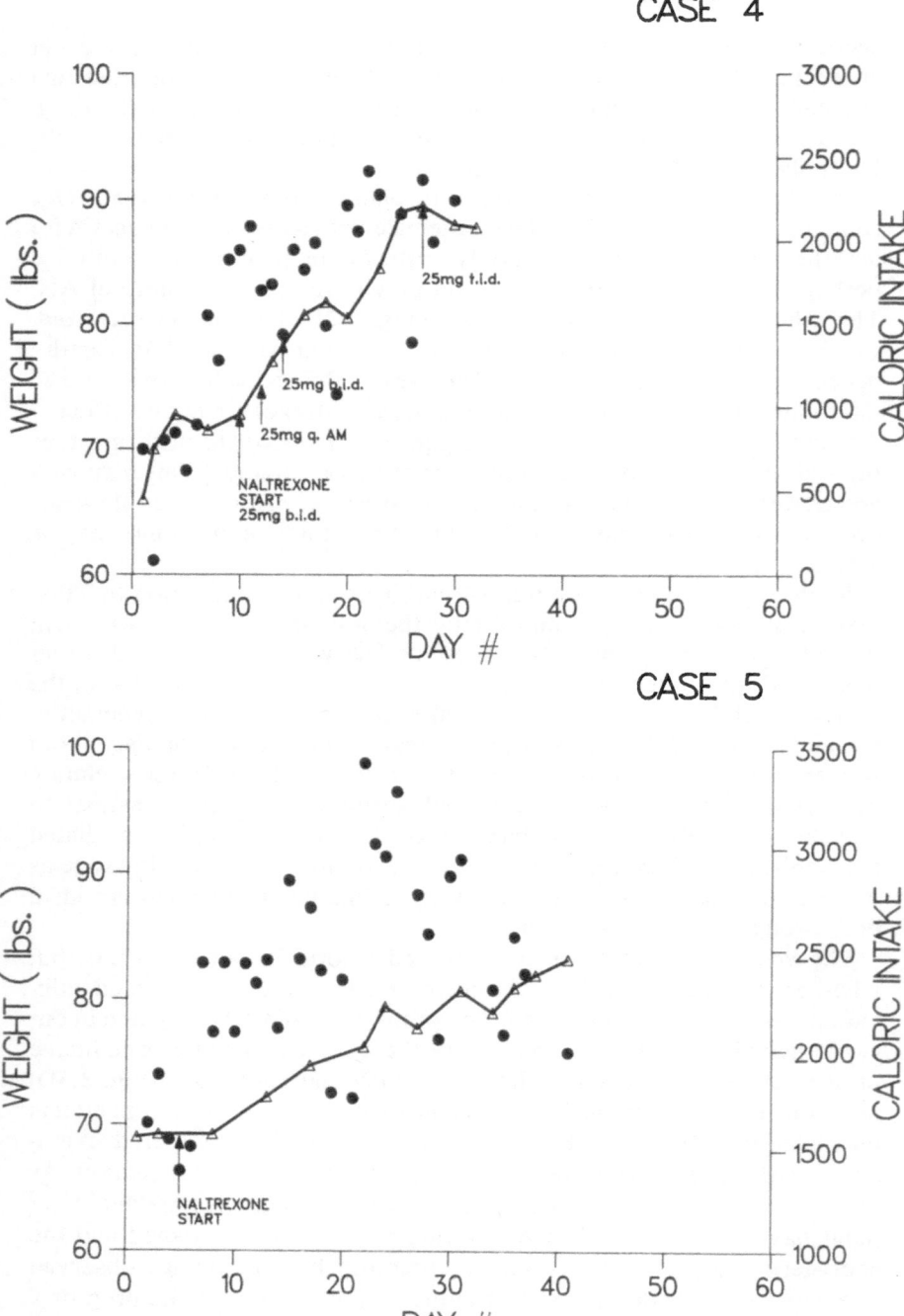

FIGURE 3.3. Caloric intake of patients with anorexia nervosa treated with naltrexone. Solid circles and squares = caloric intake. Solid lines = body weight. Cases are the same as those shown in Figure 3.2.

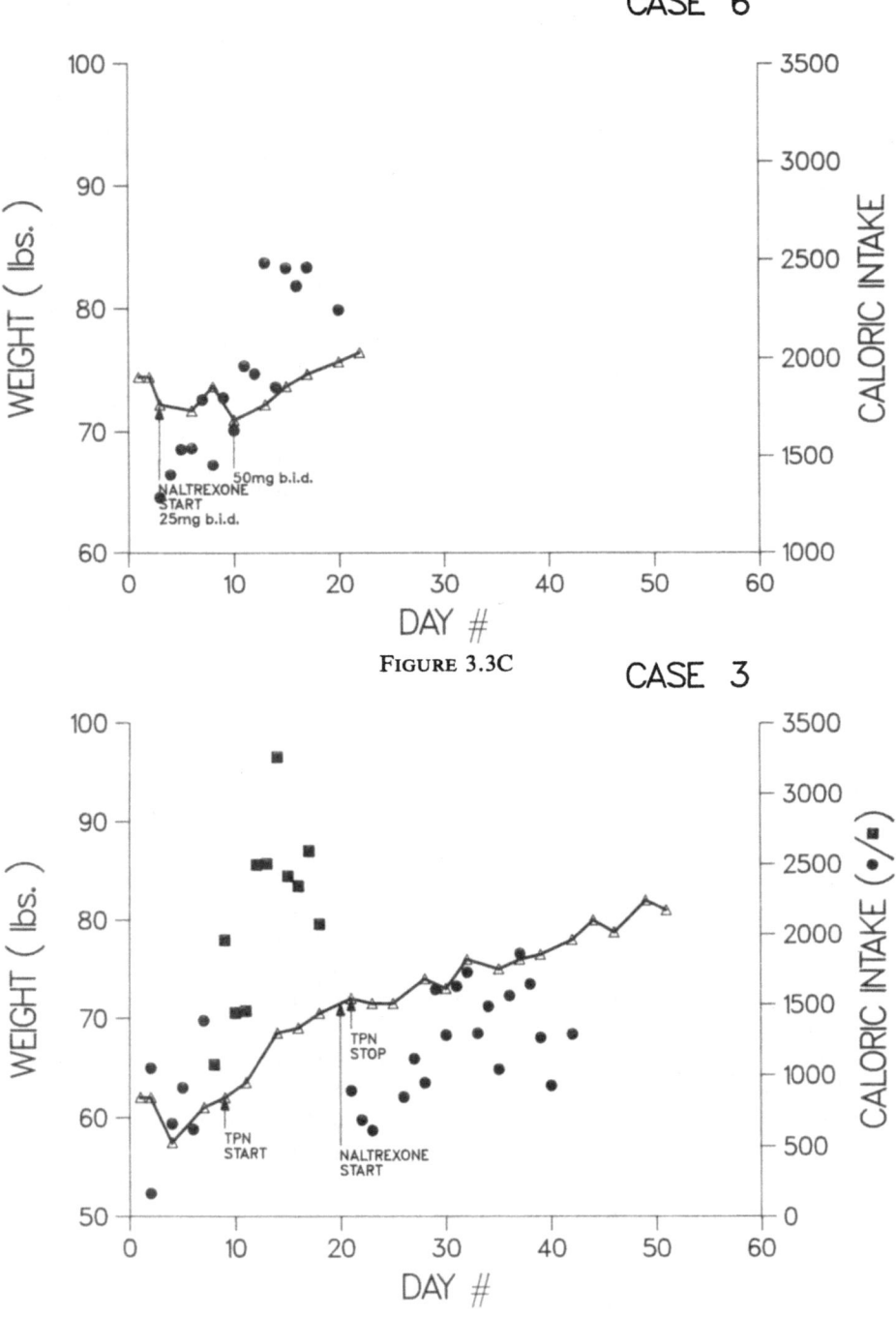

CASE 6

FIGURE 3.3C

CASE 3

FIGURE 3.3D

per week in twelve hospitalized anorectic patients and that when naloxone was stopped, the patients resumed the previous low rate of weight gain. Although the patients were treated also with the antidepressant amitriptyline (50–200 mg/day), and although it is not clear whether naloxone had altered the amount of food ingested, its utilization or both, their results were similar to ours. These investigators suggested initially that naloxone might have acted as an anti-lipolytic agent (193) or as a moderator of obsessional behavior (194), but our auto-addiction hypothesis could also explain the results.

The response of the reproductive system to naloxone is also altered in AN. For example, the rise in plasma LH and FSH concentrations following a single intravenous dose of the drug is less pronounced in anorectic patients, particularly those with amenorrhea secondary to weight loss (195). FSH did not rise and LH rose in only about half the cases. Another group reported that naloxone infusions result in a rise of LH in some but not all patients with eating disorders (196) and is more effective in normal-weight bulimic than in underweight anorectic patients. Although this difference could be due to the correction of non-opioid-mediated abnormalities, it could indicate that a given dose of naloxone is less effective when the level of endogenous opioids is high, and thus it could be interpreted as evidence for the proposed mechanism. One should add that the LH response differs in the prepubertal and pubertal states (187,188), suggesting that some of the variability observed in anorectic patients may depend upon the extent of reversion to a prepubertal pattern.

All currently available narcotic antagonists are 10-fold less effective at the κ- as opposed to the μ-opioid receptor subtypes. Our auto-addiction opioid model does not specify which endogenous opioid(s) or which receptor subtype(s) are involved. Thus, a lack of effectiveness of these drugs in some patients would not disprove the hypothesis, since the κ receptor might be preferentially mediating the effect. Alternatively, the drugs may be effective at higher doses acting through the less sensitive κ receptors. In that case, a greater potency and better therapeutic index could be expected from the use of κ receptor antagonists when they become available.

The auto-addiction hypothesis proposes that in AN the opioid drive to eat and the opioid-mediated adaptive down-regulation become uncoupled (Fig. 3.1). The patient may become addicted to the elation and/or to the down-regulation in the absence of a normal opioid induced drive to eat. Once uncoupled, patients could alternatively become addicted to the elation and/or to the drive to eat even without the down-regulation. This could result in bulimia. Thus, narcotic antagonists may also be useful in interrupting the addictive cycle of bulimia. For this reason, we have treated with naltrexone five bulimic women, ranging from late adolescence to postmenopausal age. Four had a classic binge-purge cycle, while one was only a binger. After receiving naltrexone (50 mg twice daily) for as little as two days, three of these patients became less obsessively preoccupied by food and were able to control almost completely their binging and purging. Two patients discontinued the drug soon after discharge from the hospital, and experienced a rapid recurrence of their symptoms. We have started a double-blind study using naltrexone for the treatment of bulimia. Our results are in accord with those of a recently published study of five bulimic patients, in whom intravenous naloxone significantly reduced the amount of food consumed during a binge (197).

Atypical Opioid Systems in Relation to Anorexia Nervosa

In the context of our proposed auto-addiction model, the atypical responses of the mouse to endogenous opioids become very interesting. In most species, including rats and normal humans, morphine increases food intake and blood glucose, and causes sedation. In contrast, in the mouse, morphine produces a stereotyped increase in locomotor activity. In addition, morphine decreases food intake and blood glucose over the entire effective dose response curve. These two responses, anorexia and hyperactivity, closely resemble the combination of symptoms seen in patients with AN but are opposite to those characteristic of most species. We have suggested that the endogenous opioid system as it exists in the mouse may represent that in the AN patient, in contrast to that in most species including normal humans and rats (198,199). Characterization of these differences may be useful in understanding the pathophysiology of AN.

The underlying differences in the endogenous opioid systems among individuals may be responsible for the biologic predisposition discussed in the section "Theoretical Considerations Concerning the Relationship of Anorexia Nervosa and Abnormal Opioid Function". If activated by the opioid release of dieting, the atypical responses and AN may result. Such differences may be undetectable until activated.

Our initial studies were done with female BALB/C mice. However, comparisons of various strains as well as species may show such a spectrum of opiate systems. In fact, strain differences may further support the idea that there is a range of opiate systems in humans, some of which biologically predispose the individual to AN. For example, in contrast to the BALB/C mouse, the female CF-1 mouse shows a biphasic response to morphine, increasing food intake at low doses and decreasing it at higher doses (200). Hyperactivity and a decrease in blood glucose also accompanied the decrease in food intake in this strain. Strain differences in mice are known for a number of other opiate responses, and we are currently exploring them with reference to these parameters.

Differences in the opioid systems may depend on the balance of opposing components resulting in different dominant responses. In accord with this, both species, the rat and the mouse, seem to be capable of both responses. Although an increase in food intake is the usual response to morphine in the rat, under certain conditions there is a decrease. The inverted U-shaped dose-response curve further indicates that at high doses an opposing factor comes into play (199). In the BALB/C mouse, sedation may occur at doses lower than those producing hyperactivity (199). The strain differences between CF-1 and BALB/C and the biphasic response in CF-1 further support the dual components in mice. The balance of opposing components suggests the notion that shifts in this balance could produce a spectrum of endogenous opioid systems and a range of biologic susceptibilities in different individuals. This concept also supports the uncoupling of the various opioid functions proposed in the auto-addiction model.

A possible explanation of these differences is that the various components may be mediated by different opioid receptor subtypes. This possibility is being explored by means of selective opioid agonists. A selective κ agonist, U50,488, increases food intake in both rats and mice (201). This means that morphine, a preferential μ agonist, and U50,488 have similar actions in the "typical" system of rats but opposite actions in the "atypical" systems of mice. Morphine has opposite actions, while U50,488 acts similarly in the two species. Clearly, this matter requires further exploration with other agonists.

Opioids and Other Eating Disorders in Humans

Opiate abnormalities have been observed also in other forms of appetite dysfunction. For example, the plasma levels of β-endorphin, but not of β-lipotropin or ACTH, were found to be higher in eight obese patients than in normal control subjects (202). Similarly, six obese children and six obese adolescents were found to have higher plasma β-endorphin and β-lipotropin than their age matched controls (203). A large dose of naloxone decreased food intake in a single test meal in obese but not in lean subjects (204), but chronic naltrexone given for eight weeks did not cause significant weight loss in obese subjects (205,206). However, two out of three obese patients with the special hyperphagia associated with the Prader-Willi Syndrome responded to naloxone by reducing their food intake (207). Thus, in agreement with the proposed hypothesis, these and other case studies (135,208,212) suggest that a relationship may indeed exist between high endorphin levels (measured directly or by pain sensitivity), and certain forms of hyperphagia and obesity (135,208,212).

Conversely, glucose appears to increase plasma β-endorphin, more so in obese patients than in non-obese patients (203,213,214). While this increase seems contrary to our hypothesis of opioid release with dieting, it actually occurred during the declining phase of the hyperglycemic curve. Thus, the falling caloric level may actually be the critical trigger or common denominator. On the other hand, the increase in plasma β-endorphin in response to insulin hypoglycemia was dampened in obese compared to normal subjects (203). Puberty had to be reached before the difference was evident, indicating that the sensitivity of this system changes at an age when eating disorders often manifest themselves.

Interplay of Dietary Factors in the Opioid Model

Dietary factors may influence opioid responses and hence play a role in the proposed mechanisms. Self-imposed dietary abnormalities, as well as the restrictions in AN, may be perpetuating the disorder or may represent nutritional self-selection to avoid a psychologically aversive reaction. Carbohydrates have an effect on opioid receptor binding, on the responses to exogenous opiates, and on endogenous opioid mediated functions (154,158,215,216). In diabetes, hyperglycemia may alter opioid-induced analgesia, insulin release, food intake, and even the development of physical dependence on morphine (145,217–222). Other types of dietary interplay are suggested by the effects that morphine and naloxone have on food selection by rats. Fat intake was altered in preference to carbohydrate or protein intake (130,131,223) in rats restricted to a daily 6-hour feeding period, while in free-feeding rats, protein intake was affected most and carbohydrate intake was affected least (223,224). In another study, pain threshold, presumably reflecting the action of endogenous opioids, was higher in rats fed a high-fat diet than in rats fed an isocaloric low-fat diet (225). Thus, the type of caloric restriction may also be a factor in the biologic changes induced by dieting and the resulting susceptibility to long term disturbances.

Neurotransmitters and the Auto-Addiction Hypothesis

As discussed in section "Neurotransmitter Involvement", many neurotransmitters and drugs modulating them regulate feeding behavior and have been implicated in AN. Interaction with the endogenous opiates could well be the mechanism. The proposed opioid model is compatible with other neurotransmitter hypotheses but better explains, through these interactions, the addictive-like behavior and compulsive drive in anorexia.

The best understood interaction is that between opiates and norepinephrine. Opiates inhibit the release of norepinephrine through a presynaptic mechanism, and the removal of this inhibition upon opiate withdrawal in a dependent patient results in massive norepinephrine release producing the withdrawal syndrome (135,143,144,226,227). In addition, opioids inhibit the activity of the noradrenergic receptor (228). As mentioned above, Margules suggested that the noradrenergic system mediates short-term adjustments to minimal starvation, while the endorphin system mediates long-term adaptation. Interactions with serotonin

(229,230), dopamine (135,144,227,230,231), GABA (232,233) and acetyl-choline (234,235) including those related to food intake, have been discussed elsewhere (63,135,144,229,232,236,237,238).

The Auto-Addiction Hypothesis and the Hypothalamic-Pituitary-Adrenal Axis

Disturbances in the function of the hypothalamic-pituitary-adrenal axis in AN have been discussed in section I3 and their possible interactions with the endogenous opioids proposed by the auto-addiction hypothesis are summarized in Figure 3.4.

Adenocorticotropic hormone and β-endorphin are post-translational cleavage products of pro-opiomelanocortin (85,86,88,89). They are normally released together and under the same controls and, either directly or through cortisol, exert a negative feedback secretion at the pituitary or hypothalamic level (239,240,241). Indeed morphine can be substituted for dexamethasone in the dexamethasone suppression test (242,243). In clinical depression, escape from one is usually but not always concomitant with escape from the other (242). Other opioid peptides, such as dynorphins, enkephalins and neoendorphins, derived from different precursors, prodynorphin or proenkephalin, are not released in coordination with ACTH. These other opioids may also play a role in the etiology of AN and may have actions similar or dissimilar to those of the endorphins. Thus different opioids may not be involved in some of the feedback loops described in Fig. 3.4, and a shift in the relative proportions of these peptides could significantly change the pattern of response.

Opiate drugs and opioid peptides modulate CRF and hence ACTH release (160,161,239,244–251). These effects are blocked by naloxone. Thus, the endogenous opioids may participate in the response of the hypothalamic-pituitary-adrenal axis to stress, although the details of this participation may vary with the opioid and the type of receptor involved (252).

β-Endorphin has also been shown to inhibit the stimulation of cortisol secretion by ACTH (253), suggesting that endogenous opioids may modulate the effectiveness as well as the levels of ACTH. This observation is of special interest because the stimulation of feeding by morphine seems to require cortisol, as indicated by its abolition by adrenalectomy and restoration with steroid replacement (224). If this were true also for β-endorphin-induced feeding, the two end products of CRF activity, cortisol and β-endorphin, would enhance each other's effectiveness. No information is available concerning the involvement of glucocorticoids in the opioid-mediated adaptation to starvation.

Corticotropin-releasing factor and ACTH injected directly into select brain sites decrease feeding (139,254–258) without releasing ACTH or cortisol. The relationship of this decrease to the increases produced by cortisol and by β-endorphin remains to be elucidated.

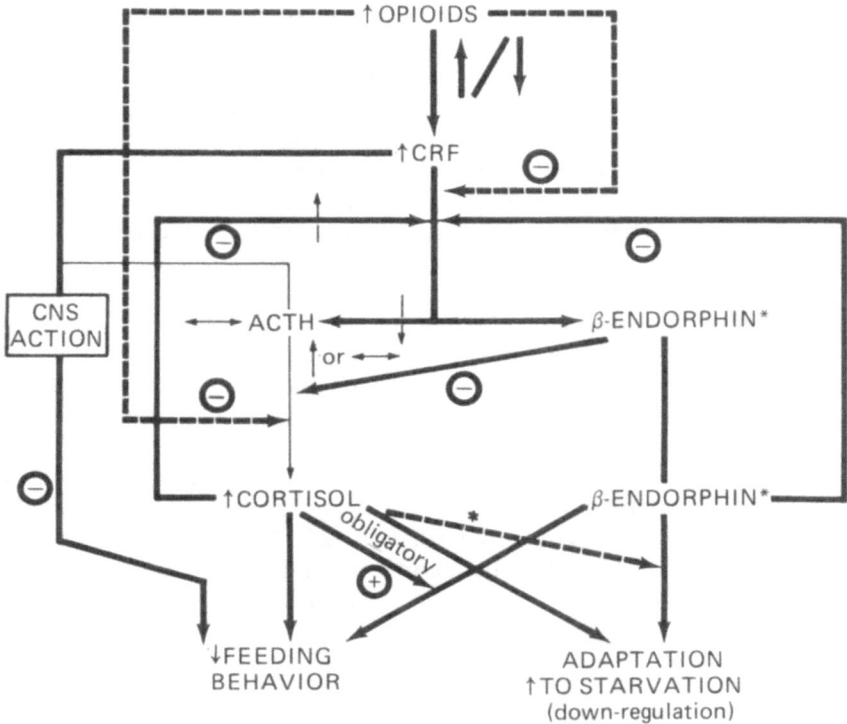

FIGURE 3.4. Hypothetical diagram of the interrelationships between opioids and the hypothalamic–pituitary–adrenal axis with reference to anorexia nervosa. The solid heavy lines and arrows indicate known relationships between opioids and the hypothalamic–pituitary–adrenal axis. Broken lines indicate other possible relationships implicated by those shown. The light arrows ($\uparrow$ or $\downarrow$) indicate changes that exist in anorexia nervosa. Other symbols are as follows $\leftrightarrow$, no change; *, undetermined; minus, inhibition, plus stimulation. The term opioid is used when any endogenous opioid may be involved. The term β-endorphin is used specifically. Hypotheses that relate anorexia nervosa to altered function of the endogenous opioids or of the hypothalamic–pituitary–adrenal axis are compatible, since disturbances of either system can generate disturbances in the other. For example, in anorexia nervosa an enhanced production of opioids would stimulate the secretion of CRF while simultaneously contributing to the feedback suppression of ACTH release.

In anorexia nervosa, there is an elevation of total endogenous opioid activity in the CSF, although the nature of the specific opioid involved is not known. In turn, this increase in opioids may bring about changes in the function of the hypothalamic-pituitary-adrenal axis that could produce the pattern in anorexia nervosa. For example, elevated opioids could increase CRF, ACTH, and hence cortisol release. Elevated cortisol could then suppress the stimulation of ACTH by CRF, resulting in increased

CRF levels but reduced ACTH release. Moreover, differences in the role of glucocorticoids in the response of the various opioids could alter the pattern and may be involved in the uncoupling of the auto-addiction. Thus, while cortisol is necessary for the stimulation of feeding by morphine, perhaps it is not necessary for feeding induced by all endogenous opioids, so that a shift in the opioid peptide might alter the need for glucocorticoids. It is fully recognized that the relationships of opioids and steroids discussed here are taken from diverse experimental systems, and that the correlations must be considered highly speculative.

Recently, it was reported that starvation reduces the CRF content of the rat hypothalamus and cerebellum, increases that of the pituitary, the midbrain, and the thalamus, but does not affect that of the pons, medulla or cerebral cortex (24). Whether these CRF changes are primary or secondary to the opioid changes induced by starvation is not known. In these animals, plasma corticosterone was elevated and plasma ACTH was decreased indicating that the feedback system was intact, as it appears to be in patients with AN (see section "Endocrine and Metabolic Function"). The actions of glucocorticoids (259,260) and opioids on food intake may also interrelate. Glucocorticoid and opioid theories are not incompatible and opioid involvement would provide an explanation for the addictive characteristics of AN.

Summary

Anorexia nervosa is a disorder of self-starvation in which patients have such a distorted body image and fear of obesity that they refuse to eat, even to the point of death. While the physical findings and the endocrine and metabolic changes characteristic of this disorder are mostly those of starvation, the behavioral changes observed in AN and in starvation are not identical. In this review we discuss clinical and experimental evidence concerning the role of hypothalamic or neurotransmitter dysfunction in the pathogenesis of AN and novel opioid auto-addiction hypothesis, according to which the release of endogenous opioids during an initial period of dieting causes elation and addiction to dieting. The hypothesis is based on (1) the addictive characteristics of the clinical behavior, (2) changes in endogenous opioid levels in AN, (3) changes in endogenous opioids and opioid systems during food deprivation, (4) relevant opioid actions, and (5) the actions of narcotic antagonists. This model suggests that interruption of the addictive cycle by opiate blockade may be therapeutically useful, and preliminary results are presented to support this concept. Moreover, we suggest that underlying differences in endogenous opioid systems may be responsible for biologic predisposition to AN. Atypical opioid systems in animals may provide useful models to investigate this possibility. The relationship of this model to opioids in other eating disorders, to dietary factors, to neurotransmitter mechanisms, and to disturbances in the hypothalamic-pituitary-adrenal axis are also discussed.

Acknowledgement. We appreciate the assistance of Ms. Carol Robinson in typing this manuscript.

References

1. Halmi KA (1978) Anorexia nervosa: recent investigations. Ann Rev Med 29:137–148.
2. Crisp AH, Palmer RL, Kalucy RS (1976) How common is anorexia nervosa? A prevalence study. Br J Psychiatry 128:549–554.
3. Hsu LKG (1980) Outcome of anorexia nervosa. A review of the literature (1954–1978). Arch Gen Psychiatry 37:1041–1046.
4. Theander S (1983) Long-term prognosis of anorexia nervosa: a preliminary report. In: Darby PL, Garfinkel PE, Garner DM, Coscina DV (eds) Anorexia Nervosa: Recent Developments in Research. Alan R. Liss, Inc., New York, pp. 441–442.
5. Diagnostic Statistical Manual of Mental Disorders, III edition, American Psychiatric Association, Washington, D.C., 1980.
6. Sein P, Searson S, Nicol AR, Hall K (1981) Anorexia nervosa and pseudo-atrophy of the brain. Br J Psychiatry 139:257–258.
7. Nussbaum M, Shenker IR, Marc J et al (1980) Cerebral atrophy in anorexia nervosa. J Pediatr 96:867–869.
8. Emrich HM, Pahl JJ, Herholz K et al (1984) PET investigation in anorexia nervosa: normal glucose metabolism during pseudoatrophy of the brain. In: Pirke KM, Ploog D (ed) The Psychobiology of Anorexia Nervosa. Springer-Verlag, Berlin, pp. 172–178.
9. Miyai K, Yamamoto T, Azukizawa M et al (1975) Serum thyroid hormones and thyrotropin in anorexia nervosa. J Clin Endocrinol Metabolism 40:334–328.
10. Vagenakis AG (1977) Thyroid hormone metabolism in prolonged experimental starvation in man. In: Vigersky RA (ed) Anorexia Nervosa. Raven Press, New York, pp. 243–253.
11. Burman KD, Vigersky RA, Loriaux DL et al (1977) Investigations concerning thyroxine deiodinative pathways in patients with anorexia nervosa. In: Vigersky RA (ed) Anorexia Nervosa. Raven Press, New York, pp. 255–261.
12. Moshang T, Utiger RD (1977). Low triiodothyronine euthyroidism in anorexia nervosa. In: Vigersky RA (ed) Anorexia Nervosa. Raven Press, New York, pp. 263–270.
13. Brown, GM (1983) Endocrine alterations in anorexia nervosa. In: Darby PL, Garfinkel PE, Garner DM, Coscina DV (Eds.), Anorexia Nervosa: Recent Developments in Research. Alan R. Liss, Inc., New York, pp. 231–248.
14. Warren MP, Vande Wiele R (1973) Clinical and metabolic features of anorexia nervosa. Am J Obstet Gynecol 117:435–449.
15. Wachslight-Rodbard H, Gross HA, Rodbard D et al (1979) Increased insulin binding to erythrocytes in anorexia nervosa. N Engl J Med 300:882–887.
16. Mecklenberg RS, Loriaux DL, Thompson RH et al (1974) Hypothalamic dysfunction in patients with anorexia nervosa. Medicine 53:147–159.

17. Vigersky RA, Loriaux DL, Anderson AE et al (1976) Anorexia nervosa: behavioral and hypothalamic aspects. J Clin Endocrinol Metabol 5:517–535.
18. Gold PW, Kaye W, Robertson GL et al (1983) Abnormalities plasma and cerebrospinal fluid arginine vasopressin in patients with anorexia nervosa. N Engl J Med 308:1117–1123.
19. Ebert MH, Kaye WK, Gold PW (1984) Neurotransmitter metabolism. In: Pirke KM, Ploog D (eds) The Psychobiology of Anorexia Nervosa. Springer-Verlag, Berlin, pp. 58–72.
20. Weiner H, Katz, JL (1983) The hypothalamic-pituitary-adrenal axis in anorexia nervosa: a reassessment. In: Darby PL, Garfinkel PE, Garner DM, Coscina DV, (eds) Anorexia Nervosa: Recent Developments in Research. Alan R. Liss, Inc., New York, pp. 249–270.
21. Gold PW, Gwirtsman H, Avgerinos PC et al (1986) Abnormal hypothalamic-pituitary-adrenal function in anorexia nervosa. New Engl J Med 314:1335–1342.
22. Hotta M, Shibasaki T, Masuda A et al (1986) The response of plasma adrenocorticotropin and cortisol to corticotropin-releasing hormone (CRH) and cerebrospinal fluid immunoreactive CRH in anorexia nervosa patients. J Clin Endocrinol Metab 62:319–324.
23. Cavagnini F, Invitti C, Passamonti M et al (1986) Response of ACTH and cortisol to corticotropin-releasing hormone in anorexia nervosa. N Engl J Med 314:184–185.
24. Suemaru S, Hishimoto K, Hattori T et al (1986) Starvation-induced changes in rat brain corticotropin-releasing factor (CRF) and pituitary-adrenocortical response. Life Sci 39:1161–1166.
25. Nillius SJ, Wide L (1977) The pituitary responsiveness to acute and chronic administration of gonadotropin-releasing hormone in acute and recovering stages of anorexia nervosa. In: Vigersky RA (ed) Anorexia Nervosa. Raven Press, New York, pp 225–242.
26. Boyar RM, Katz, J (1977) Twenty-four hour gonadotropin secretory patterns in anorexia nervosa. In: Vigersky RA (ed) Anorexia Nervosa. Raven Press, New York, pp. 177–188.
27. Brown, GM (1977) Endocrine profiles in anorexia nervosa. In: Vigersky RA (ed) Anorexia Nervosa. Raven Press, New York, pp. 123–126.
28. Casper RC, Davis JM, Pandey GN (1977) The effect of nutritional status and weight changes on hypothalamic function tests in anorexia nervosa. In: Vigersky RA (ed) Anorexia Nervosa. Raven Press, New York, pp. 137–148.
29. Fisch, RE (1977) Food intake, fatness, and reproductive ability. In: Vigersky RA (ed) Anorexia Nervosa. Raven Press, New York, pp. 149–162.
30. Sherman BM, Halmi KA (1977) Effect of nutritional rehabilitation on hypothalamic-pituitary function in anorexia nervosa. In: Vigersky RA (ed) Anorexia Nervosa. Raven Press, New York, pp. 211–224.
31. Vigersky RA, Loriaux, DL (1977) Anorexia vervosa as a model of hypothalamic dysfunction. In: Vigersky RA (ed) Anorexia Nervosa. Raven Press, New York, pp. 109–123.
32. Wakeling A, DeSouza VFA (1983) Differential endocrine and menstrual response to weight change in anorexia nervosa. In: Darby PL, Garfinkel PE, Garner DM, Coscina DV (eds) Anorexia Nervosa: Recent Developments in Research. Alan R. Liss, Inc., New York, pp. 271–278.

33. Warren, MP (1977) Weight loss and responsiveness to LH-RH In: Vigersky RA (ed) Anorexia Nervosa. Raven Press, New York, pp. 189–198.
34. Lupon T, Simon L, Barry V et al (1976) Biological aspects of anorexia nervosa. Life Sci 18:1341–1348.
35. Crisp AH, Blendis LM, Pawan GLS (1968) Aspects of fat metabolism in anorexia nervosa. Metabolism 17:1109–1118.
36. Klinefelter HF (1965) Hypercholestrolemia in anorexia nervosa (letter). J Clin Endocrinol Metab 25:1520–1521.
37. Nestel P (1974) Cholesterol metabolism in anorexia nervosa and hypercholestrolemia. J Clin Endocrinol Metab 38:325–328.
38. Mordasini R, Klose G, Greten H (1978) Secondary type II hyperlipoproteinemia in patients with anorexia nervosa. Metabolism 27:71–79.
39. Fichter MM, Pirke KM (1984). Hypothalamic-pituitary function in starving healthy subjects. In: Pirke KM, Ploog D (eds) The Psychobiology of Anorexia Nervosa. Springer-Verlag, Berlin, pp. 124–135.
40. Fichter MM, Pirke KM, Holsboer F (1986) Weight loss causes neuroendocrine disturbances: experimental study in healthy starving subjects. Psych Res 17:61–72.
41. Pirke KM, Spyra B, Warnhoff M et al (1984) Effect of starvation on central neurotransmitter systems and on endocrine regulation. In: Pirke KM, Ploog D (eds) The Psychobiology of Anorexia Nervosa. Springer-Verlag, Berlin, pp. 47–57.
42. Kaye WH, Gwirtsman H, George T et al (1986) Caloric consumption and activity levels after weight recovery in anorexia nervosa: a prolonged delay in normalization. Int J Eating Disorders 5:489–502.
43. Walker J, Roberts SL, Halmi KA et al (1979) Caloric requirements for weight gain in anorexia nervosa. Am J Clin Nutr 32:1396–1400.
44. Stordy BJ, Marks V, Kalucy RS et al (1977) Weight gain, thermic effect of glucose, and resting metabolic rate during recovery from anorexia nervosa. Am J Clin Nutr 30:138–146.
45. Dempsey DT, Crosby LO, Pertschuk MJ et al (1984) Weight gain and nutritional efficacy in anorexia nervosa. Am J Clin Nutr 39:236–242.
46. Russell GFM, Mezey AG (1962) An analysis of weight gain in patients with anorexia nervosa treated with high calorie diets. Clin Sci 23:449–461.
47. Kaye WH, Gwirtsman HE, Obarzanek E et al (1986) Caloric intake necessary for weight maintenance in anorexia nervosa: nonbulimics require greater caloric intake than bulimics. Am J Clin Nutr 44:435–443.
48. Pertschuk MJ, Crosby LO, Mullen, JL (1983) Nonlinearity of weight gain and nutrition intake in anorexia nervosa. In: Darby PL, Garfinkel PE, Garner DM, Coscina DV (eds) Anorexia Nervosa: Recent Developments in Research. Alan R. Liss, Inc., New York, pp. 301–310.
49. Armstrong S, Coleman G, Singer G (1980) Food and water deprivation: changes in rat feeding, drinking, activity and body weight. Neurosci Biobehav Rev 4:377–402.
50. Boyle PC, Storlien LH, Keesey RE (1978) Increased efficiency of food utilization following weight loss. Physiol Behav 21:261–264.
51. Coscina CV, Dixon LM (1983) Body weight regulation in anorexia nervosa: insights from an animal model. In: Darby PL, Garfinkel PE, Garner DM, Coscina DV (eds) Anorexia Nervosa: Recent Developments in Research. Alan R. Liss, Inc., New York, pp. 207–219.

52. Hill JO, Fried SK, DiGirolamo, M (1984) Effects of fasting and restricted refeeding on utilization of ingested energy in rats. Am J Physiol 247:R318–R327.
53. Kanarek R, Collier, GH (1983) Self-starvation: a problem of overrriding the satiety signal? Physiol Behav 30:307–311.
54. Levitsky DA, Faust I, Glassman, M (1976) The ingestion of food and the recovery of body weight following fasting in the naive rat. Physiol Behav 17:575–580.
55. Stricker EM, Andersen AE (1980) The lateral hypothalamic syndrome: comparison with the syndrome of anorexia nervosa. Life Sci 26:1927–1934.
56. Blundell JE (1977) Is there a role for serotonin (5-hydrotryptamine) in feeding? Int J of Obes 1:15–42.
57. Blundell JE (1984) Systems and interactions: an approach to the pharmacology of eating and hunger. In: Stunkard AJ, Stellar E (eds) Eating and Its Disorders. Raven Press, New York, pp. 39–65.
58. Hoebel BG (1976) Satiety: Hypothalamic stimulation, anorexic drugs and neurochemical substrates. In: Novin D, Wyrwicka W, Bray G (eds) Hunger: Basic Mechanisms and Clinical Implications. Raven Press, New York, pp. 33–50.
59. Hoebel BG (1977) Pharmacological control of feeding. Ann Rev Pharmacol 17:605–621.
60. Hoebel BG (1984) Neurotransmitters in the control of feeding and its rewards: monoamines, opiates, and brain-gut peptides. In: Stunkard AJ, Stellar E (eds) Eating and Its Disorders. Raven Press, New York, pp. 15–38.
61. Grossman SP (1978). Correlative analyses of ingestive behavior and regional amine depletions after surgical transections of neural pathways in the mesncephalon, diencephalon, and striatum. In: Garratini S, Samanin R (eds) Central Mechanisms of Anorectic Drugs. Raven Press, New York pp. 1–38.
62. Leibowitz SF (1980) Neurochemical systems of the hypothalamus in control of feeding and drinking behavior and water and electrolyte excretion. In: Morgane P, Panksepp J (eds) Handbook of the Hypothalamus vol. 3. Decker, New York, pp 299–437.
63. Leibowitz SF (1986) Brain monoamines and peptides: role in the control of eating behavior. Fed Proc 45:1396–1403.
64. Morley JE (1980) The neuroendocrine control of appetite: the role of the endogenous opiates, cholecystokinin, TRH, gamma-amino-butyric acid and the diazepam receptor. Life Sci 27:355–368.
65. Morley JE, Levine AS (1985) Pharmacology of eating behavior. Ann Rev Pharmacol Toxicol 25:127–146.
66. Gross HA, Lake CR, Ebert MH et al (1979) Catecholamine metabolism in primary anorexia nervosa. J Clin Endocrinol Metabol 49:805–809.
67. Halmi KA, Dekirmenjian AH, Davis JM et al (1979) Catecholamine metabolism in anorexia nervosa. Arch Gen Psychiatry 35:458–460.
68. Kaye WH, Ebert MH, Raleigh M et al (1984) Abnormalities in CNS monoamine metabolism in anorexia nervosa. Arch Gen Psych 41:350–355.
69. Kaye WH, Jimerson DC, Lake R et al (1985) Altered norepinephrine metabolism following long-term weight recovery in patients with anorexia nervosa. Psychiatry Res 14:333–342.
70. Heufelder A, Warnhoff M, Pirke KM (1985) Platelet α-adrenoceptor and

adenylate cyclase in patients with anorexia nervosa and bulimia. J Clin Endocrinol Metab 61:1053–1060.

71. Kaye WH, Ebert MH, Gwirtsman HE et al (1984) Differences in brain serotonergic metabolism between nonbulimic and bulimic patients with anorexia nervosa. Am J Psychiat 141:1598–1601.

72. Riederer P, Toifl K, Kruzik P (1982) Excretion of biogenic amine metabolites in anorexia nervosa. Clin Chim Acta 123:27–32.

73. Barry VC, Klawans HL (1976) On the role of dopamine in the pathophysiology of anorexia nervosa. J Neural Transm 38:107–122.

74. Mawon AR (1974) Anorexia nervosa and the regulation of intake: a review. Psychol Med 4:280–308.

75. Crisp AH (1966). A treatment regime for anorexia nervosa. Br J Psychiatr 112:505–512.

76. Dally PJ, Sargent W (1966) Treatment and outcome of anorexia nervosa. Br Med J 2:93–95.

77. Johanson AJ, Knorr NJ (1977) L-DOPA as treatment for anorexia nervosa. In: Vigersky RA (ed) Anorexia Nervosa. Raven Press, New York, pp. 363–372.

78. Marrazzi M, Luby ED (1986) An auto-addiction opioid model of chronic anorexia nervosa. Int J Eating Disorders 5:191–208.

79. World Health Organization, Experimental Committee on Addiction Producing Drugs, Thirteenth Report. WHO Technical Report Series No. 273. World Health Organization, Geneva 1964.

80. Rakoff, V (1983) Multiple determinants of family dynamics in anorexia nervosa. In Darby PL, Garfinkel PE, Garner DM, Coscina DV (Eds) Anorexia Nervosa: Recent Developments in Research. Alan R. Liss, Inc., New York, pp. 29–40.

81. Szmukler GI, Tantam D (1984) Anorexia nervosa: Starvation dependence. Br J Med Psychol 57:303–310.

82. Margules DL (1979) Beta-endorphin and endoloxone: Hormones of the autonomic nervous system for the conservation or expenditure of bodily resources and energy in anticipation of famine or feast. Neursci Biobehav Rev 3:155–162.

83. Kaye WH, Pickar DM, Naber D et al (1982) Cerebrospinal fluid opioid activity in anorexia nervosa. Am J Psychiatr 139:643–645.

84. Gerner RH, Sharp B (1982) CSF β-endorphin immunoreactivity in normal, schizophrenic, depressed, manic and anorexic subjects. Brain Res 237:244–247.

85. Herbert E. Birnberg N, Civelli O, Lissitzky JC et al (1982) Regulation of genetic expression of pro-opiomelanocortin in pituitary and extrapituitary tissues of mouse and rat. In: Costa E, Trabucchi M (eds) Regulatory Peptides: From Molecular Biology to Function. Raven Press, New York, pp. 9–18.

86. Hosobuchi Y, Bloom, FE (1983) Analgesia induced by brain stimulation in man: its effect on release of β-endorphin and adrenocorticotropin into cerebrospinal fluid. In: Wood JH (ed) Neurobiology of Cerebrospinal Fluid 2. Plenum Press, New York, pp. 97–105.

87. Jeffcoate WJ, McLoughlin L, Hope J et al (1978). β-endorphin in human cerebrospinal fluid. Lancet 1:119–121.

88. Kraft K, Lang RE, Kirilow G et al (1983) Differential regulation of β-endor-

phin in the anterior pituitary, intermediate lobe, hypothalamus and brain stem. Life Sci 33 (Suppl. 1):491–494.

89. Krieger DT, Liotta AS, Nicholsen G et al (1979) Brain ACTH and endorphin reduced in rats with monosodium glutamate-induced arcuate nuclear lesions. Nature 278:562–563.

90. Facchinetti F, Petraglia F, Sances G et al (1986) Dissociation between CSF and plasma β-endorphin in major depressive disorders: Evidence for a different regulation. J Endocrinol Invest 9:11–14.

91. Knuth UA, Friesen HG (1983) Changes of beta-endorphin and somatostatin concentrations in different hypothalamic areas of female rats after chronic starvation. Life Sci 33:827–833.

92. Gambert SR, Garthwaite TL, Pontzer CH et al (1980) Fasting associated with decrease in hypothalamic β-endorphin. Science 210:1271–1272.

93. Vaswani KK, Tejwani GA (1986) Food deprivation induced changes in the level of opioid peptides in the pituitary and brain of rat. Life Sci. 38:197–201.

94. Majeed NH, Lason W, Przewlocki B et al (1986) Brain and peripheral opioid peptides after changes in ingestive behavior. Neuroendocrinol 42:267–272.

95. Takahashi H, Motomatsu T, Nobunaga M (1986) Influences of water deprivation and fasting on hypothalamic, pituitary and plasma opioid peptides and prolactin in rats. Physiol Behav 37:603–608.

96. Scallet AC, Della-Fera MA, Baile CA (1985) Satiety, hunger and regional brain content of cholecystokinin/gastrin and met-enkephalin immunoreactivity in sheep. Peptides 6:937–943.

97. Bodnar RJ, Kelly DD, Spiaggia A et al (1978) Biphasic alterations of nociceptive thresholds induced by food deprivation. Physiol Psychol 6:391–395.

98. McGivern R, Berka C, Bernston GC et al (1979) Effect of naloxone on analgesia induced by food deprivation. Life Sci 25:885–888.

99. Hamm RJ, Kniseley JS, Watson A et al (1985) Hormonal mediation of the analgesia produced by food deprivation. Physiol Behav 35:879–882.

100. Hamm RJ, Knisely JS (1986) The analgesia produced by food deprivation in 4-month old, 14-month old, and 24-month old rats. Life Sci 39:1509–1515.

101. Einhorn D, Young JB, Lansberg L (1982) Hypotensive effect of fasting: possible involvement of the sympathetic nervous system and endogenous opiates. Science 217:727–729.

102. Hamm RJ, Lyeth BG (1984) Nociceptive thresholds following food restriction and return to free-feeding. Physiol Behav 33:499–501.

103. Davis JM, Lowy MT, Yim GKW et al (1982) Relationship between plasma concentrations of immunoreactive beta-endorphin and food intake in rats. Peptides 4:79–83.

104. Yim GKW, Lowy MT, Davis JM et al (1982) Opiate involvement in glucoprivic feeding. In: Hoebel B, Novin D (eds) The Neural Basis of Feeding and Reward. Haer Institute, Brunswick, Maine, pp 485–498.

105. Bodnar RJ, Kelly DD, Brutus M et al (1978) Chronic 2-deoxy-D-glucose treatment adaptation of its analgesic, but not hyperphagic properties. Pharmacol Biochem Behav 9:763–768.

106. Bodnar RJ, Kelly DD, Brutus M et al (1978) 2-Deoxy-D-glucose induced decrements in operant and reflex pain thresholds. Pharmacol Biochem Behav 9:543–9.

107. Bodnar RJ, Kelly DD, Glusman M (1979) 2-Deoxy-D-glucose analgesia: influence of opiates and non-opiate factors. Pharmacol Biochem Behav 11:297–301.
108. Bodnar RJ, Kelly DD, Mansour A et al (1979) Differential effects of hypophysectomy upon analgesia induced by two glucoprivic stressors and morphine. Pharmacol Biochem Behav 11:303–308.
109. Bodnar RJ, Kramer E, Simone DN et al (1983) Dissociation of analgesic and hyperphagic responses following 2-deoxy-D-glucose Int J Neurosci 21:225–236.
110. Spiaggia A, Bodnar RJ, Kelly DD et al (1979) Opiate and non-opiate mechanisms of stress-induced analgesia: cross-tolerance between stressors. Pharmacol Biochem Behav 10:761–765.
111. Kramer E, Sperber ES, Bodnar RJ (1986) Age-related decrements in the analgesic and hyperphagic responses to 2-deoxy-D-glucose. Physiol Behav 35:929–934.
112. Si ECC, Bryant HU, Yim GKW (1986) Opioid and non-opioid components of insulin-induced feeding. Pharmacol Biochem Behav 24:899–903.
113. Thompson DA, Welle SL, Lilavivat U et al (1982). Opiate receptor blockade in man reduces 2-deoxy-D-glucose induced food intake but not hunger, thirst, and hypothermia. Life Sci 31:847–852.
114. Atrens DM, Marfaing-Jallat P, LeMagnen J (1983) Ethanol preference following hypothalamic stimulation: relation to stimulation parameters and energy balance. Pharmacol Biochem Behav 19:571–575.
115. Carroll ME, Boe IN (1984) Effect of dose on increased etonitazene self-administration by rats due to food deprivation. Psychopharmacology 82:151–152.
116. Carroll ME, France CP, Meisch RA (1981) Intravenous self-administration of etonitazene, cocaine and phencyclidine in rats during food deprivation and satiation. J Pharmacol Exper Therap 217:241–247.
117. Meisch RA, Kliner, DJ (1979) Etonitazene as a reinforcer for rats: increased etonitazene-reinforced behavior due to food deprivation. Psychopharmacology 63:97–98.
118. Oei TPS (1983) Effect of body weight reduction and food deprivation on cocaine self-administration. Pharmacol Biochem Behav 19:453–455.
119. Carroll ME, Lac ST, Walker MJ, Kragh R, Newman T (1986). Effects of naltrexone on intravenous cocaine self-administration in rats during food satiation and deprivation. J Pharmacol Exp Ther 238:1–7.
120. Carroll, ME (1985) The role of food deprivation in the maintenance and reinstatement of cocaine-seeking behavior in rats. Drug Alcoh Depend 16:85–109.
121. Snell D, Feller D, Bylund D et al (1982) Sensitization produced by repeated administration of naloxone is blocked by food deprivation. J Pharmacol Exper Ther 221:444–452.
122. Takahashi RN and Singer G (1979) Self-administration of tetrahydrocannabinol by rats. Pharmacol Biochem Behav 11:737–740.
123. Carroll ME, Pederson MC, Harrison RG (1986) Food deprivation reveals strain differences in opiate intake of Sprague-Dawley and Wistar Rats. Pharmacol Biochem Behav 24:1095–1099.
124. Sanger DJ and McCarthy PS (1980) Differential effects of morphine on food

and water intake in food deprived and freely-feeding rats. Psychopharmacology 72:103–106.

125. Brown DR, Holtzman SG (1979) Suppression of food and water intake in rats and mice by naloxone. Pharmacol Biochem Behav 11:567–573.

126. Sanger DJ, McCarthy, PS (1982) The anorectic action of naloxone is attenuated by adaptation to a food deprivation schedule. Psychopharmacology 77:336–338.

127. Gosnell BA, Levine AS, Morley, JE (1983) N-Allylnormetazocine (SKF-10,047): the induction of feeding by a putative sigma agonist. Pharmacol Biochem Behav 19:737–742.

128. Levine AS, Morley, JE (1983) Butorphanol tartrate induces feeding in rats. Life Sci 32:781–785.

129. McLean S, Hoebel, BG (1983) Feeding induced by opiates injected into the paraventricular hypothalamus. Peptides 4:287–292.

130. Marks-Kaufman, R (1982) Increased fat consumption induced by morphine administration in rats. Pharmacol Biochem Behav 16:949–955.

131. Marks-Kaufman R, Kanarek RB (1981) Modifications of nutrient selection induced by naloxone in rats. Psychopharmacology 74:321–324.

132. Morley JE, Levine, AS (1981) Dynorphin (1–13) induces spontaneous feeding in rats. Life Sci 29:1901–1903.

133. Morley JE, Levine, AS (1983) Involvement of dynorphin and the kappa opioid receptor in feeding. Peptides 4:797–800.

134. Morley JE, Levine AS, Kneip J et al (1983) The effect of peripherally administered satiety substance on feeding induced by butorphanol tartrate. Pharmacol Biochem Behav 19:577–582.

135. Morley JE, Levine AS, Yim GK et al (1983) Opioid modulation of appetite. Neurosci Biobehav Rev 7:281–305.

136. Sanger DJ (1981) Endorphinergic mechanisms in the control of food and water intake. Appetite: J Intake Res 2:193–208.

137. Sanger DJ, MaCarthy, PS (1981) Increased food and water intake produced in rats by opiate receptor agonists. Psychopharmacol 74:217–220.

138. Tannenbaum MG, Pivorun EB (1984) Effect of naltrexone on food intake and hoarding in white-footed mice. Pharmacol Biochem Behav 20:35–37.

139. Morley JE, Levine AS, Gosnell BA et al (1985) Peptides and feeding. Peptides 6 (Suppl. 2):181–192.

140. Reid LD (1985) Endogenous opioid peptides and regulation of drinking and feeding. Am J Clin Nutr 42:1099–1132.

141. Baile CA, McLaughlin CL, Della-Fera MA (1986) Role of cholecystokinin and opioid peptides in control of food intake. Physiol Rev 66:172–234.

142. Yim, GK and Lowy, MT (1984) Opioids, feeding and anorexias. Fed Proc 43:2893–2897.

143. Jaffe JH, Martin WR (1980) Opioid analgesics and antagonists. In: Gilman AG, Goodman LS, Gilman A (eds) The Pharmacological Basis of Therapeutics, 6th ed. Macmillan Publishing Co, New York, pp. 494–534.

144. Morley JE (1981) The endocrinology of the opiates and opioid peptides. Metabol 30:195–209.

145. Giugliano E, Ceriello A, DiPinto P et al (1982) Impaired insulin secretion in human diabetes mellitus: the effect of naloxone-induced opiate receptor blockade. Diabet 31:367–370.

146. Helman AM, Giraud P, Nicolaidis S et al (1983) Glucagon release after

stimulation of the lateral hypothalamic area in rats: Predominant β-adrenergic transmission and involvement of endorphin pathways. Endo 113:1–6.

147. Ipp, E (1984) Central and peripheral endorphins: Their role in the control of glucose homeostasis. In: Muller EE, Genazzani AR (eds) Central and Peripheral Endorphins: Basic and Clinical Aspects. Raven Press, New York, pp 251–257.

148. Ipp E, Dobbs RE, Unger, RH (1978) Morphine and β-endorphin influence the secretion of the endocrine pancreas. Nature 276:190–191.

149. Ipp E, Schusdziarra V, Harris V et al (1980) Morphine induced hyperglycemia: role of insulin and glucagon. Endo 107:461–463.

150. Kanter RA, Ensinck JW, Fujimoto WY (1980) Disparate effects of enkephalin and morphine upon insulin and glucagon secretion by islet cell cultures. Diabet 29:84–86.

151. Morley JE, Baranetsky NG, Wingert TD et al (1980) Endocrine effects of naloxone-induced opiate receptor blockade. J Clin Endocrinol Metab 50: 251–257.

152. Reid RL, Yen SSC (1981) β-Endorphin stimulates the secretion of insulin and glucagon in humans. J Clin Endocrinol Metab 52:592–594.

153. Schusdziarra V, Henrichs I, Holland A et al (1981) Evidence for an effect of exorphins on plasma insulin and glucagon levels in dogs. Diabet 30:362–364.

154. Schusdziarra V, Rewes B, Lenz N et al (1983) Carbohydrates modulate opiate receptor mediated mechanisms during postprandial endocrine function. Reg Peptides 7:243–252.

155. Schusdziarra V, Rewes B, Lenz N et al (1983) Evidence for a role of endogenous opiates in postprandial somatostatin release. Reg Peptides 6:355–361.

156. Schusdziarra V, Schick R, DeLaFuente A et al (1983) Effect of β-casomorphins on somatostatin release in dogs. Endocrinology 112:1948–1951.

157. Schusdziarra V, Schick R, DeLaFuente A et al (1983) Effect of β-casomorphins and analogs on insulin release in dogs. Endo 112:885–889.

158. Schusdziarra V, Schick R, Holland A et al (1983) Effect of opiate-active substances on pancreatic polypeptide levels in dogs. Peptides 4:205–210.

159. Schusdziarra V, Specht J, Schick R et al (1983) Effect of morphine, leu-enkephalin and β-casomorphins on basal somatostatin release in dogs. Horm Metabol Res 15:407–408.

160. El-Tayer KMA, Gauthier CJT, Brubaker PL et al (1986) Hormonal and metabolic responses to intracarotid and intrajugular infusion of β-endorphin in normal dogs. Can J Physiol Pharmacol 64:306–310.

161. El-Tayer KMA, Brubaker PL, Lickley HLA et al (1986) Effect of opiate-receptor blockade on normoglycemic and hypoglycemic glucoregulation. Am J Physiol 250:E236–E242.

162. Hussain MN, Kikuchi K, Cukerman E et al (1986) The effect of β-endorphin on biogenic amines, insulin, and glucagon levels in the hepatic portal circulation of normal and pancreatectomized dogs. Endo 119:685–690.

163. Levin ER, Yamada T, Levin S et al (1986) Endogenous opioid modulation of pancreatic hormone secretion: studies in dogs. Metabol 35:59–63.

164. El-Tayer KMA, Brubaker PL, Vranic M et al (1985) Beta-endorphin modula-

tion of the glucoregulatory effects of repeated epinephrine infusion in normal dogs. Diabet 34:1293–1300.

165. Knudtzon J (1986) Effects of pro-opiomelanocortin-derived peptides on plasma levels of glucagon, insulin and glucose. Horm Metabol Res 18:579–583.

166. Werther GA, Joffe S, Artal R et al (1984) Opiates modulate insulin action in vivo in dogs. Diabetologia 26:65–69.

167. Meites J (1984) Effects of opiates on neuroendocrine functions in animals? An Overview. In: Delitala G, Motta M, Serio M (eds) Opioid Modulation of Endocrine Function, Raven Press, New York, pp. 53–63.

168. Millard WJ, Martin JB (1984) Opioid modulation of human growth hormone secretion. In: Delitala G, Motta M, Serio M (eds) Opioid Modulation of Endocrine Function, Raven Press, New York pp. 111–124.

169. Morley JE, Willenbring ML, Krahn DD et al (1984) Opioid control of thyroid function. In: Delitala G, Motta M, Serio M (eds) Opioid Modulation of Endocrine Function, Raven Press, New York, pp. 267–276.

170. Jean-Baptiste E, Rizack, MA (1980) In vitro cyclic AMP-mediated lipolytic activity of endorphins, enkephalins and naloxone. Life Sci 27:135–141.

171. Schwandt P, Richter W, Morley, JS (1981) β-Lipotropin contains two lipolytic sequences. Neuropept 1:211–216.

172. Mandenoff A, Fumeron F, Appelbaum M et al (1982) Endogenous opiates and energy balance. Science 215:1536–1538.

173. Adler BA, Crowley WR (1984) Modulation of luteinizing hormone release and catecholamine activity by opiates in the female rat. Neuroendocrinology 38:248–253.

174. Ferin M, vanVugt D, Wardlaw S (1984) Hypothalamic control of the menstrual cycle and the role of endogenous opioid peptides. Rec Prog Horm Res 40:441–485.

175. Van Vugt DA, Meites J (1980) Influence of endogenous opiates on anterior pituitary function. Federation Proceedings 39:2533–2538.

176. Van Vugt DA, Sylvester PW, Alysworth CF et al (1982). Counteraction of gonadal steroid inhibition of luteinizing hormone release by naloxone. Neuroendocrinol 34:274–278.

177. Piva F, Limonta P, Maggi R et al (1984) Dual effects of the opioids in the control of gonadotropin secretion. In: Delitala G, Motta M, Serio M (eds) Opioid Modulation of Endocrine Function. Raven Press, New York, pp. 155–170.

178. Kalra SP, Leadem CA (1984) Control of luteinizing hormone secretion by endogenous opioid peptides. In: Delitala G. Motta M, Serio M (eds) Opioid Modulation of Endocrine Function, Raven Press, New York, pp. 171–184.

179. Knuth UA, Sikand GS, Casanueva FF et al (1983) Changes in beta-endorphin content in discrete areas of the hypothalamus throughout proestrus and diestrus of the rat. Life Sc 33:1443–1450.

180. Hong JS, Lowe C, Squibb R et al (1981) Monosodium glutamate exposure in the neonate alters hypothalamic and pituitary neuropeptide levels in the adult. Reg Peptides 2:347–352.

181. Hong JS, Yoshikawa K, Lamartiniere CA (1982) Sex-related difference in the rat pituitary (Met)-enkephalin level altered by gonadectomy. Brain Res 25 1:380–383.

182. Mueller G (1980) Attenuated pituitary β-endorphin release in estrogen-treated rats. Proc Soc Exper Biol Med 165:75–81.
183. Wilkinson M, Bhanot R, Wilkinson DA et al (1983) Prolonged estrogen treatment induces changes in opiate, benzodiazepine and β-adrenergic binding sites in female rat hypothalamus. Brain Res Bull 11:279–281.
184. Jirikowski GF, Merchenthaler I, Rieger GE et al (1986) Estradiol target sites immunoreactive for β-endorphin in the arcuate nucleus of rat and mouse hypothalamus. Neurosci Lett 65:121–126.
185. Margules DL (1978) Molecular theory of obesity, sterility and other behavioral and endocrine problems in genetically obese mice. Neurosci Biobehav Rev 2:231–233.
186. Genezzani AR, Facchinetti F, Pintor C et al (1983) Proopiocortin-related peptide levels throughout prepuberty and puberty. J Clin Endocrinol Metab 57:56–61.
187. Mauras N, Veldhuis JD, Rogol AD (1986) Role of endogenous opiates in pubertal maturation: opposing actions of naltrexone in prepubertal and late pubertal boys. J Clin Endocrinol Metab 62:1256–1263.
188. Petraglia F, Bernasconi S, Iughetti L et al (1986) Naloxone-induced luteinizing hormone secretion in normal, precocious and delayed puberty. J Clin Endocrinol Metab 63:1112–1116.
189. Crisp AH, Kalucy RS, Lacey JH et al (1977) The long-term prognosis in anorexia nervosa: Some factors predictive of outcome. In: Vigersky RA (ed) Anorexia Nervosa. Raven Press, New York, pp 55–65.
190. Halmi KA, Goldberg SC, Eckert E et al (1977) Pretreatment evaluation in anorexia nervosa. In: Vigersky RA (ed) Anorexia Nervosa. Raven Press, New York, pp. 43–54.
191. Bernstein IL, Borson S (1986) Learned food aversion: A component of anorexia syndromes. Psychol Rev 93:462–472.
192. Luby ED, Marrazzi MA, Kinzie J (1987) Case reports - treatment of chronic anorexia nervosa with opiate blockade. J Clin Psychopharmacol 7:52–53.
193. Moore R, Mills IH, Forster A (1981) Naloxone in the treatment of anorexia nervosa: effect on weight gain and lipolysis. J Roy Soc Med 74:129–131.
194. Mills IH, Medicott, L (1984) The basis of naloxone treatment in anorexia nervosa and the metabolic responses to it. In: Pirke KM, Ploog D (eds) The Psychobiology of Anorexia Nervosa. Springer-Verlag, Berlin, pp 161–171.
195. Baranowska B, Rozbicka G, Jeske W et al (1984) The role of endogenous opiates in the mechanism of inhibited luteinizing hormone (LH) secretion in women with anorexia nervosa: the effect of naloxone on LH, follicle stimulating hormone, prolactin and β-endorphin secretion. J Clin Endocrinol Metab 59:412–416.
196. Baraban JM, Walsh BT, Gladis M et al (1986) Effects of naloxone on luteinizing hormone secretion in eating disorders: A pilot study. Int J Eating Disord 5:149–155.
197. Mitchell JE, Laine DE, Morley JE et al (1986) Naloxone but not CCK-8 may attenuate binge-eating behavior in patients with the bulimia syndrome. Biol Psychiatr 21:1399–1406.
198. Buck M, Efthyvoulidis S, Marrazzi MA et al (1986) An animal model of chronic anorexia nervosa? Neurosci Abst 11:344.
199. Buck M, Marrazzi MA (1987) An animal model of chronic anorexia nervosa? Life Sci 41:765–773.

200. Marrazzi MA, Powers RJ, Buck M (1986) Mouse opiate systems as a model of chronic anorexia nervosa. Appetite 7:278—279 (full manuscript in preparation).
201. Marrazzi MA, L'Abbe D (1986) Characterization of the mouse opiate system in relation to anorexia nervosa. Pharmacologist 28:2441 (full manuscript in preparation).
202. Facchinetti F, Giovannini C, Barletta C et al (1986) Hyperendorphinemia in obesity and relationships to affective state. Physiol Behav 36:937–940.
203. Genazzani AR, Facchinetti F, Petraglia F et al (1986) Hyperendorphinemia in obese children and adolescents. J Clin Endocrinol Metab 62:36–40.
204. Atkinson RL (1982) Naloxone decreases food intake in obese humans. J Clin Endocrinol Metab 55:196–198.
205. Atkinson RL, Berke LK, Drake CR et al (1985) Effects of long-term therapy with naltrexone on body weight in obesity. Clin Pharmacol Ther 38:419–422.
206. Malcolm R, O'Neil PM, Sexauer JD et al (1985) A controlled trial of naltrexone in obese humans. Int J Obes 9:347–353.
207. Kyriakides M, Silverstone T, Jeffcoate W et al (1980) Effect of naloxone on hyperphagia in Prader-Willi Syndrome. Lancet 1:876–877.
208. Givens JR, Wiedemann E, Andersen RN et al (1980) β-Endorphin and β-lipotropin plasma levels in hirsute women: Correlation with body weight. J Clin Endocrinol Metab 50:975–976.
209. McKendall MJ, Haier, RJ (1983) Pain sensitivity and obesity. Psychiatr Res 8:119–125.
210. Pradalier A, Willer JC, Boureau F et al (1981) Relationship between pain and obesity: An electrophysiological study. Physiol Behav 27:961–964.
211. Dunger DB, Leonard JV, Wolff OH et al (1980) Effect of naloxone in a previously undescribed hypothalamic syndrome. Lancet 1:1277–1281.
212. Fraioli F, Fabbri A, Moretti C et al (1981) Endogenous opioid peptides and neuroendocrine correlations in a case of congenital indifference to pain. Endo 108:238A.
213. Getto CJ, Swift WJ, Carlson IH et al (1986) Immunoreactive beta-endorphin increases after IV glucose in obese human subjects. Brain Res Bull 17:435–437.
214. Getto CJ, Fullerton DT, Carlson IH (1984) Plasma immunoreactive beta-endorphin response to glucose ingestion in human obesity. Appetite 5:329–335.
215. Davis WM, Miya TS, Edwards LD (1956) The influence of glucose and insulin pretreatment upon morphine analgesia in the rat. J Am Pharm Assoc 45:60–62.
216. Werther GA, Hogg A (1984) Opiate receptor modulation by glucose and insulin. Presented at Symposium, Neural and Metabolic Bases of Feeding held in Napa Valley, California.
217. Levine AS, Morley JE, Brown DM et al (1982) Extreme sensitivity of diabetic mice to naloxone-induced suppression of food intake. Physiol Behav 28:987–989.
218. Levine AS, Morley JE, Wilcox G et al (1982) Tail pinch behavior and analgesia in diabetic mice. Physiol Behav 28:39–43.
219. Simon GS, Dewey WL (1981) Narcotics and Diabetes: I. The effects of streptozotocin-induced diabetes on the antinociceptive potency of morphine. J Pharmacol Exper Therap 218:318–323.

220. Simon GS, Gorzelleca J., Dewey WL (1981) Narcotics and diabetes. II. Streptozotocin-induced diabetes selectively alters the potency of certain narcotic analgesics. J Pharmacol Exper Therap 218:324–329.

221. Shook JE, Dewey WL (1986) Morphine dependence and diabetes I. The development of morphine dependence in streptozotocin-diabetic rats and spontaneously diabetic C57BL/KsJ mice. J Pharmacol Exper Therap 237:841–847.

222. Shook JE, Kachur JF, Brase DA et al (1986) Morphine dependence and diabetes II. Alterations of normorphine potency in the guinea pig ileum and mouse van deferens and of ileal morphine dependence by changes in glucose concentration. J Pharmacol Exper Therap 237:848–852.

223. Shor-Posner G, Azar AP, Filart R et al (1986) Morphine-stimulated feeding: analysis of macronutrient selection and paraventricular nucleus lesions. Pharmacol Biochem Behav 24:931–939.

224. Bhakthavatsalam P, Leibowitz SF (1986) Morphine-elicited feeding: Diurnal rhythm, circulating corticosterone and macronutrient selection. Pharmacol Biochem Behav 24:911–917.

225. Yehuda S, Leprohon-Greenwood CE, Dixon LM et al (1986) Effects of dietary fat on pain threshold, thermoregulation and motor activity in rats. Pharmacol Biochem Behav 124:1775–1777.

226. Palmer MR, Seiger A, Hoffer BJ et al (1983) Modulatory interactions between enkephalins and catecholamines: anatomical and physiological substrates. Fed Proc 42:2934–2945.

227. Mulder AH, Frankhuyzen AL, Stof JC et al (1984) Catecholamine receptors, opiate receptors, and presynaptic modulation of neurotransmitter release in the brain. In: Neuropharmacology and Central Nervous System— Theoretical Aspects. Alan Liss, New York, pp. 47–58.

228. Tsang D, Tan AT, Henry JL et al (1978) Effect of opioid peptides on noradrenaline-stimulated cyclic AMP formation in homogenates of rat cerebral cortex and hypothalamus. Brain Res 152:521–527.

229. Harsing LG, Yang HYT, Costa E (1982) Accumulation of hypothalamic endorphins after repeated injections of anorectics which release serotonin. J Pharmacol Exper Therap 223:689–694.

230. Sapun-Malcolm D, Farah JM Jr., Mueller GP (1986) Serotonin and dopamine independently regulate pituitary β-endorphin release in vivo. Neuroendocrinol 42:191–196.

231. Wood, PL (1983) Opioid regulation of CNS dopaminergic pathways, a review of methodology, receptor types, regional variations and species differences. Peptides 4:595–601.

232. Cooper SJ (1983) Benzodiazepine-opiate antagonist interactions in relation to feeding and drinking behavior. Life Sci 32:1043–1051.

233. Sivam SP, Hong JS (1986) GABAergic regulation of enkephalin in rat striatum: alterations in met-enkephalin level, precursor content, and pre-proenkephalin messenger RNA abundance. J Pharmacol Exper Therap 237:326–331.

234. Green PG, Kitchen J (1986) Antinociception opioids and the cholinergic system. Prog Neurobiol 26:119–146.

235. Domino EF (1979) Opiate interaction with cholinergic neurons. Adv Biochem Pharmacol 20:339–355.

236. Morley JE, Levine AS (1982) Opiates, dopamine and feeding. In: Hoebel BG

(ed) The Neural Basis of Feeding and Reward. Haer Institute, Brunswick, Maine, pp. 499–506.

237. Morley JE, Levine AS, Grace M et al (1982) Dynorphin (1–13), dopamine and feeding in rats. Pharmacol Biochem Behav 16:701–705.

238. Tepperman FS, Hirst M, Gowdey CW (1981) A probable role for norepinephrine in feeding after hypothalamic injection of morphine. Pharmacol Biochem Behav 15:555–558.

239. Taylor T, Dluhy DG, Williams GH (1983) β-Endorphin suppresses adrenocorticotropin and cortisol levels in normal human subjects. J Clin Endocrinol Metab 57:592–596.

240. Krantz DE, Brown WA (1985) Dexamethasone suppresses beta-endorphin in humans. Psychoneuroimmunol 10:211–214.

241. Beyer HS, Parker L, Li CH et al (1986) β-Endorphin attenuates the serum cortisol response to exogenous adrenocorticotropin. J Clin Endocrinol Metab 62:808–811.

242. Zis AP, Haskett RF, Albala A et al (1985) Opioid regulation of hypothalamic-pituitary-adrenal function in depression. Arch Gen Psych 42:383–386.

243. Zis AP, Haskett RF, Albala A et al (1985) Cortisol escape from morphine suppression. Psychiatr Res 15:91–95.

244. Buckingham JC (1982) Secretion of corticotropin and its hypothalamic releasing factor in response to morphine and opioid peptides. Neuroendocrinol 35:111–116.

245. Buckingham JC, Cooper TA (1984) Differences in hypothalamic-pituitary-adrenocortical activity in the rat after acute and prolonged treatment with morphine. Neuroendocrinol 38:411–417.

246. Buckingham JC, Cooper TA (1986) Effects of naloxone on hypothalamo-pituitary-adrenocortical activity in the rat. Neuroendocrinol 42:421–426.

247. Yajima F, Suda T, Tomori N et al (1986) Effects of opioid peptides on immunoreactive corticotropin-releasing factor release from the rat hypothalamus in vitro. Life Sci 39:181–186.

248. Desouza EB, van Loon GR (1983) D-Ala-met-enkaphalinamide, a potent opioid peptide, alters pituitary-adrenocortical secretion in rats. Endo 111:1483–1490.

249. Munson PL (1973) Effects of morphine and related drugs on the corticotrophin (ACTH)-stress reaction. Prog Brain Res 39:361–372.

250. Eisenberg RM (1980) Effect of naloxone on plasma corticosterone in the opiate-naive rat. Life Sci 20:935–943.

251. Siegel RA, Chowers I, Conforti N et al (1982) Effects of naloxone on basal and stress-induced ACTH and corticostrone secretion in the male rat-site and mechanism of action. Brain Res 249:103–109.

252. Buckingham JC, Cooper TA (1986) Pharmacological characterization of opioid receptors influencing the secretion of corticotrophin releasing factor in the rat. Neuroendocrinol 44:36–40.

253. Beyer HS, Parker L, Li CH et al (1986) β-Endorphin attenuates the serum cortisol response to exogenous adrenocorticotropin. J Clin Endocrinol Metab 62:808–811.

254. Britton DR, Koob G, Rivier J et al (1982) Intraventricular corticotropin-releasing factor enhances behavioral effects of novelty. Life Sci. 31:363–367.

255. Morley JE, Levine AS (1982) Corticotropin releasing factor, grooming and ingestive behaviors. Life Sci 31:1459–1664.

256. Gosnell BA, Morley JE, Levine AS (1983) A comparison of the effects of corticotropin-releasing factor and sauvagine on food intake. Pharmacol Biochem Behav 19:771–775.
257. Levine AS, Rogers B, Kneip J et al (1983) Effect of centrally administered corticotropin releasing factor (CRF) on multiple feeding paradigms. Neuropharmacol 22:337–339.
258. Vergoni AV, Poggiolo R, Bertolini A (1986) Corticotropin inhibits food intake in rats. Neuropeptides 7:153–158.
259. Dallman MF (1984) Viewing the ventromedial hypothalamus from the adrenal gland. Am J Physiol 248:R1–R12.
260. Freedman MR, Horwitz BA, Stern JS (1986) Effect of adrenalectomy and glucocorticoid replacement on development of obesity. Am J Physiol 250:R595–R607.

4

The Pineal and Its Indole Products: Basic Aspects and Clinical Applications

RUSSEL J. REITER

Introduction

After decades of conflicting findings concerning the functions of the pineal, research within the last 20 years has finally begun to unravel the complex interactions of this ubiquitously acting gland. Indeed, this complexity has accounted in part for the long delay in solving the mysteries of this important organ of internal secretion. Additionally, the seeming unconventional nature of its interactions and the species variations in response to the pineal and its constituents have made elucidation of the functions of the pineal gland difficult and sometimes frustrating. It is, however, yielding to the steady onslaught of investigation; whereas the last two decades have proven productive in partially clarifying the biochemistry and physiology of the pineal, research in the years ahead will undoubtedly be even more rewarding.

What is summarized herein relates to the mechanisms of control of pineal indole metabolism in mammals and to the physiologic interactions of pineal hormones with other endocrine and nonendocrine structures. Although the presentation is geared to data garnered from mammals including man, it should be pointed out that studies on nonmammalian vertebrates have also been very instrumental in advancing the state of knowledge in this field (1–4).

Anatomic Connections of the Visual System with the Pineal Gland

The photoperiod has a major role in synchronizing the circadian rhythms of pineal indole metabolism. While in many nonmammalian vertebrates the effects of light on the pineal are direct, in mammals the perception of light by the retina is essential for curbing the synthesis of pineal hormonal products. The retinas are physiologically connected to the pineal gland by a circuitous, multineuronal pathway which involves axons in both the central and peripheral nervous systems (Fig. 4.1) (5,6). After photic energy is transduced into a neural message in the photoreceptor cells, action potentials are transferred over the bipolar neurons to the ganglion cells of the retina, some of whose axons form the retinohypothalamic tract (RHT) which terminates in both the ipsilateral and especially the contralateral suprachiasmatic nucleus. (SCN) (7,8). Remarkably, the retina–pineal connection remains functional even in the presence of very few anatomically intact photoreceptor elements (9,10). Information arriving at the SCN via the RHT serves to entrain the activity of the nuclei by suppressing the signal between the SCN and the pineal gland (11). During darkness, when the inhibitory effect of light on the SCN is lifted, the number of neural action potentials reaching the pinealocyte is increased resulting in augmented enzyme activity relating to melatonin production (12).

The exact location of the neurons that connect the SCN with the intermediolateral cell column of the thoracic cord is a matter of debate. While the paraventricular nuclei (PVN) of the hypothalamus receive projections from the SCN and may be relay sites for the descending information (13–15), there is not total agreement on this point (6). Conveniently, axons whose cell bodies are in the PVN do project, among other places, to the spinal cord (16–17). On the other hand, there are several other direct descending fiber pathways which could carry neural information from the hypothalamus to the preganglionic sympathetic neurons of the thoracic cord (17,18). Undoubtedly, bilateral lesions in the hypothalamus caudal to the SCN in Syrian hamsters (19) and transection of the cervical cord in humans (20) interfere with the endocrine and melatonin forming ability of the pineal gland.

Regardless of the specific route the descending fibers take to arrive at the preganglionic sympathetic perikarya in the spinal cord, those carrying information destined for the pineal gland probably end on the intermediolateral cell column of spinal segments C8–T5 (Fig. 4.1), since axons of these cell bodies project to the superior cervical ganglia (SCG) (21). The axons of the preganglionic cells exit the cord through the ventral root, pass up the sympathetic chain, and synapse on postganglionic sympathetic neurons in the SCG. From here, the course that the fibers take to the pineal is again circuitous; the bulk of the axons travel with blood vessels but eventually diverge to form a discrete bundle of neurons in the tentorium cerebelli. These bundles, usually paired, emerge from the tentorium to enter the apex of the pineal gland (5) where they terminate primarily within perivascular spaces but occasionally between adjacent pinealocytes (22). In those species in which the pineal complex consists of both a superficial and deep component, the sympathetic fibers are abundantly distributed to both portions of the gland (23).

Although the neural pathways connecting the retina to the pineal gland have been described almost exclusively in nonhuman material, there are

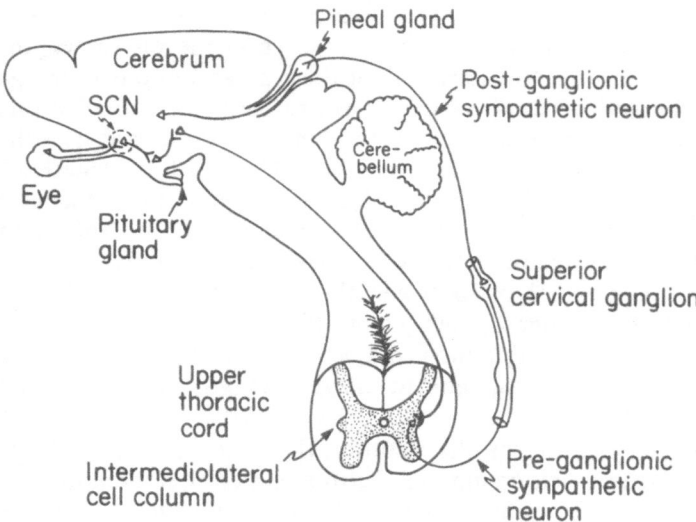

FIGURE 4.1. Photic information detected by the retina of mammals influences the pineal gland by means of a circuitous neuronal network which involves cells in both the central and peripheral nervous systems. Synapses in this pathway have been identified in the suprachiasmatic nuclei (SCN) of the hypothalamus, in the intermediolateral cell column of C8-T5 and in the superior cervical ganglia. Another possible synaptic site is within the paraventricular nuclei of the hypothalamus. A central innervation is represented by neurons in the brain whose axons enter the pineal stalk and terminate in the pineal gland. From Reiter (22).

strong indications that the human brain has evolved similar neural connections to relay information about the ambient photoperiod to the epithalamus. In this regard, it is important to note that SCN have been tentatively identified in human brain (24). These nuclei had been overlooked for many years in the human brain because of their diffuse nature and short anteroposterior dimension.

Besides this well-documented sympathetic innervation to the mammalian pineal gland, a parasympathetic cholinergic input has been described in several mammals, e.g., the rabbit and monkey. In these species the preganglionic fibers course in the superficial petrosal nerves and eventually enter the pineal to synapse on intramural parasympathetic cells (25). The short postganglionic fibers terminate among the parenchymal elements of the gland. Whether the parasympathetic innervation to the pineal gland is in fact unique to so few species is unknown, as is its functional significance.

In addition to the catecholaminergic sympathetic and possible cholinergic parasympathetic innervations to the mammalian pineal gland, a third type of innervation has been described and in fact is being aggressively investigated by several groups. This central innervation, defined as direct, consists of axons that pass from the neurons in the central nervous system through the pineal stalk and into the gland (Fig. 4.1). Recent anatomic studies using the horseradish peroxidase (HRP) technique have shown pinealopetal fibers to arise from thalamic, hypothalamic, epithalamic, and mesencephalic areas to enter the pineal gland (26,27). In several species including the guinea pig (28), Mongolian gerbil (29), and the rat (30,31) the central fibers have been shown in part to have their parikarya in the PVN of the hypothalamus. These fibers in the rat project through the stria medullaris thalami to the level of the medial habenular nuclei and habenular commissure; from here they enter the deep pineal gland and the pineal stalk (32). Terminals of this central projection could not, however, be identified in the superficial pineal gland by this group of investigators. The authors point out that the number of fibers in the stria medullaris thalami that actually enter the pineal gland is very small. In the human, nerve fibers from the posterior and habenular commissures also enter the pineal gland.

In reference to the central innervation of the pineal gland, the anatomic findings are supported by electrophysiologic evidence as well. In both the guinea pig (33,34) and the rat (32,34), electric stimulation of the PVN reportedly influences unit activity in the pineal gland via the monosynaptic PVN-pineal projection. Likewise, hypothalamic stimulation was found to decrease serotonin N-acetyltransferase (NAT) activity and melatonin content of the rat superficial pineal (35). The neurotransmitter involved in this pathway remains unknown but, considering the large number of presumptive neurally active substances found in the pineal (36), there are many candidates available for this role.

The only pineal innervation for which a function has been unequivocally demonstrated is the sympathetic component arising from the SCG. Destruction of these postganglionic sympathetic axons was shown many years ago to disrupt rhythmic melatonin production (37) and the ability of the pineal gland to influence the neuroendocrine–reproductive axis (38). The findings have been repeatedly confirmed and the functional importance of the sympathetic innervation is no longer questioned. This, however, is not true for the parasympathetic and the direct central projections whose physiologic importance remains the subject of future studies. In reference to the central innervation, if photic information from the retinas is relayed through the PVN (13), there are at least two pathways by which photoperiodic influences could be exerted at the level of the pineal gland, i.e., via the direct central innervation and via the peripheral sympathetic input.

Transynaptic Regulation of Pineal Indoleamine Metabolism

The neurotransmitter regulating the nocturnal rise in pineal melatonin production is norepinephrine (NE) (39). It is found in high concentrations

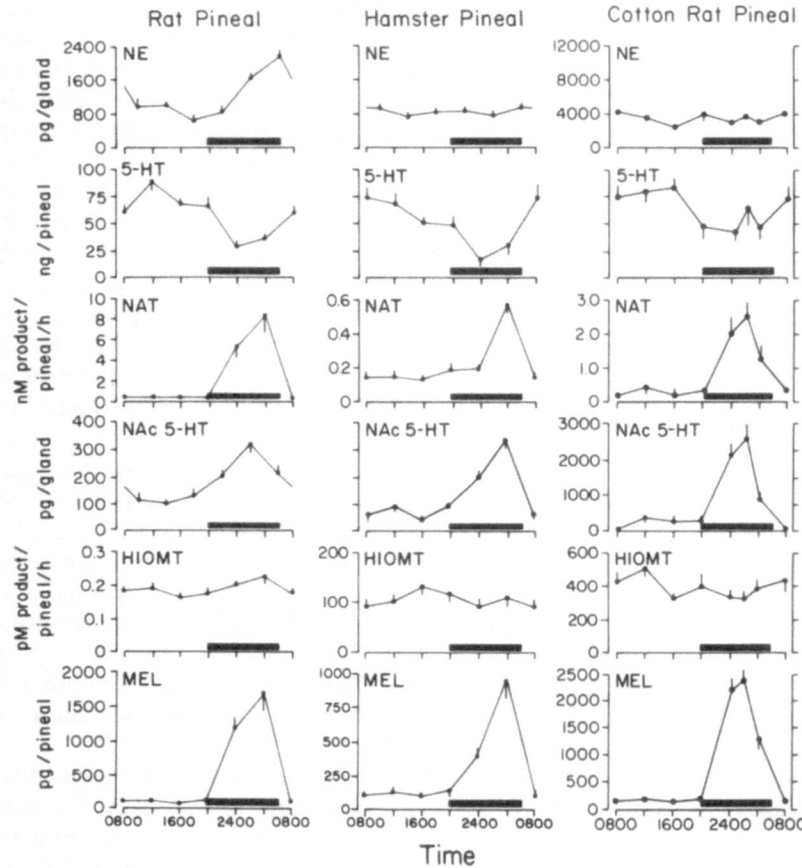

FIGURE 4.2. Pineal rhythms in three rodent species. Norepinephrine (NE) exhibits a nocturnal increase in the rat pineal but not in the Syrian hamster or the cotton rat. Serotonin (5-HT) levels drop during the daily dark period while N-acetyltransferase (NAT) activity, N-acetylserotonin (NAc 5-HT) and melatonin (MEL) levels rise at night. Hydroxyindole-O-methyltransferase (HIOMT) exhibits a uniform activity throughout the daily light/dark cycle. These changes appear to occur in the pineal gland of all mammals. Black bar along abscissa represents the daily period of darkness. From Reiter (22).

in the sympathetic nerve endings of the pineal (40) and its content in the gland is greatly diminished after bilateral removal of the superior cervical ganglia (41). NE levels in the rat pineal gland exhibit a nyctohemeral increase (42,43) but this is not the case in a number of other species, including the Syrian hamster (44), Mongolian gerbil (43), and cotton rat (45) (Fig. 4.2). Even in those species that lack a 24-hour rhythm in pineal NE content, however, a nocturnal increase in the activity of tyrosine hydroxylase, the rate-limiting enzyme in catecholamine synthesis is seen (46). This is also true for the rat (47) and presumably for all mammals. The neuronal vesicles which store NE in the pineal sympathetic nerve terminals also contain serotonin which is produced in the pinealocytes and then is shunted to the nerve endings (48). The function of this monoamine, if any, in influencing melatonin production after its simultaneous release with NE remains unknown.

NE is a mixed α- and β-receptor agonist and it is capable of interacting with α_1-, α_2-, β_1- and β_2 receptors. The α_1- and β_1-adrenoreceptors are located on pinealocyte membranes (49–51). At least in the rat, the number of β-receptors on the pinealocyte membrane increases at night, presumably allowing for the maximal action of NE and the resultant stimulation of melatonin production (52,53).

Species differ in their reliance on β- and α-receptors to mediate pineal melatonin synthesis. In the best studied model, the rat, released NE primarily interacts with β_1-receptors leading to a rapid rise in the conversion of serotonin to melatonin (48,54); the NE also acts on α_1-receptors causing a potentiation of the β-receptor effect (55) although stimulation of α-receptors by themselves does not promote the cascade of events which leads to an increased melatonin synthesis. This latter phenomenon was observed in dispersed cultured rat pinealocytes and its applicability to the in vivo situation remains to be investigated.

As first described in the eastern chipmunk (56), there are species in which the nighttime rise in melatonin production cannot be blocked by propranolol, a β-adrenergic receptor blocker. Most recently this has been shown to be the case for the sheep as well (57). In the sheep, the nocturnal rise in serum melatonin is blocked by intravenously administered prazosin (an α-receptor blocker), but, as in the chipmunk, not by propranolol. Binding studies in ovine pineal membranes using a specific ligand for α_1-adrenoreceptors indicated the binding is rapid, reversible, saturable, and stereospecific. The control of melatonin production in the sheep pineal seems to be under the control of an α_1-subclass of receptor at a point beyond N-acetylation of 5-hydroxytryptamine (5-HT, serotonin). In very few species have such detailed studies been done regarding the interaction of NE with pinealocyte receptors. Considering this, it seems likely that, as more species are tested, animals other than sheep, which utilize primarily α-adrenoreceptors to control the nocturnal rise in melatonin production, will be discovered.

Even during the day the rat pineal gland responds with increased melatonin production after either the exogenous administration of NE, especially when it is preceded by a NE uptake inhibitor, or by isoproterenol, which is not taken up into nerve endings (54). In contrast, numerous attempts to stimulate 5-HT metabolism in the Syrian hamster pineal during the day with either the natural transmitter or its agonist have yielded negative results (58–60). The only exception to this was a report in which a weak melatonin rise followed two injections of isoproterenol (61), an effect that could not be duplicated in a subsequent study (58). Although β-agonists cannot induce melatonin production in the Syrian hamster pineal during the day, the nocturnal rise in pineal melatonin content is curtailed by β-receptor antagonists (62). Furthermore, isoproterenol administration delays the depression in melatonin at night when the animals are exposed to a short pulse of light (60). These findings suggested a unique situation wherein the β-receptors in the Syrian hamster pinealocyte membrane are insensitive to stimulation during the day and responsive at night.

Experiments to check the differential sensitivity according to time of day followed quickly and showed that, indeed, a 24-hour rhythm in the responsivity of the hamster pineal to either exogenously administered NE or isoproterenol existed (63–65). Both in vivo and in vitro studies clearly revealed that the hamster pineal will respond to β-agonists only during the latter half of the dark phase of the light/dark cycle (65). As will be detailed in a subsequent section of this review, this is precisely when the normal nocturnal rise in melatonin production occurs in the Syrian hamster (66,67). This means that in the hamster as in the rat, the β-adrenoceptors mediate the nocturnal rise in the pineal melatonin production and that the increase may be controlled as much by receptor sensitivity as by the release of NE from the sympathetic nerve endings within the pineal (65). The number of β-receptors on hamster pinealocyte membranes as a function of the light/dark cycle is currently being investigated.

What makes these results from the Syrian hamster particularly interesting is the fact that a rise in serum melatonin in the human also does not follow isoproterenol infusion during the day (68). It seems probable, however, that duplicating this infusion protocol at night, as was done in the Syrian hamster (65), may reveal that the human pineal, like that of other species, is sensitive to isoproterenol and that the normal nocturnal rise in human pineal indoleamine metabolism depends upon the release of NE and its subsequent action on the β-adrenoceptors.

The β-receptors on the pinealocyte membrane are linked to the production of cyclic AMP (Fig. 4.3). NE administration leads to roughly a 10-fold rise in adenylate cyclase activity in the rat pineal gland (69–71). This activation is mediated by a GTP-binding adenylate cyclase stimulatory protein which is present in the membrane. The stimulation is rapid with the rise in enzyme activity being detectable within 10 min; thereafter the effect disappears as a consequence of desensitization (55). NE also acts simultaneously on α-adrenoceptors further stimulating cyclic AMP production. NE potentiation of adenylate cyclase activity seems to occur via a mechanism involving phosphotidylinositol turnover (72). The prevalence among species of α-adrenergic potentiation of β-receptor stimulation of cyclic AMP production in the pineal gland remains unknown. The bulk of currently available evidence suggests that cyclic AMP is involved

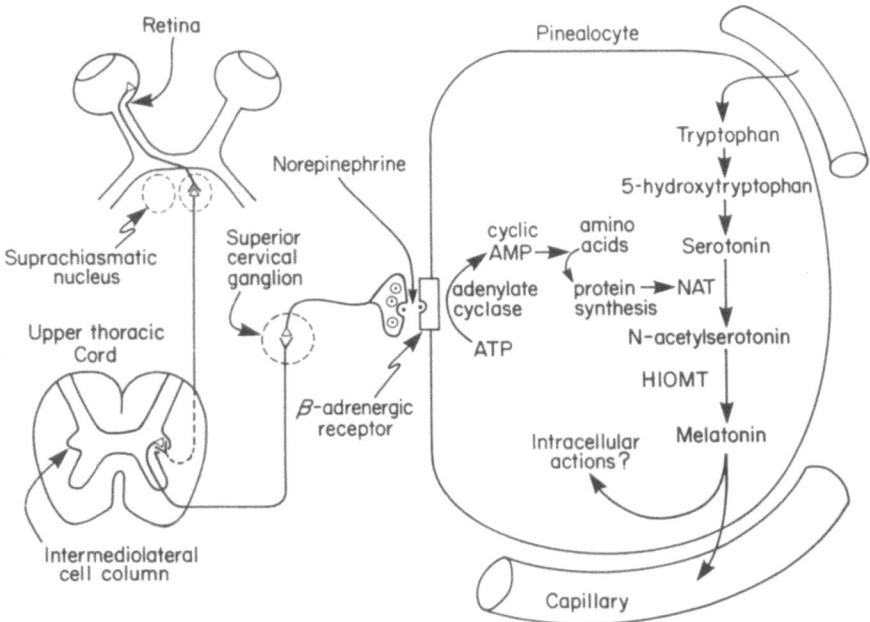

FIGURE 4.3. Diagrammatic representation of the neural connections between the eyes and the pinealocytes. The neurotransmitter released from the postganglionic terminals in the pineal gland is norepinephrine (NE); after its release NE acts via β- and in some cases α-receptors (or both) to stimulate melatonin production via a mechanism which involves cyclic AMP as a second messenger. After melatonin is formed it seems to be quickly released into the systemic circulation. From Reiter (75).

in melatonin production by virtue of its stimulation of NAT activity (73). This enzyme reduces the quantity of serotonin and serotonin oxidation products within the pineal gland, while increasing the levels of N-acetylserotonin, the immediate precursor of melatonin (48).

Pineal Serotonin Metabolism

Melatonin production within the pineal gland has been thoroughly studied (48,74–76) (Fig. 4.4). Melatonin is a product of tryptophan metabolism, an amino acid taken up by the pinealocytes from systemic circulation. The injection of either tryptophan or its metabolic product, 5-hydroxytryptophan (5-HTP), stimulates, at least under some circumstances, plasma melatonin levels presumably by increasing the production and release of the indole from the pineal gland (56,77). Tryptophan is converted to

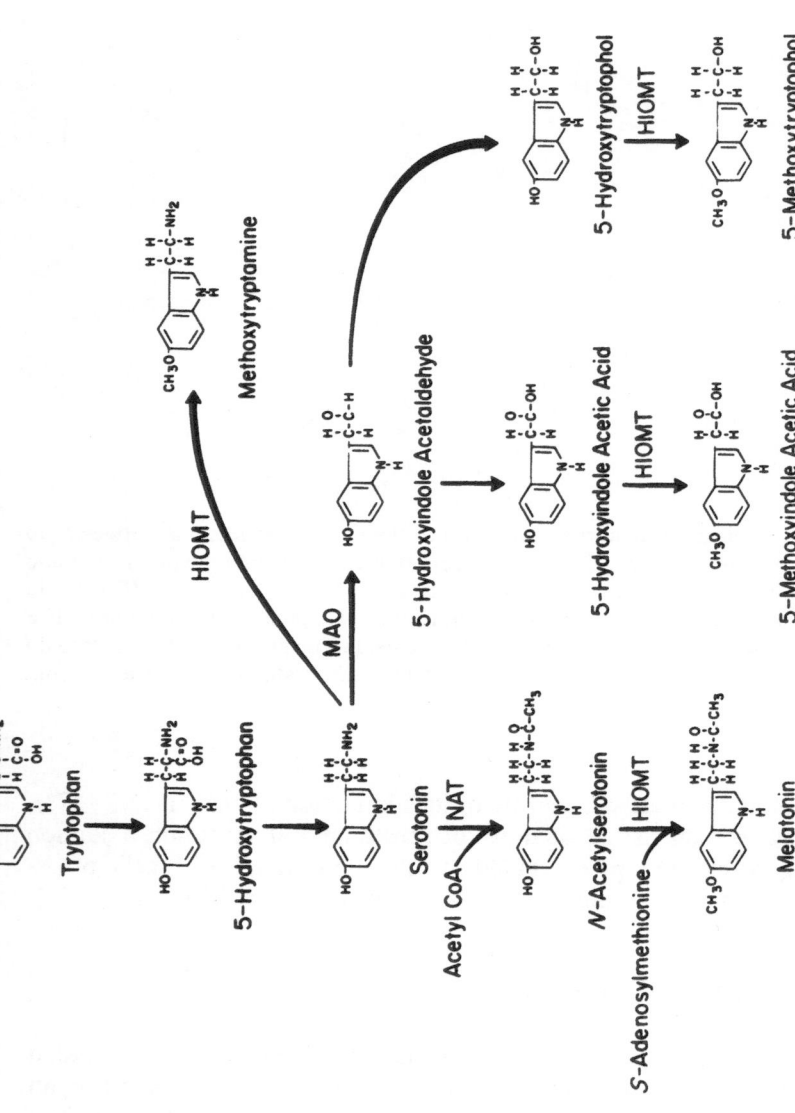

FIGURE 4.4. Tryptophan metabolism in the pineal gland. Melatonin is the pineal hormone released into the systemic circulation. Other possible pineal hormones are methoxytryptamine, N-acetylsero-tonin, and 5-methoxytryptophol. MAO = mono-amine oxidase; NAT = N-acetyltransferase; HIOMT = hydroxyindole-O-methyltransferase.

5-HTP in the presence of tryptophan hydroxylase; 5-HTP is then metabolized to 5-HT by aromatic amino acid decarboxylase (48,73). The serotonin content of the pineal gland is normally very high, exceeding its concentration in other organs (78,79).

Serotonin can be converted to a variety of indoles within the pineal gland (Fig. 4.4). At night its metabolism to melatonin is particularly active. A major reason for this seems to be the marked increase in the activity of NAT, which N-acetylates serotonin to N-acetylserotonin (Fig. 4.2). The activity of this enzyme in the rat pineal is increased up to 100-fold at night (80,81) while in other species the nocturnal rise is less robust (82). Because of this conversion, N-acetylserotonin also is higher at night than during the day (83). The final step in the synthesis of melatonin requires the action of hydroxyindole-O-methyltransferase (HIOMT) (84). Whereas initial reports claimed that this enzyme activity also increases at night (85,86), more recent studies discount this rise. It appears that HIOMT either exhibits no or a very weak 24-hour rhythm within the pineal gland of a variety of mammals (87–89). Nonetheless, mass action is sufficient to convert large quantities of N-acetylserotonin to melatonin at night, resulting in a large increase in melatonin content of the pineal gland in all mammals studied to date (90–95). Although several factors can influence the actual quantity of melatonin formed, the activity of the enzyme which N-acetylates serotonin seems to be a primary process in controlling melatonin formation.

Once melatonin is formed within the pineal it seems to be rapidly released (Fig. 4.3); as a consequence the blood levels of melatonin are typically highest during the night (95–98). Little is known concerning the mechanism of melatonin release, but considering the highly lipophilic nature of the molecule it is generally thought that, once produced, it diffuses passively out of the pinealocyte. In any case, there seems to be no prolonged storage of large amounts of melatonin within the pineal gland such as is characteristic of hormones produced by many other endocrine organs.

Pineal 5-HT can also be metabolized to several other potentially active compounds. As in other tissues, 5-HT is acted upon by monoamine oxidase (MAO) with eventual conversion, among other metabolites, to 5-methoxytryptophol (Fig. 4.4), a compound that reportedly exhibits a 24-hour rhythm in the blood (99), and that, according to a number of publications, has endocrine effects similar to those of melatonin (100–105).

Research on the functions of 5-methoxytryptophol will likely benefit from a recently developed improved radioimmunoassay (106). Skene and colleagues found that the daytime pineal concentrations of 5-methoxytryptophol varied greatly among species. Thus, afternoon levels were found to be 0.052 pmol/pineal gland in the rat, 0.539 pmol/gland in the Syrian hamster, 1.73 pmol/pineal in the sheep, and 7.15 pmol/gland in the tortoise. The 5-methoxytryptophol/melatonin ratio also exhibited great interspecies variations with values of 0.02 in rat, 0.22 in sheep, 2.7 in hamster, and 17 in the tortoise. The presence of 5-methoxytryptophol in the mammalian pineal gland, albeit in different amounts, has been repeatedly confirmed (107–110).

The immediate precursor of 5-methoxytryptophol, 5-hydroxytryptophol, has also been examined in terms of potential endocrine actions. In a few studies this compound has been shown to have biologic activity, but in general the information is sparse and not especially convincing (111). If 5-hydroxytryptophol is indeed released from the pinealocyte and is biologically active, its mechanism of action remains unidentified.

HIOMT, the melatonin forming enzyme, can also act directly on 5-HT to convert it to 5-methoxytryptamine (112). Although this compound is produced in the pineal gland, its secretory processes, if any, remain unknown. The action of exogenously administered 5-methoxytryptamine on the reproductive system is similar to that of melatonin (113,114). Its potency, however, seems to be one-tenth that of melatonin (115–118).

Factors Influencing Pineal and Plasma Melatonin Levels

Regardless of whether the locomotor activity of mammalian species is diurnal, nocturnal, or crepuscular, peak pineal melatonin production and plasma levels of the hormone are greatest at night. In general, nighttime pineal melatonin values exceed those measured during the day by a factor of 2 to 12. The pattern of nocturnal melatonin production, however, seems to vary among mammals. Three different patterns of nocturnal melatonin production have been described (119). This classification is arbitrary and, for any given mammal, the nocturnal pineal melatonin pattern may vary either according to the previous environmental history of the animal, or to some yet unknown factors (120). However, when the 24-hour melatonin rhythm for a species is compared in different laboratories in which the animals are kept under similar environmental conditions, it seems to be more-or-less reproducible.

We have made an observation concerning the 24-hour pineal melatonin rhythm, particularly in the Syrian hamster, which is still unexplained. Typically, when this species is kept under a light/dark (LD) environment of 14:10, peak melatonin production usually occurs about 8 hours after the onset of darkness with the peak being of rather short duration. However, in a few cases we have noted that the peak may be delayed for 1 hour, occurring 9 hours after lights are out. What is interesting is that on any given night, the melatonin rhythms among different animals seem to be closely synchronized; thus, the rhythm seems to be delayed in all

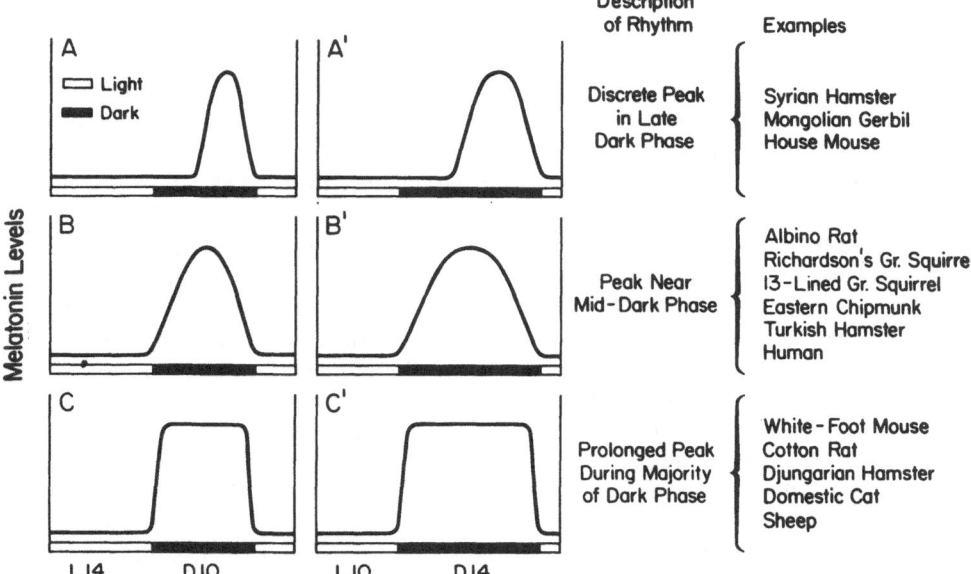

FIGURE 4.5. Provisional classification of the patterns of melatonin production (A, B, and C) and the animals in which they occur (right). Regardless of the specific nocturnal pattern of melatonin production, prolongation of the dark period increases the duration of the melatonin surge (A' B', and C').

animals. This judgment is based on the small standard errors when group means are compared. The implication is that animals, at least highly inbred Syrian hamsters, kept under essentially identical conditions may possess a means of synchronizing their pineal melatonin production rhythms. Whether this phenomenon will stand up to additional scrutiny and what the signalling process (pheromonal?) may be must await further experimentation. Aside from this casual observation, there are no reports of a presumed synchronization phenomenon in other mammals.

The Syrian hamster has what is referred to as a Type A melatonin rhythm (119) (Fig. 4.5). Other species in which a similar melatonin pattern is observed are the Mongolian gerbil (*Meriones unguiculatus*) (121,122) and the house mouse (*Mus domesticus*, C57B1/J6 strain) (123). If these species are kept under LD cycles of 14:10, increased melatonin production after the onset of darkness is delayed for several hours; during this time pineal melatonin concentrations remain equivalent to those during the day. Following this interval, melatonin levels increase rapidly to reach a peak that is sustained only briefly (<1 hour). Thus, in these species melatonin levels above those measured during the day are observed only for 4 to 5 hours.

The Type B (Fig. 4.5) melatonin pattern seems to be the most common and is observed in the following species: Turkish hamster (*Mesocricetus brandti*) (124), albino rat (125), eastern chipmunk (*Tamias striatus*) (56), Richardson's ground squirrel (*Spermophilus richardsonni*) (89), Mexican ground squirrel (*Spermophilus mexicanus*) (126), 13-lined ground squirrel (*Spermophilus tridecemlineatus*) (120), and in the human (68,127). In the human, of course, the nocturnal pineal melatonin rhythm has been most often deduced from studying the level of this constituent in the blood. Also, the human, more than any other species, seems to exhibit greater variation in both the nature of the 24-hour melatonin rhythm and the magnitude of the nighttime increase (128–130). Attempts to compare pineal melatonin concentrations in humans who expired during the day or the night generally support high nocturnal pineal melatonin production (131–133); the same is true when enzyme activity is measured in postmortem human pineal tissue (134,135). The Type B pattern is characterized by a gradual increase in melatonin levels in the pineal, beginning at the onset of darkness and reaching a peak near the middle of the dark phase. Melatonin levels fall during the latter half of the dark phase (98,119).

A third pattern (Type C) of pineal melatonin production is observed in the white-footed mouse (*Peromyscus leucopus*) (136), the domestic sheep (97,137), the cotton rat (*Sigmodon hispidus*) (138), the Djungarian (dwarf) hamster (*Phodopus sungorus*) (124), and the domestic cat (*Felix domesticus*) (96). In these species melatonin peak levels are reached shortly after the lights are turned off, remain high for the duration of the dark period, and then fall just prior to the onset of light or are truncated by turning the lights on. Regardless of the type of the 24-hour melatonin rhythm, lengthening the daily park period prolongs the duration of elevated melatonin levels, up to a limit. Thus, if the nights are increased to 16 hours in the case of the cotton rat, for example, melatonin values will only remain high for about 10 hours (139), furthermore, melatonin levels do not remain persistently elevated under continued darkness (140,141).

The pattern of melatonin secretion cannot be predicted on the basis of the animal species. Thus, nocturnal and diurnal mammals may have similar melatonin rhythms; but rodents cannot be distinguished from ungulates. Finally, whether blood melatonin levels over a 24-hour period correspond to the amount or rhythm present in the pineal gland has not been determined for many species. In the case of the albino rat (140) and Syrian hamster (Fig. 4.6), however, there seems to be a strong correlation between pineal and blood melatonin concentrations.

The subsequent paragraphs describe the impact of a number of extrinsic and intrinsic factors on the conversion of serotonin to melatonin

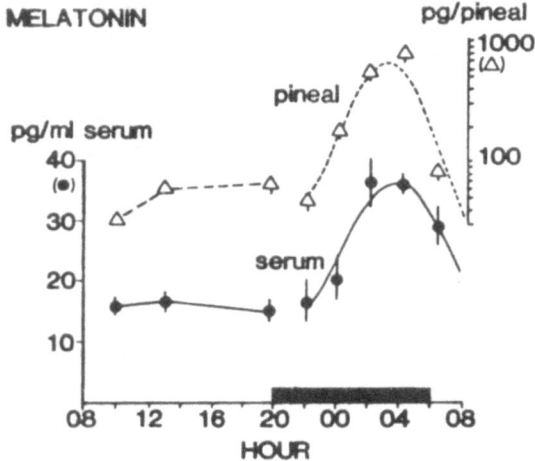

FIGURE 4.6. The close correlation between pineal and serum melatonin levels in the Syrian hamster is depicted in this figure. Nighttime pineal melatonin production is accompanied by a similar rise in serum levels of the indole. Black bar on abscissa represents the daily dark period. Figure provided by G.M. Vaughan.

in the mammalian pineal gland. Some of these factors clearly influence, in a dramatic way, the ability of the pineal gland to produce melatonin while others have only a slight modulatory influence.

Photoperiod

Perhaps the single most important factor regulating indole synthesis by the pineal gland is the LD cycle to which the animals are exposed. It has already been pointed out that pineal melatonin production is maximal during the normal nighttime, while intervals of light are associated with low levels of melatonin synthesis. Characteristically, increasing the duration of daily dark exposure also leads to longer intervals of high melatonin production. The imposition of light during the night abruptly curtails the conversion of serotonin to melatonin, causing a rapid decline in both blood and pineal melatonin concentrations in mammals (Fig. 4.7) (89,141) including man (142). At least in the pineal gland, melatonin plummets rapidly with a half-life (t_1) of slightly less than 10 min; in the cotton rat, the t_1 seems to be somewhat faster, on the order of 2 minutes (143). Under any circumstance, 20 to 25 minutes after acute light exposure during the night, basal daytime levels of melatonin are achieved, provided the light is sufficiently bright and intense (144).

The sensitivity of the pineal gland of different mammals to light intensity (irradiance) varies widely. If Syrian hamsters or albino rats are

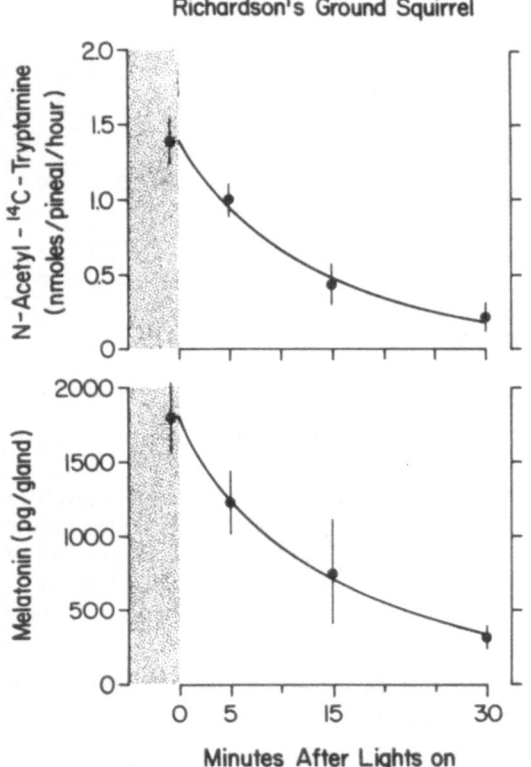

FIGURE 4.7. When animals are acutely exposed to a brief period of light during the night, pineal NAT activity (top) and melatonin levels (bottom) drop precipitiously as indicated here for the Richardson's ground squirrel. This occurs in all mammalian species with a t_A of less than 10 min. The shaded area represents darkness. From Reiter et al. (89).

merely exposed to normal laboratory light (which has an irradiance of roughly 50 to 150 μW/cm^2) during the normal nighttime period, pineal melatonin levels never rise (67,141). In some other species this is clearly not the case; hence, in the diurnally active human (145), eastern chipmunk (56), Richardson's ground squirrel (89), and 13-lined ground squirrel (120) usual room light is of insufficient brightness to inhibit pineal melatonin production. Even when Richardson's ground squirrels are exposed to continual room light for 7 days the pineal melatonin rhythm does not disappear, although it is somewhat dampened (146).

Although the problem has not been adequately investigated, an irradiance of roughly 2 to 3 times the normal room light appears to inhibit plasma melatonin levels in psychologically normal humans (147). In rodents the inhibitory irradiance for pineal melatonin production varies greatly. For example, the albino rat pineal responds to a very low light irradiance (0.0005 μW/cm^2) at night (148), while the minimal light irradiance required to suppress pineal indoleamine activity in the Syrian hamster is about 0.1 μW/cm^2 (Fig. 4.8) (149). Finally, both the

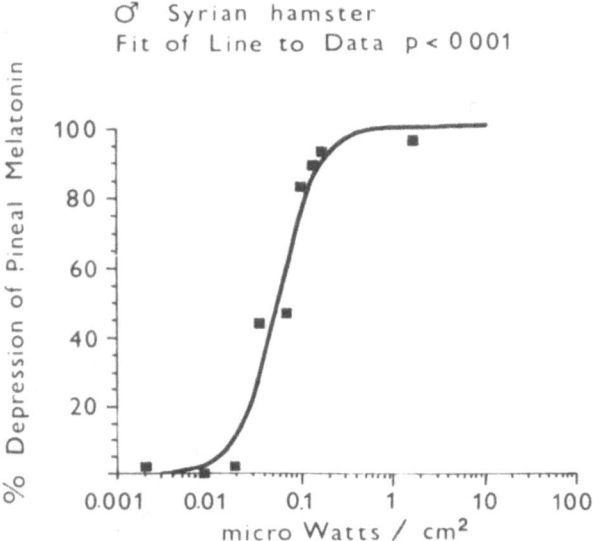

FIGURE 4.8. Dose–response curve illustrating percent depression of pineal melatonin in the Syrian hamster after exposure to a number of different light irradiances (in $\mu W/cm^2$). An irradiance of roughly 0.1 $\mu W/cm^2$ totally inhibits melatonin production in this species. [Reprinted with permission from ref. 149, © The Endocrine Society, 1983.]

Richardson's ground squirrel (89) and the eastern chipmunk (56) must be exposed to much brighter light (up to 1,850 $\mu W/cm^2$) before melatonin levels drop. Simple calculations reveal that the albino rat pineal is more sensitive to light inhibition than is that of the ground squirrel by a factor of 3.6×10^7. In these studies the light was provided by fluorescent bulbs.

To put these light irradiances into perspective for the reader, full sunlight on a clear day can have a brightness of up to 50,000 $\mu W/cm^2$, normal indoor lighting is of the order of 50 to 150 $\mu W/cm^2$, while moonlight usually does not have an irradiance greater than 0.2 $\mu W/cm^2$.

Regardless of the light irradiance required to suppress pineal melatonin production, the inhibition seems to be an all-or-nothing phenomenon, at least in rodents. Thus, if the light is sufficiently bright pineal melatonin production falls to basal levels; conversely, if it does not reach a brightness equivalent to the minimal irradiance required for inhibition, pineal melatonin content remains at high levels. Thus, there seems to be no partial inhibition of pineal melatonin production (149,150), although one unconfirmed study indicates that in humans, graded intensities of light cause partial suppression of circulating melatonin values (147).

The studies described above indicate also that the pineal gland of nocturnally active mammals seems to be more sensitive to light inhibition than the gland of diurnally active animals. This may relate to the structure of the retina in these species. Typically nocturnal mammals have rod-dominated retinas while diurnal animals have predominantly cone retinas. Because of the visual pigments associated with these photoreceptor types, light effects on the pineal may be influenced accordingly. This issue will be discussed more extensively below.

Another interesting feature concerning the acute inhibition of nocturnal pineal melatonin production by light is the duration of the exposure required. Using NAT activity as an index of melatonin production, a 1-minute light exposure at night was found to totally arrest the melatonin synthesis activity of the rat pineal gland (151,152). Five seconds are enough in the Richardson's ground squirrel (89) while a 1-second light pulse suffices in the cotton rat (153) and the Syrian hamster (154). Once melatonin levels begin to diminish usually they must return to low daytime values before than can start up again. In the Syrian hamster (154) and in the albino rat (152) the timing of the light exposure during the dark period is instrumental in determining whether the return to darkness will reinitiate pineal melatonin production. Thus, if the brief light exposure occurs early in the dark phase, the reexposure to darkness is followed by elevated melatonin production; conversely, later in the dark period a return to darkness (after brief light exposure) is not associated with the increased conversion of serotonin to melatonin (154).

Whether moonlight can suppress pineal melatonin synthesis, especially in nocturnally active animals, is an important issue. Indeed, it was proposed that for nocturnal fossorial rodents, such as the Syrian hamster, the pineal melatonin rhythm in the wild may be 180° out of phase with that generally seen in the laboratory (155). This speculation is based on the following rationale. These animals are usually in underground darkened burrows during the day and they emerge above ground at night, when moonlight might be sufficient to suppress pineal melatonin synthesis. Thus, maximal production of the indole really occurs during the day when the animals are primarily under ground. The effect of natural moonlight has been checked in two nocturnal rodents, the Syrian hamster (156) and the cotton rat (153). In neither, however, was moonlight capable of significantly restraining the ability of the pineal to produce melatonin, even though the irradiance of the moon was sufficient to suppress pineal activity. This suggested the idea that some other property of light may also be important in determining its ability to act on the pineal gland of these species and the need to investigate the effect of different wave-lengths (157,158).

Indeed, in Syrian hamsters, wavelengths in the blue range (around 500 nm) are most effective in reducing melatonin production in the pineal gland, while other visible wavelengths are minimally active or not at all (158). More recently, it was shown that near ultraviolet irradiation (UV, 360 nm), generally considered to be outside the visible range, is also capable of suppressing the pineal melatonin content in the Syrian hamster

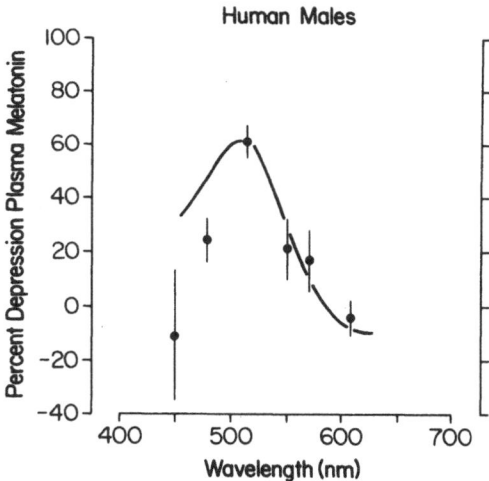

FIGURE 4.9. Depression of circulating melatonin levels in humans after exposure to different wavelengths of light. Wavelengths at 509 nm were most inhibitory to plasma melatonin titers. The curved line represents the spectral sensitivity of human rods. These data indicate that the light induced suppression of plasma melatonin in humans is mediated by a retinal photopigment. However, it is premature to identify this pigment as either rhodopsin or any other known cone photopigment. Redrawn from Brainard et al. (160).

(159). This did not happen if the animals had been surgically blinded, suggesting that the eyes are involved in mediating this inhibitory effect. Rhodopsin can absorb photons in the range of 300 to 400 nm, although the efficiency of absorption is much greater at wavelengths in the 450 to 550 nm range. The ability of UV light to suppress melatonin synthesis in the hamster pineal is a novel observation that should be studied in other species.

The ability of blue light to inhibit hamster pineal synthesis and the inhibitory effect of a 509 nm light on humans (Fig. 4.9) (160) are consistent with, but certainly do not prove that the visual pigment rhodopsin may be part of the mechanism involved in light inhibition of the pineal gland. If this were true, it could explain why the pineal gland of nocturnal animals, whose retinas are rich in rhodopsin-containing rods, seem to be more sensitive to light inhibition than the gland of diurnally active mammals.

Extremely Low Frequency (ELF) Electrical Fields

The biologic effects of ELF electric fields (60-Hz) remain undefined. Several reports claim that such fields have behavioral effects which may be related to basic alterations in central nervous system (CNS) physiol-

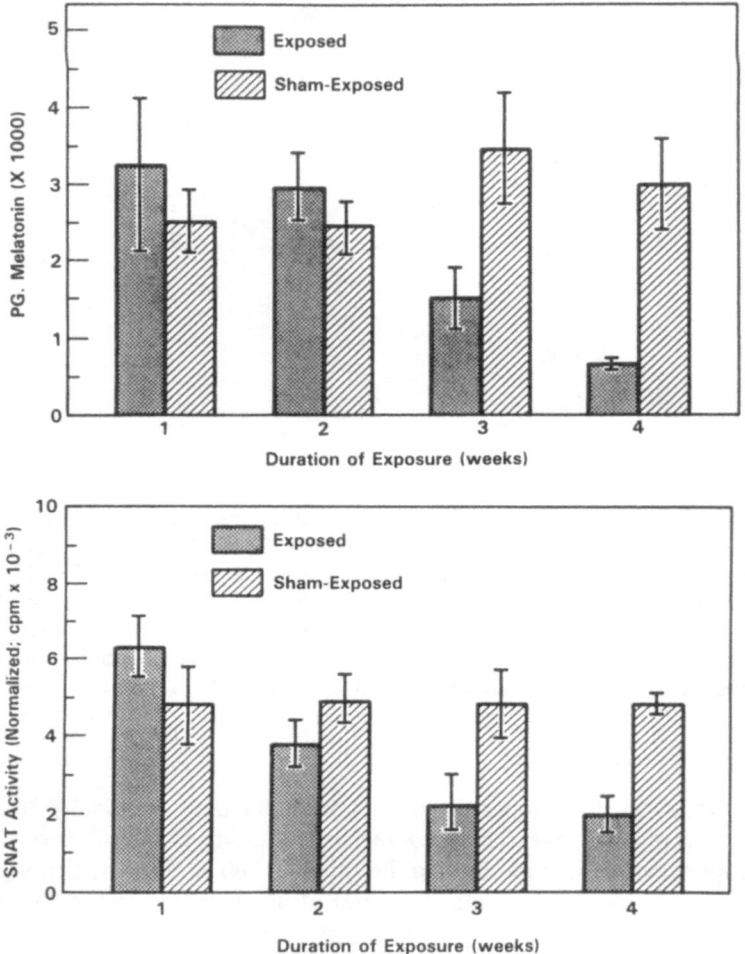

FIGURE 4.10. Pineal serotonin N-acetyltransferase (top) and melatonin levels (bottom) in the pineal gland of control rats (sham exposed) and of rats exposed to extremely low frequency electric fields (60-Hz) (exposed). Pineals were collected after either 1, 2, 3, or 4 weeks. With prolonged exposure to these fields, pineal NAT and melatonin levels fell gradually. From Wilson et al. (162).

ogy. Pineal function is one endpoint that has been used to identify neuroendocrine changes associated with chronic exposure to 60-Hz electric fields. Male rats exposed to an ELF electric field (65 Kv/m) for 30 days were found to have depressed nocturnal increases in NAT activity and melatonin content while 5-methoxytryptophol levels were increased (161). The authors interpret the changes in pineal synthetic activity as a reflection of alterations in CNS function. Although the authors do not specify the possible neuronal site involved, it may be that the SCN, which control the circadian production of melatonin, are influenced by the electric fields. It would have been of interest to determine whether other circadian rhythms that depend on the integrity of the SCN were affected by the treatment.

In a second report, the same group confirmed the suppression of the nocturnal rise in pineal melatonin in rats exposed to a 60-Hz electric field for 30 days (162). By sampling pineal tissue at weekly intervals after the onset of exposure to the ELF field, they observed a progressive reduction in both the acetylating activity and the melatonin content of the gland (Fig. 4.10). They also proved that the deficit was not permanent, since within 3 days after removing the rats from the 60-Hz field the normal nocturnal rise in pineal NAT activity and melatonin levels were re-established (162). Again, the authors strongly imply that the alterations in pineal physiology are likely due to changes in the neural message that the CNS transfers to the pineal gland. It is also possible that the suppression of the augmented pineal levels of melatonin normally measured at night is due to field-induced firing of neurons in the visual system which could mimic the effect of light exposure.

Animals and humans are exposed to ELF electric and magnetic fields, many times greater than normal ambient levels, when they are in the vicinity of energy generating, storage, and transmission systems. It is possible that long-term exposure to such fields could disrupt CNS physiology, thereby altering the melatonin synthesis activity of the pineal gland. It is conceivable that potentially deleterious consequences of ELF fields could be monitored by examining the 24-hour urinary excretion rate of either melatonin or its major metabolites.

Age

In some animals, advanced age is associated with substantial nocturnal reductions in the ability of the pineal gland to produce melatonin while low daytime levels seem to be minimally affected. In 18-month-old male and female Syrian hamsters, for example, the night-time increase in pineal melatonin is severely blunted compared to that observed in 2-month-old control animals (Fig. 4.11) (163). The old female but not the male hamsters used in this study were well beyond their reproductive age (these animals rarely live more than 22 to 24 months). The reduction in the nocturnal melatonin increase may be related to a relative insensitivity of the β-receptors on the pinealocyte membrane, although other possible explanations have not been excluded (164). Remarkably, despite the greatly reduced pineal and plasma melatonin rhythms in these animals, if the old hamsters are kept under short daylight conditions, their neuro-endocrine–reproductive system exhibits pineal-induced regression, simi-

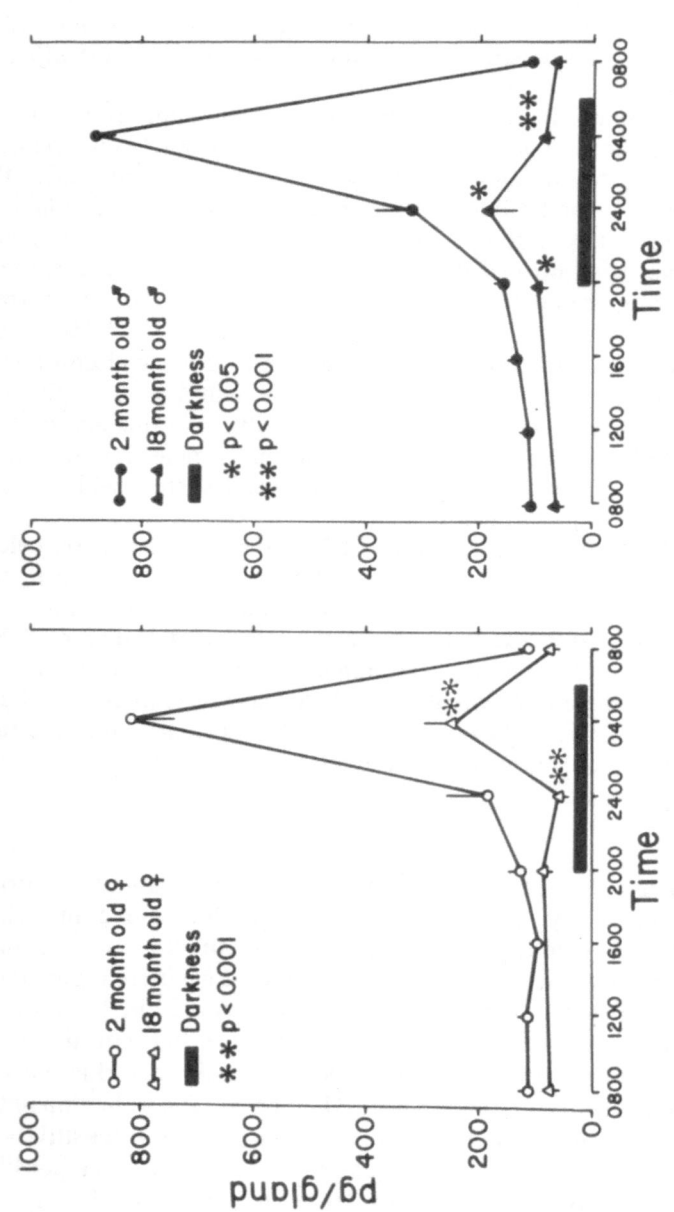

FIGURE 4.11. Pineal melatonin levels in female (left) and male (right) Syrian hamsters aged 2 or 18 months. Since hamsters rarely live beyond two years of age, the 18-month-old animals are quite old. The drop in pineal malatonin production as a function of increasing age is common in the species in which it has been adequately tested. * $p < 0.05$; ** $p < 0.001$. [From Reiter et al. (163). Copyright 1980 by the AAAS.]

lar to that observed in young hamsters (165). Thus, although the melatonin rhythm may diminish with advanced age, the endocrine capabilities of
the pineal gland apparently do not.

Similar reductions in melatonin production with advanced age have
been observed in rats (165) and gerbils (167,168). In the rat, pineal NAT
and melatonin levels appear to behave independently; while melatonin
levels fall gradually with age (166), NAT activity remains at the level
characteristic of young animals. This reduction in pineal melatonin
concentration is accompanied by a fall in the blood levels of this
constituent and of its immediate precursor, *N*-acetylserotonin (169,170).

In the human, effects of the aging process on plasma melatonin levels
and on the urinary excretion of its metabolites have been described, but
the findings are not uniform. A number of reports claim that age has no
effect on the level of melatonin or its metabolites in body fluids (171–174),
while other reports (175–178) suggest that these levels decline with age.
This discrepancy may be due to the age of the individuals used in these
studies, to their health, to the medications that they were taking, to the
time of year when the studies were conducted, and to other factors. The
report by Iguchi et al. (175) seems to be one of the most convincing; this
group found that the nocturnal rise in plasma melatonin was significantly
smaller in old (mid-80s) than in young (mid-20s) males. This is in
agreement with a recent report (178) that urinary levels of 6-hydroxymela-
tonin, the principal hepatic metabolite of melatonin (179), decrease
gradually between 20 and 100 years of age.

If the data from the animal and human studies are evaluated col-
lectively, it appears that advancing age is, in fact, related to a reduced
ability of the pineal gland to produce and perhaps secrete melatonin. With
the exception of one study (165), the consequences of this reduction have
not been explored.

Calcification

The deposition of calcium within the pineal gland is well known but is not
a universal feature among mammals (180,181). It is commonly observed in
humans, although its significance remains unclear. Corpora arenacea
(brain sand, concretions, acevuli) were initially thought to be related
strictly to aging, although this is currently debated (182); they are,
however, more frequent in the pineal glands of older animals. In humans,
it has been speculated that pineal calcification may be more frequent in
patients with tumors (183,184) or with various endocrinopathies
(185,186).

The corpora arenacea consist of needle-shaped crystals which may
form bundles, these are usually embedded in an amorphous material of
moderate electron density (187). They are composed of either hydroxyap-
atite or carbon apatite (188,189) with small amounts of strontium and

other elements (190,191). The organic matrix of the corpora arenacea is made up of glycoproteins and proteoglycans (180,181,192). These structures are either intracellular or extracellular (187,193) and, when they enlarge intracellularly, the cell usually degenerates.

The study of pineal calcifications is limited by the fact that they occur more commonly in large animals, e.g., monkey, cow, donkey, horse, etc. and are rarely seen in the rat (194,195) and guinea pig pineal gland (196). One exception is the Mongolian gerbil, *Meriones unguiculatus* (182,193,197) and virtually everything that is known concerning the origin, distribution, and factors controlling the formation of corpora arenacea comes from the study of this animal.

The number of corpora arenacea within the pineal gland of the gerbil can be experimentally manipulated. Thus, if the gerbil's superior cervical ganglia are removed at a young age, before corpora arenacea are present in the pineal gland, their formation is prevented, suggesting that it is related to some metabolic process of the pineal gland and dependent upon the sympathetic innervation of the gland itself (198). Furthermore, the exposure of gerbils to reduced amounts of light enhance concretion formation (199), while the daily treatment of gerbils with propranolol, a β-receptor blocking agent, retards the formation of corpora arenacea (200). Remarkably, it was recently demonstrated the sympathetic denervation not only prevents the formation of corpora arenacea, but if the superior cervical ganglia are removed from adult gerbils with heavily calcified pineal glands the concretions are markedly reduced within 12 weeks, further emphasizing the lability of these deposits (201). In general, the reported findings strongly suggest a relationship between the deposition of calcium in the pineal gland and the functional status of the organ, i.e., highly active pineal glands exhibit an increased number of corpora arenacea whereas an inactive pineal gland is likely to contain few calcified deposits.

A specific functional relationship between the corpora arenacea and the metabolism of the pineal gland has yet to be firmly established. It has been speculated that the concretions may form as a result of the secretion of pineal polypeptides (202), that they may be related to altered lymph flow within the gland (180), or that they may be associated with hypertension (182,203) or stress (204).

In this regard, Vaughan et al. (205) found that the subcutaneous placement of melatonin-beeswax pellets into gerbils reduced the incidence of pineal concretion formation. It may also be of interest that the gerbil, which has the greatest number of pineal calcium deposits reported for any species, also has the weakest of all known 24-hour pineal melatonin rhythms (122,167). The implication is that pineal calcification may be associated with a reduced ability of the pineal gland to convert serotonin to melatonin. This supposition, however, is based on circumstantial evidence and, given the marked variability in the apparent magnitude of the 24-hour melatonin rhythm among mammals, the weak melatonin cycle in gerbils may represent a characteristic of the species.

It has been claimed that, in the human, the enzyme activities associated with pineal indoleamine metabolism do not change with age; however, specific studies correlating the activities with the degree of pineal calcification are few (182,187). Pineal calcification in the human varies greatly among individuals and can be identified with various radiologic and nonradiologic methods and it would be of considerable interest to correlate the degree of pineal concretion with the 24-hour plasma melatonin pattern.

Genetics

Genetic factors that determine the rhythmic production of melatonin have received very little attention. However, several recent reports suggest that selective breeding can substantially change the production of melatonin by the pineal gland. This is important to investigators who use highly inbred strains of animals, obtained from commercial suppliers who select their breeding stock to produce the largest litters. Considering the influence of melatonin on reproduction, such selective procedures could well play a role in determining the type or magnitude of the melatonin rhythm in commercially available animals. This emphasizes the importance of studying species captured in the wild in conjunction with those that have been commercially bred. Genetic factors may also explain the occasional interspecies variations in the response to pinealectomy.

The work of Ebihara and colleagues (123) illustrates how artificial selection may markedly change the 24-hour pineal melatonin rhythm. These investigators compared radioimmunoassayable pineal melatonin levels in various strains of mice over a 24 hour period and found that several inbred strains, e.g., C57BL/J6, AKR/J, BALB/c, and NZB/ BLNJ, always had pineal melatonin contents (during either day or night) below 10 pg/gland, which was the lower limit of detection in their assay. Interestingly, each of these strains of mice had been inbred for over 40 years. Other strains of mice that had been inbred for a shorter time (<30 years), and one strain bred in the laboratory for only 5 years, exhibited a robust nocturnal increase in pineal melatonin. The 24-hour melatonin levels in the pineal gland of the C57BL/6J and the field-derived strain (FDS) are shown in Fig. 4.12.

To determine whether genetic factors were involved in the observed pineal melatonin differences between these strains, C57BL/J6 were crossbred with FDS mice and the pineal melatonin content of the F_1 hybrids was estimated during the night when melatonin levels would be expected to be high. The pineal gland of the hybrids was found to contain measurable quantities of melatonin. Measurements of the segregation patterns of pineal melatonin in the offspring of three crosses, $F_1 \times F_1$,

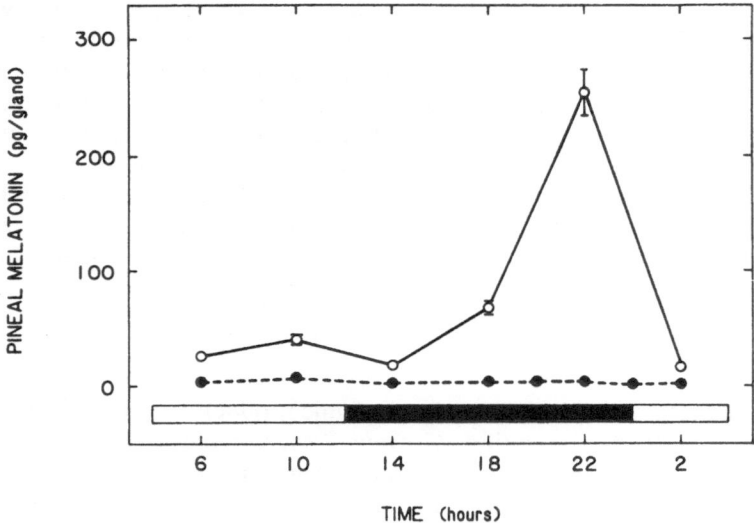

FIGURE 4.12. Pineal melatonin content in the C57BL/J6 (dots) and in the FDS field derived mouse strain (circles) over a 24-hour period. The bar along the abscissa represents the light/dark cycle. The absence of a melatonin rhythm in the C57BL/J6 mice is due to a genetic absence of NAT and HIOMT. Each point represents the pineals of at least 10 males and 10 females (mean ± SEM). [Reprinted from Ebihara et al. (123). Copyright 1986 by the AAAS.]

F_1 × C57BL/J6, and F_1 × FDS, allowed the prediction that the pineal melatonin deficiency in the C57BL/J6 mice is a consequence of mutations in two independently segregating, autosomal recessive genes. In the pineals of this strain both NAT and HIOMT activities are absent.

Ebihara et al. (123) concluded that the absence of the melatonin rhythm in the pineal of some mouse strains may be an important factor in the successful breeding of these animals since melatonin is capable, at least under certain conditions, of suppressing reproduction in rodents. However, in the specific case of the strains of mice that were used in these studies, the stimulatory or inhibitory effects, of melatonin on reproduction are essentially unknown.

Shortly after this report, Maestroni and co-workers (206), in a study designed to investigate melatonin interactions with the immune system, determined the 24-hour levels of melatonin in the serum of female C57BL/J6 mice and observed a nyctohemeral rise (roughly 4× greater than basal daytime values) with a short peak (<1 hour) occurring 4 to 5 hours after the lights were turned off. A similar nocturnal rise in serum melatonin in this strain of mouse was observed also by G. M. Brown (personal communication).

The apparently opposing results reported for pineal (123) and serum melatonin (206) in the C57BL/J6 mouse suggest the interesting possibility that serum melatonin in the C57BL/J6 mouse may derive from nonpineal sources, such as the retina (207,208) and the Harderian gland (209–211), even though there is no evidence that these organs can release it, especially in a cyclic manner, and even though, in the C57BL/J6 mouse there is a genetically determined deficiency of the melatonin forming enzymes, NAT and HIOMT (123).

A difference in the melatonin content of the pineal glands of different strains of rats has also been reported (212). In this study, six strains and/or stocks of rats were acclimated for at least 2 weeks under a LD cycle of 12 : 12 (lights on 0500 hours) and their pineals were then taken for melatonin analyses 5.5 to 6.5 hours after the lights were turned on. Remarkably, pineal melatonin content from these rats varied from 30 to 5,000 pg/gland, with large individual and strain differences. Twelve of the 65 glands that were assayed had pineal melatonin levels greater than 900 pg and all came from two outbred rat stocks. The remaining 53 glands had melatonin values of 25 to 500 pg/gland; some of these were higher than the mid-light levels reported by other workers (213–215).

In general the results of this study indicate that inbred strains of rats have more uniform daytime melatonin levels than outbred stocks, even though individual strains of animals and/or stocks tended to have different daytime melatonin levels possibly for genetic reasons.

In humans, also, there are marked individual variations in the 24-hour level and pattern of serum melatonin, suggesting substantial variations in the ability of the pineal gland to either synthesize or secrete the indole. Some individuals reportedly exhibit no nighttime rise in circulating melatonin (216,217) possibly due to genetic factors or to specific disease states. Examples of disease states in which a plasma melatonin cycle is reportedly absent are psoriasis vulgaris, Klinefelter's syndrome, Turner's syndrome, and spina bifida occulta (217). These observations, however, have not been independently confirmed.

Stress

The effects of aversive stimuli on pineal melatonin production are undoubtedly more complex than originally envisaged (218). It has been assumed that the large amounts of catecholamines released from the adrenal medulla and other sources in stressful situations would stimulate pineal melatonin production. However, Parfitt and Klein (219) have argued that this cannot occur because the sympathetic nerve endings in the pineal gland sequester excess catecholamines by active neuronal uptake, thereby preventing stimulation of the pinealocyte (Fig. 4.13). This theory is based on the observation that certain stressors, i.e., swimming for 2.5 hours in 29–32°C water, do not induce a large rise in pineal NAT

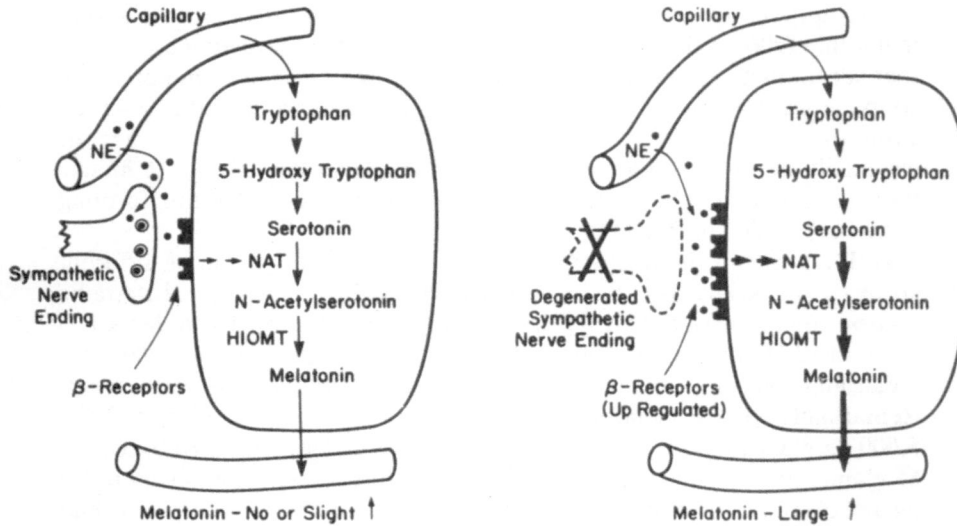

FIGURE 4.13. Proposed mechanism for the failure of circulating catecholamines to stimulate pineal melatonin production, suggesting that the sympathetic nerve endings in the pineal gland sequester excess circulating catecholamines (left). If the superior cervical ganglia are removed or if the animals are treated with a catecholamine uptake inhibitor (right), the circulating NE can act on the pinealocyte to stimulate melatonin formation. This hypothesis does not explain why some types of stress associated with high plasma catecholamines levels stimulate pineal melatonin production. From Reiter (218).

activity unless a catecholamine uptake inhibitor, e.g., imipramine, is given before the stress occurs (219).

In other cases, the pineal gland does respond to circulating catecholamines. For example, when rats are kept under continuous light for 48 hours, a procedure known to increase β-receptor density on the pinealocyte membranes (51), either insulin-induced hypoglycemia or immobilization stress dramatically increases the NAT activity (300- and 100-fold, respectively) of the pineal gland and, presumably the synthesis of melatonin. This response is not observed when hyperinsulinism is produced by a high carbohydrate diet or when the adrenal glands have been removed, the latter indicating that something of adrenal origin, presumably medullary catecholamines, is responsible.

Subsequent studies showed that immobilization of rats for 3 hours during the day led to a 50% rise in pineal NAT activity (220) and that pineal NAT and melatonin rise in the rats with daytime insulin-induced hypoglycemia (221), or following daytime exposure to a variety of stressful stimuli, such as immersion in cold water, noise, novel environment, exposure to ether vapor or to cold temperature, or immobilization (222,223). For unexplained reasons others (224) were unable to confirm these results in rats, although in the Syrian hamster, insulin-induced hypoglycemia and the associated rise in plasma catecholamines actually decrease pineal HIOMT activity and melatonin content (221). Thus, although the number of studies in which the Syrian hamster was used is limited, it appears that rat and hamster pineals respond very differently to circulating catecholamines, which might be expected in view of the fact that the rat pineal gland can be stimulated during the day by exogenously administered isoproterenol or NE (when preceded by a NE uptake inhibitor) (54,75) whereas the Syrian hamster pineal gland cannot (58,59).

The effect of chronic stress on the pineal melatonin rhythm has been investigated rather sparingly. On a number of occasions, the pineal NAT and melatonin cycles have been described in rodents captured in the wild and held in captivity for 7 to 14 days. Judging from their behavior, these animals were under a great deal of stress, yet they had a 24-hour melatonin rhythm comparable to that of domestic species kept under similar conditions (56,89,126,225). Thus, chronic stress, at least that associated with capture and caging, does not noticeably impair the 24-hour rhythm of melatonin production.

The bulk of the studies enumerated above describes the responses of the pineal gland to aversive stimuli during the day, when melatonin synthetic activity is at basal levels. This was obviously done because it would be expected that if stress has any influence on melatonin production it would likely be of a stimulatory nature. In studies conducted at night, when melatonin production is already high, the results are dramatically different. For example, Joshi et al. (226) observed that 1.5 cc saline injected into the hind leg of rats at night when both pineal NAT activity and melatonin levels were elevated, unexpectedly led to a substantial decline in the activity of the acetylating enzyme and in melatonin levels (Fig. 4.14). The changes followed a time course similar to that seen following acute light exposure at night (10,141), and were clearly adrenal-mediated since they were prevented by adrenalectomy.

The drop in pineal melatonin induced by a hind leg injection of saline is not unique to this stressor. More recently, Troiani and colleagues (227) found that swimming for 10 min in 22 ± 2°C water at night led to a precipitous decline in pineal melatonin levels in the rat; again, the time course of this change was very rapid as basal indole values were reached within 15 to 20 minutes after the onset of swimming. This particular response differs substantially from that induced by the hind leg injection of saline. In the latter both NAT and melatonin drop, while in the case of the night-time swimming melatonin levels fall but NAT activity continues to increase. Furthermore, whereas the hind leg injection response is prevented by bilateral adrenalectomy, the melatonin drop induced by swimming is not. Indeed, to date we have found no manipulation that will prevent the swimming-induced decline in melatonin levels in the rat pineal gland (unpublished observations).

One other finding in reference to this phenomenon is worthy of

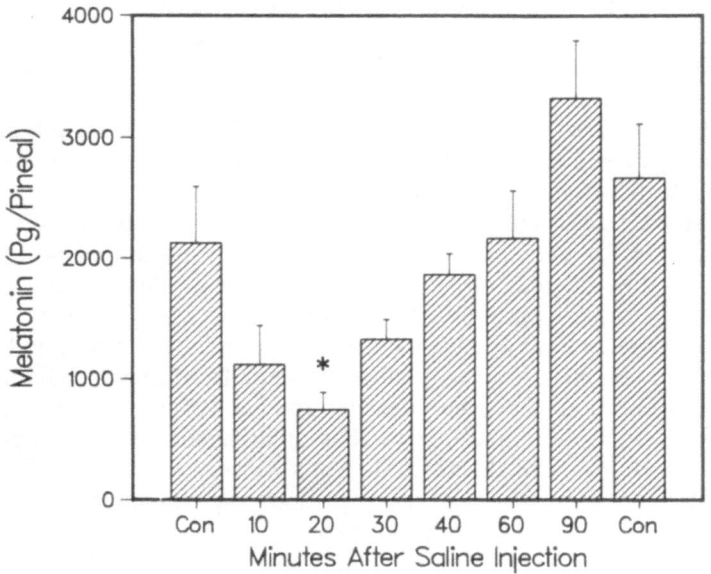

FIGURE 4.14. Rapid drop in the pineal melatonin level of rats following the injection of 1.5 cc saline during the dark phase of the light/dark cycle. The unexpected drop in pineal melatonin is prevented by adrenalectomy, but what adrenal constituent mediates the decline remains unknown. [Reprinted with permission from (226), Copyright 1986, Pergamon Press p 7c.]

mention. Whereas 10 minutes of swimming induce a decline in elevated night-time melatonin levels in the rat, during the same interval plasma concentrations of indole rise. The implication of these findings is that stored pineal melatonin is being rapidly released.

Observations in humans are few. One group of investigators found that 30 or 60 min after either pneumoencephalography or an insulin tolerance test, there was no change in daytime serum melatonin levels in spite of an obvious increase in blood cortisol levels (229,230). Under conditions of chronically elevated circulating catecholamines due to burn injury, the 24-hour serum melatonin rhythm in humans was found to be slightly attenuated (231), a small effect reminiscent of that seen in wild-captured animals described above (126,225).

The most provocative finding concerning the effect of stress on blood melatonin in humans comes from the work of Carr and associates (232). They observed that basal daytime melatonin levels in the blood of normal, healthy women could be increased by 200% by a 1-hour work-out on a bicycle ergometer sufficient to raise the pulse rate to 85% of maximum. These findings are noteworthy in view of the suppressive role of melatonin on the reproductive system (32,75,137,233,234) and of the reproduc-

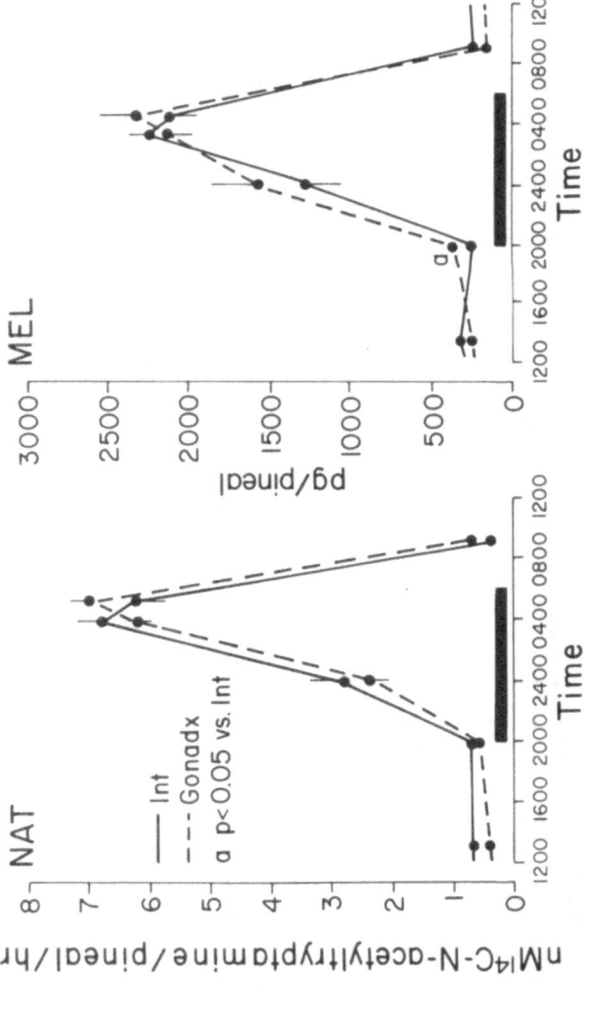

FIGURE 4.15. In adult male rats gonadectomy (Gonadx) has little influence on the 24-hour pineal rhythms of either N-acetyltransferase (NAT) activity or melatonin (MEL). The rhythms of intact (Int) rats of the same age are also shown. Black bar along the abscissa represents the daily period of darkness. [From Reiter et al. (236), by permission of S. Karger AG, Basel.]

tive difficulties commonly associated with chronic exercise, e.g., long distance running, in women (235).

Hormones

The conspicuous effects of pineal secretory products on the neuro-endocrine–reproductive axis suggest that there may be a reciprocal interaction of the pineal gland and the reproductive system. Orchiectomy in adult male rats, however, has essentially no effect on either the 24-hour pattern or magnitude of the pineal NAT or melatonin rhythms (Fig. 4.15) (236). On the other hand, a number of authors have shown that the pineal melatonin content (237,238) and associated enzyme activities (239–241) change during the course of the estrous cycle in the female rat. Usually, melatonin production is lowest at the time of expected ovulation, presumably because estradiol suppresses its synthesis by the pineal gland (242). The observations fit the expectation that production of melatonin, a hormone whose exogenous administration suppresses ovulation (243), should be suppressed when the ova are shed, although it has not been conclusively demonstrated that this is required for ovulation. Indeed in the Syrian hamster, the pineal melatonin content exhibits no statistically significant changes during the estrous cycle (244).

The role of the menstrual cycle in regulating pineal function in women remains unclear. Wetterberg and co-workers (245) and Arendt (246)

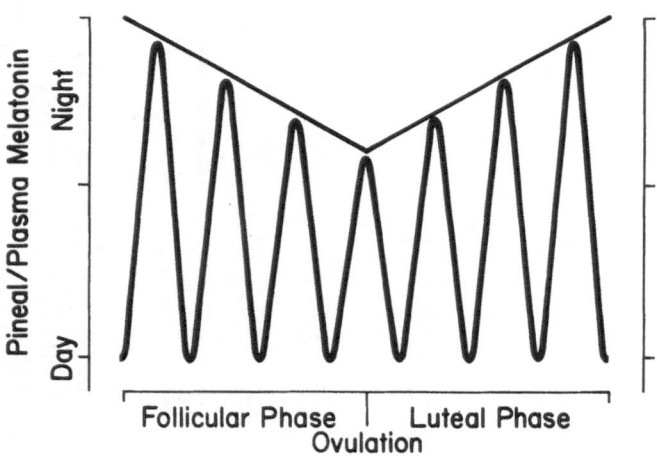

FIGURE 4.16. Presumed changes in plasma melatonin levels in human females during the menstrual cycle; several reports claim that the nocturnal rise in circulating melatonin values are lowest at the time of expected ovulation. If this were so, the restricted melatonin peaks could have a permissive effect on the hormonal surges required for ovulation. [From Reiter (234), with permission.]

presented preliminary data indicating that the nocturnal rise in plasma melatonin was lowest at the time of expected ovulation. Another study indicated that the nocturnal rise in young women was lower during the midmenstrual period (days 13–17 of the cycle) than during the premenstrual period (days 25–30 of the cycle) (247). These changes are diagrammatically represented in Fig. 4.16.

Other endocrine glands do not seem to be of paramount importance in regulating the rhythmic production of melatonin in experimental animals. In rodents neither thyroidectomy (236,248) nor adrenalectomy (236,249) greatly changes the nocturnal rise in pineal melatonin, although hypophysectomy, which eliminates a variety of hormones, does reduce both the NAT rhythm as well as that of melatonin (236). These rather meager effects are somewhat surprising in view of the many hormone receptors tentatively identified within the pineal gland (25). Perhaps there receptors are involved in processes not related to tryptophan metabolism.

Similarly, endocrine dysfunctions in humans usually cause only minor variations in the normal melatonin rhythms (127,251), with the exception of panhypopituitarism which may be associated with weak nocturnal surges of melatonin (217) reminiscent of those seen in hypophysectomized rats (236).

Alcohol

Data suggesting a possible effect of the pineal gland on ethanol consumption in rats (252,253) and Syrian hamsters (254) were published over a decade ago. In general, they indicate that animals increase alcohol consumption in preference to water when they receive no light via their visual systems, and that pinealectomy prevents psychologic dependence on ethanol.

More recently, Moss and colleagues (255) have demonstrated that rats kept in a state of continuous ethanol intoxication for 4 days had a blunted nighttime rise in pineal melatonin content. In this study pineal melatonin levels were measured at only a single time point (0300 hours) during the night. The drop in melatonin was accompanied by a commensurate rise in pineal serotonin suggesting a deficit in the synthesis of melatonin. No changes in either pineal melatonin or serotonin levels were noted during the alcohol withdrawal syndrome. A similar decrease in the amplitude of the nighttime melatonin rise was observed in a group of alcoholics regardless of whether they were consuming alcohol at the time of the test or not (217).

These observations suggest that the reportedly higher incidence of breast cancer in women who consume alcohol (256) may be related to depressed melatonin production and consequent development of endocrine-dependent tumors. Indeed, several studies have shown that pinealectomy leads to a higher incidence and faster growth of induced mammary carcinomas in rats (257,258) while melatonin retards the initiation and growth of such tumors (258). Finally, an association between an increased incidence of estrogen-dependent breast tumors and low nocturnal melatonin surges in human females has been reported (259).

Melatonin in Other Body Fluids

Cerebrospinal Fluid

The route of melatonin release in mammals has been a matter of debate (260). Thus, there is evidence that both melatonin (261) and AVT (262) may be released directly into the CSF, but most workers believe that the blood receives the bulk of the melatonin produced in the pineal gland. However, Pavel (262) argues that AVT is not only discharged into the CSF but that melatonin serves as the stimulus for such a release.

In one study (263), measurements of plasma and CSF melatonin levels in calves exposed to a LD cycle of 14 : 10 (Fig. 4.17) revealed a robust CSF melatonin rhythm, characterized by a rapid increase shortly after the onset of darkness and by a plateau which was maintained until the lights were turned on. Indeed, the melatonin rhythm of the CSF was more clearly defined than that of the blood and the actual levels were higher throughout the 24 hours. These findings suggest that melatonin is discharged first into the CSF and secondarily into the blood. Nevertheless, most investigators are of the opinion that CSF melatonin derives from the blood rather than directly from the pineal gland (260,263,264).

Reppert and colleagues found, in the rhesus monkey, that the CSF exhibits a large and distinct daily melatonin rhythm (Fig. 4.18), strongly coupled to the prevailing environmental photoperiod (265,266), that the nighttime rise in CSF melatonin could be suppressed by constant light exposure and that under persistent darkness, the rhythm free-ran (267,268).

No clear day/night rhythm in CSF has been documented in human subjects (127,269), although observations in other species suggest that it likely exists. Elevated daytime concentrations of CSF melatonin have been found in leukemic children (270) and in adults suffering from schizophrenia (271), but it is not known if this elevation is specific for these diseases. In general, serum melatonin levels in man have been reported to be higher than those in the CSF (127,272,273), at least in the daytime.

As suggested in an earlier report (260), a direct release of melatonin into the CSF would have the advantage of being delivered near its primary sites of action in the CNS, undiluted by previous delivery into the blood and less available for destruction by the liver. Thus, the total amount of melatonin needed to achieve effective concentrations would be greatly reduced. In spite of this argument and even though the necessary

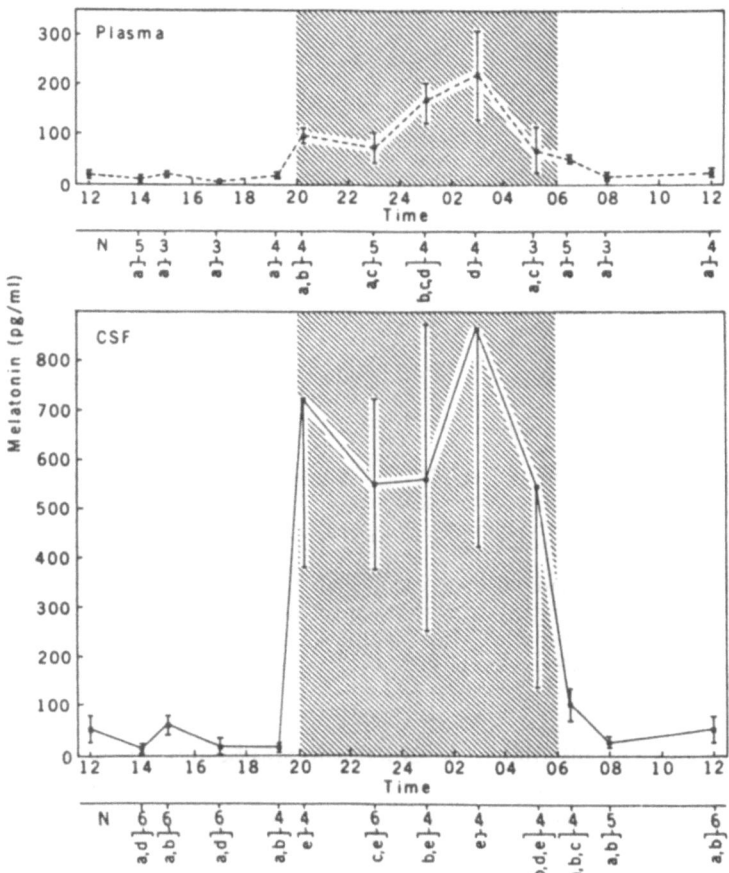

FIGURE 4.17. Plasma (top) and CSF (bottom) concentrations of melatonin in calves over a 24-hour period with darkness extending from 2000 h to 0600 h. Both day-time and night-time immunoreactive melatonin concentrations were greater in the CSF than in the blood and the rhythm was also clearly more robust. N = number of calves; points represent mean ± SEM; the difference between two points is significant (p < 0.05) when the points do not share the same letter. [From Hedlund et al. (263). Copyright 1977 by the AAAs.]

anatomic structures exist, the direct release of melatonin into the CSF is not considered significant.

Saliva

Recently, melatonin was found to be secreted in human saliva (274–277) in a pattern that closely reflected that found in the blood, although at much lower concentrations. Since about 70% of the blood melatonin is

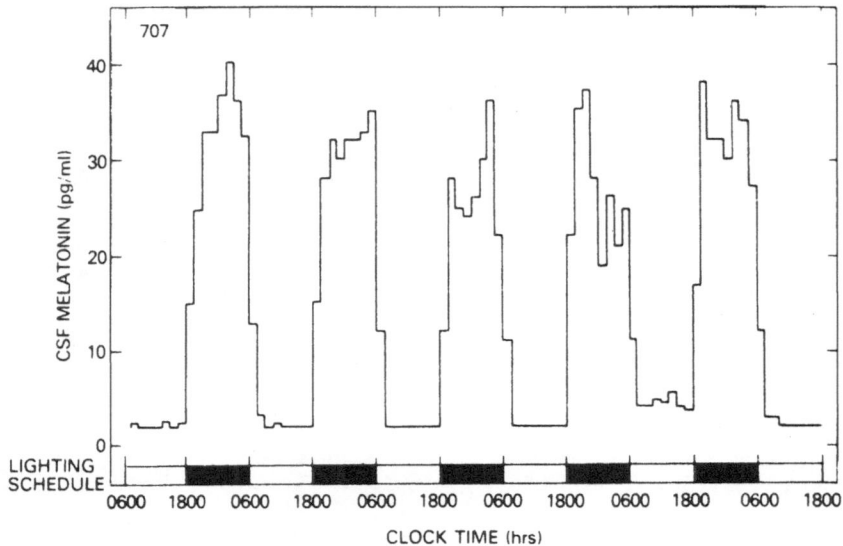

FIGURE 4.18. Nyctohemeral increase in CSF melatonin in a rhesus monkey in which the cycle was studied for 6 consecutive days. CSF was collected every 90 minutes. [From Reppert et al. (265), with permission.]

bound to plasma albumin and since albumin-bound hormones are not secreted in the saliva, the concentration of melatonin in the saliva may actually represent the free, biologically available hormone.

Since there is no evidence that the salivary glands have the enzymatic machinery to produce melatonin, it seems likely that salivary melatonin derives from the blood and, therefore represents a general index of pineal function. This notion is confirmed indirectly by the observation that an oral dose of 100 mg of melatonin in humans results in peak salivary concentrations 60 minutes later (278) and that its elimination follows a first-order kinetics with a half-life of 38 minutes (compared to a half life of 41 min for serum melatonin).

Urine

Melatonin is rapidly metabolized in the liver and brain and only less than 1% of it is excreted in the urine. Lynch and colleagues (279,280) have found that the urinary excretion of melatonin in humans is about 3 to 5 times higher at night than during the day and frequently reflects the concentration in the plasma (281,282). However, since the amount is small and obviously profoundly affected by even slight changes in the rate of metabolic breakdown, measurements of the 24-hour urinary melatonin

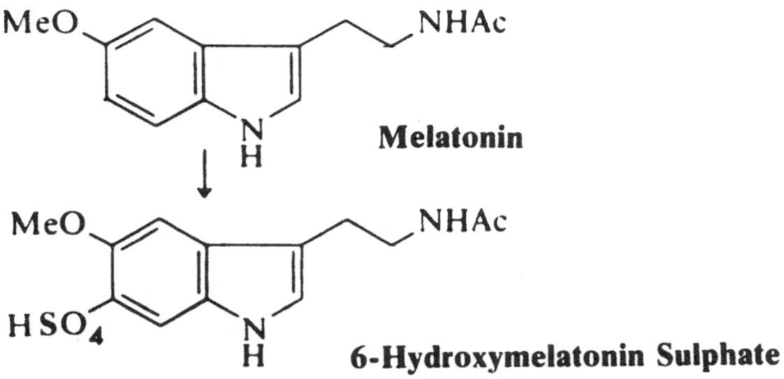

FIGURE 4.19. Melatonin and its chief hepatic metabolite, 6-hydroxymelatonin sulfate. The latter is excreted in the urine, more abundantly during the night than during the day.

excretion are difficult to evaluate. Thus, measurement of its chief hepatic metabolite, 6-hydroxymelatonin (6-OHM) sulfate is often preferred (Fig. 4.19) (283,284). Tetsuo et al (205) reported that normal men excrete an average of 16 μg 6-OHM per 24 hour, most of it during the period of darkness. Similar day/night differences have been found in women (286) and in the urinary excretion of 6-OHM sulfate (287). All indications are that the concentration of these metabolites of melatonin in the urine of humans parallels the synthesis and secretion of pineal melatonin and, therefore adequately reflects pineal activity, provided of course that nothing significant alters melatonin metabolism.

Final Comment

The pineal gland of all mammals, and perhaps of all vertebrates, synthesizes melatonin from tryptophan with the largest amount being formed during periods of darkness. The gland receives information about the daily cycle of light and darkness though a circuitous nervous pathway which connects it to the retina via the superior cervical ganglia and postganglionic sympathetic neurons. The mechanisms of the nocturnal rise in melatonin production have been at least partially identified. In addition to melatonin, the pineal contains other products of serotonin metabolism which may prove to have endocrine functions.

A variety of perturbations can influence the 24-hour pattern of melatonin synthesis and secretion. Among them, the most extensively studied by far is the cycle of light and darkness. Under all circumstances light, provided it is of sufficient irradiance (brightness) and of the proper wavelength, inhibits melatonin production. Another environmental variable to which some individuals may be exposed is represented by ELF electric fields which at least in rats inhibit the nighttime melatonin increase, possibly mediated by acting through the sympathetic innervation of the pineal gland.

The effects of age and calcification on pineal melatonin metabolism seem to be a result of different mechanisms since in only a few mammalian species does the pineal gland exhibit significant calcium deposition, whereas advanced age is always associated with a substantial reduction in pineal gland and blood melatonin levels. A genetic absence of a melatonin rhythm has been described in the mouse but there are reasons to believe that in other mammals, especially highly inbred strains of animals, a similar deficiency may exist.

The influence of aversive stimuli on pineal melatonin production varies, especially in relation to the time when the stimulus is applied. During the day when melatonin is at basal levels, a stress-induced surge of catecholamines may stimulate melatonin secretion and increase pineal and plasma melatonin levels. On the other hand, at night, when the conversion of serotonin to melatonin is already maximal, a similar stimulus causes a precipitous decline. The reason for this difference is unknown. Other hormones appear to have rather minor, albeit possibly important, influences on pineal melatonin production, whereas alcohol consumption may prove to be a significant factor.

Melatonin and its metabolites are present in the CSF, saliva, and urine, in amounts that follow the daily fluctuations seen in blood and in the pineal itself, and seem to reflect the activity of the pineal gland. Indeed, their measurement can be used as a reliable index of pineal activity.

In a number of nonhuman mammalian species, the major role of the pineal is to mediate seasonal changes in reproductive physiology (13,75,93,137,288,289). Although this role has been proven beyond reasonable doubt, the pineal may have other physiologic functions. Indeed, some investigators believe that there may not be an organ system which escapes its influence (290–294).

In humans, as well, the pineal is very active endocrine organ and, although its specific functions have not been totally clarified, there are strong indications that its malfunction may be linked to a number of clinical syndromes (50,97,127,216,234,295,296).

References

1. Binkley SA (1981) Pineal biochemistry: Comparative aspects and circadian rhythms. In: Reiter RJ (ed) The Pineal Gland, Vol. I, Anatomy and Biochemistry. CRC Press, Boca Raton, pp 155–172.
2. Ralph CL (1983) Evolution of pineal control of endocrine function in lower vertebrates. Am Zool 23:597–605.
3. Underwood H (1984) The pineal and circadian rhythms. In: Reiter RJ (ed) The Pineal Gland. Raven, New York, pp 221–251.

4. Menaker M (1985) Eyes—the second (and third) pineal glands? In: Photoperiodism, Melatonin and the Pineal. Pitman, London. pp 78–86.
5. Kappers JA (1965) Survey of the innervation of the epiphysis cerebri and the accessory pineal organs of vertebrates. Progr Brain Res 10:87–153.
6. Moore RY (1978) The innervation of the mammalian pineal gland.In: Reiter RJ (ed) The Pineal and Reproduction. Karger, Basel, pp 1–29.
7. Moore RY, Lenn NJ (1972) A retino-hypothalamic projection in the rat. J Comp Neurol 146:1–14.
8. Eichler VB, Moore RY (1974) The primary and accessory optic system in golden hamster, *Mesocricetus auratus*. Acta Anat 89:359–371.
9. Reiter RJ, Klein DC (1971) Observations on the pineal gland, the Harderian glands, the retinas and the reproductive organs of adult female rats exposed to continuous light. J Endocrinol 51:117–125.
10. Webb SM, Champney TH, Lewinski AK et al (1985) Photoreceptor damage and eye pigmentation: Influence on the sensitivity of rat pineal N-acetyltransferase activity and melatonin levels to light at night. Neuroendocrinology 40:205–209.
11. Korf H-K, Møller M (1984) The innervation of the mammalian pineal gland with special reference to central pinealopetal projections. Pineal Res Rev 2:41–86.
12. Moore RY, Klein DC (1974) Visual pathways and the central neural control of a circadian rhythm in pineal serotonin *N*-acetyltransferase activity. Brain Res 71:17–33.
13. Bittman EL (1984) Melatonin and photoperiodic time measurement: Evidence from rodents and ruminants. In: Reiter RJ (ed) The Pineal Gland. Raven, New York, pp 155–192.
14. Swanson LW, Cowan WM (1975) The efferent connections of the suprachiasmatic nucleus of the hypothalamus. J Comp Neurol 160:1–12.
15. Swanson LW, Kuypers HGJM (1980) The paraventricular nucleus of the hypothalamus: Cytoarchitectonic subdivisions and organization of projections to the pituitary, dorsal vagal complex, and spinal cord as demonstrated by retrograde fluorescence double-labeling method. J Comp Neurol 194:555–570.
16. Berk ML, Finkelstein JA (1981) An autoradiographic determination of the efferent projections of the suprachiasmatic nucleus of the hypothalamus. Brain Res 226:1–13.
17. Swanson LW, Kuypers HGJM (1980) A direct projection from the ventromedial nucleus and retrochiasmatic area of the hypothalamus to the medulla and spinal cord of the rat. Neurosci Lett 17:307–312.
18. Berk ML, Finkelstein JA (1982) Efferent connections of the lateral hypothalamic area of the rat: An autoradiographic investigation. Brain Res Bull 8:511–526.
19. Reiter RJ (1972) Surgical procedures involving the pineal gland which prevent gonadal degeneration in adult male hamsters. Ann d'Endocr 33:571–582.
20. Kniesley LW, Moskowitz MA, Lynch HJ (1978) Cervical spinal cord lesions disrupt the rhythm in human melatonin excretion. J Neural Transm (Suppl) 13:311–323.
21. Rando TA, Bowers CW, Zigmond RE (1981) Localization of neurons in the

rat spinal cord which project to the superior cervical ganglion. J Comp Neurol 196:73–83.

22. Reiter RJ (1981) The mammalian pineal gland: Structure and function. Am J Anat 162:287–313.

23. Reiter RJ, Hedlund L (1976) Peripheral sympathetic innervation of the deep pineal gland of the golden hamster. Experientia 32:1071–1072.

24. Lydic R, Shoene WC, Czeisler CA et al (1980) Suprachiasmatic region of the human hypothalamus: Homolog to the primate circadian pacemaker? Sleep 2:355–361.

25. Romijn HJ (1973) Parasympathetic innervation of the rabbit pineal gland. Brain Res 55:431–436.

26. Korf H-W, Møller M (1985) The innervation of the mammalian pineal gland with special reference to the central pinealopetal projections. Pineal Res Rev 2:41–86.

27. Korf H-W, Møller M (1985) The central innervation of the pineal organ. In: Mess B, Ruzsas C, Tima L, Pevet P (eds). The Pineal Gland. Akademiai Kiado, Budapest, pp 48–69.

28. Korf H-W, Wagner U (1980) Evidence for a nervous connection between the brain and the pineal organ in the guinea pig. Cell Tissue Res 209:505–510.

29. Møller M, Korf H-W (1983) The origin of central pinealopetal nerve fibers in the Mongolian gerbil as demonstrated by the retrograde transport of horseradish peroxidase. Cell Tissue Res 230:273–287.

30. Buijs RM, Pevet P (1980) Vasopressin- and oxytocin-containing fibers in the pineal gland and subcommissural organ of the rat. Cell Tissue Res 205:11–17.

31. Guerillot C, Pfister A, Müller J et al (1982) Recherche de l'origine des fibres nerveuses extraorthosympathiques innervant l'epiphyse du rat (etude du transport retrograde de la peroxydase de raifort). Reprod Nutr Dev 22:371–378.

32. Reuss S, Møller M (1986) Direct projections to the rat pineal gland via the stria medullaris thalami. Cell Tissue Res 244:691–694.

33. Semm P, Schneider T, Vollrath L (1981) Morphological and electrophysiological evidence for habenular influence on the guinea pig pineal gland. J Neural Transm 50:247–266.

34. Reuss S, Semm P, Vollrath L (1984) Electrophysiological investigations on the central innervation of the rat and guinea pig pineal gland. J Neural Transm 60:31–43.

35. Reuss S, Olcese J, Vollrath L (1985) Electrical stimulation of the hypothalamic paraventricular nuclei inhibits pineal melatonin synthesis in male rats. Neuroendocrinology 41:192–196.

36. Ebadi M, Govitrapong P (1986) Orphan transmitters and their receptor sites in the pineal gland. Pineal Res Rev 4:1–54.

37. Wurtman RJ, Axelrod J, Fischer JE (1964) Melatonin synthesis in the pineal gland: Effect of light mediated by the sympathetic nervous system. Science 143:1328–1329.

38. Reiter RJ, Hester RJ (1966) Interrelationship of the pineal gland, the superior cervical ganglia, and the photoperiod on the regulation of the endocrine systems of hamsters. Endocrinology 79:1168–1170.

39. Shein HM, Wurtman RJ (1971) Stimulation of (^{14}C) tryptophan 5-hydroxylation by norepinephrine and dibutyryl adenosine 3′,5′-monophosphate in rat pineal organ cultures. Life Sci 10:935–940.

40. Pellegrino de Iraldi A, Zieher LM, De Robertis E (1963) The 5-hydroxy-tryptamine content and synthesis of normal and denervated pineal gland. Life Sci 2:691–696.
41. Morgan WW, Hansen J (1978) Time course of the disappearance of pineal noradrenaline following superior cervical ganglionectomy. Exp Brain Res 32:429–434.
42. Wurtman RJ, Axelrod J (1974) A 24-hour rhythm in the content of norepi-nephrine in the pineal and the salivary glands of the rat. Life Sci 5:665–670.
43. Morgan WW, Reiter RJ (1977) Pineal noradrenaline levels in the Mongolian gerbil and in different strains of laboratory rats over a lighting regimen. Life Sci 21:555–558.
44. Morgan WW, Reiter RJ, Pfeil KA (1976) Hamster pineal noradrenaline levels over a regulated lighting period and the influence of superior cervical ganglionectomy. Life Sci 19:437–440.
45. Matthews SA, Evans KL, Morgan WW et al (1982) Pineal indoleamine metabolism in the cotton rat, *Sigmodon hispidus:* Studies on norepinephrine serotonin *N*-acetyltransferase and melatonin. In: Reiter RJ (ed) The Pineal and Its Hormones. Alan R. Liss, New York, pp 35–44.
46. Craft CM, Morgan WW, Reiter RJ (1983) 24-hour changes in catecholamine synthesis in rat and hamster pineal glands. Neuroendocrinology 38:193–198.
47. Brownstein M, Axelrod J (1974) Pineal gland: 24-hour rhythm in norepineph-rine turnover. Science 184:163–165.
48. Axelrod J (1974) The pineal gland: A neurochemical transducer. Science 184:134–1348.
49. Auerbach DA, Klein DC, Woodward B et al (1981) Neonatal rat pinealo-cytes: Typical and atypical characteristics of ^{125}I-iodohydroxybenzylpindolol binding and adenosine 3′,5′-monophosphate accumulation. Endocrinology 108:559–567.
50. Lewy, AJ (1983) Biochemistry and regulation of mammalian melatonin production. In: Relkin R (ed). The Pineal Gland. Elsevier, New York, pp 77–129.
51. Craft CM, Morgan WW, Jones PJ et al (1985) Hamster and rat pineal gland β-adrenoceptor characterization with iodocyanopindolol and the effect of decreased catecholamine synthesis on the receptor. J Pineal Res 2:51–66.
52. Reiter RJ, Esquifino AI, Champney TH et al (1985) Pineal melatonin production in relation to sexual development in the male rat. In: Gupta D, Borrelli P, Attanasio A (eds). Pediatric Endocrinology. Croom & Helm, London, pp 190–202.
53. Vaughan GM, Pruitt BA Jr, Mason AD Jr (1987) Nytohemeral rhythm in melatonin responses to isoproterenol *in vitro*: Comparison of rats and Syrian hamsters. Comp Biochem Physiol, 87C:71–74.
54. Zatz M (1981) Pharmacology of the rat pineal gland. In: Reiter RJ (ed) The Pineal Gland, Vol. 1, Anatomy and Biochemistry. CRC Press, Boca Raton, pp 229–242.
55. Vanecek J, Sugden D, Weller J et al (1985) Atypical synergistic alpha$_1$- and β_1-adrenergic regulation of adenosine 3′5′-monophosphate in cultured rat pinealocytes. Endocrinology 116:2167–2173.
56. Reiter RJ, King TS, Richardson BA et al (1982) Pineal melatonin levels in a diurnal species, the eastern chipmunk (*Tamias striatus*): Effects of light at

night, propranolol administration or superior cervical ganglionectomy. J Neural Transm 54:275–284.

57. Sugden D, Namboodiri MAA, Klein DC et al (1985) Ovine pineal α_1-adrenoceptors: Characterization and evidence for a functional role in the regulation of serum melatonin. Endocrinology 116:1960–1967.

58. Lipton JS, Petterborg LJ, Steinlechner S et al (1982) *In vivo* responses of the pineal gland of the Syrian hamster to isoproterenol or norepinephrine. In: Reiter RJ (ed) The Pineal and Its Hormones. Alan R. Liss, New York, pp 107–115.

59. Steinlechner S, King TS, Champney TH et al (1984) Comparison of the effects of β-adrenergic agents on pineal serotonin N-acetyltransferase activity and melatonin in two species of hamsters. J Pineal Res 1:23–30.

60. Steinlechner S, King TS, Champney TH et al (1985) Pharmacological studies on the regulation of N-acetyltransferase activity and melatonin content of the pineal gland of the Syrian hamster. J Pineal Res 2:109–119.

61. Tamarkin L, Reppert SM, Klein DC (1979) Regulation of pineal melatonin in the Syrian hamster. Endocrinology 104:385–389.

62. Lipton JS, Petterborg LJ, Reiter RJ (1981) Influence of propranolol, phenoxybenzamine or phentolamine on the *in vivo* nocturnal rise of pineal melatonin levels in the Syrian hamster. Life Sci 28:2377–2382.

63. Vaughan GM, Lasko J, Coggins SH et al (1986) Rhythmic melatonin response of the Syrian hamster pineal gland to norepinephrine *in vitro* and *in vivo*. J Pineal Res 3:235–250.

64. Reiter RJ, Oaknin S, Troiani ME et al (1986) Factors controlling melatonin rhythmicity in the pineal gland of the Syrian hamster. Annu Rev Chronopharmacol 3:41–44.

65. Reiter RJ, Vaughan, GM Oaknius et al (1987) Norepinephrine or isoproterenol stimulation of pineal N-acetyltransferase activity and melatonin content in the Syrian hamster is restricted to the second half of the daily dark phase. Neuroendocrinology 45:249–256.

66. Panke ES, Reiter RJ, Rollag MD et al (1978) Pineal serotonin N-acetyltransferase activity and melatonin concentrations in prepubertal and adult Syrian hamsters exposed to short daily photoperiods. Endocr Res Commun 5:311–374.

67. Panke ES, Rollag MD, Reiter RJ (1979) Pineal melatonin concentrations in the Syrian hamster. Endocrinology 104:194–197.

68. Vaughan GM, Pelham RW, Pang SF et al (1976) Nocturnal elevation of plasma melatonin and urinary 5-hydroxyindoleacetic acid in young men: Attempts at modification by brief changes in environmental lighting and sleep and by autonomic drugs. J Clin Endocrinol Metab 42:752–764.

69. Weiss B, Costa E (1967) Adenyl cyclase activity in rat pineal gland: Effects of chronic denervation and norephinephrine. Science 156:1750–1752.

70. Weiss B (1969) Effects of environmental lighting and chronic denervation in the activation of adenylcyclase of rat pineal gland by norepinephrine and sodium fluoride. J Pharmacol Exp Ther 168:146–152.

71. Weiss B, Strada SJ (1972) Neuroendocrine control of the cyclic AMP system of brain and pineal gland. J Cyclic Nucleotide Res 1:357–374.

72. Smith TL, Eichberg J, Hauser G (1979) Postsynaptic localization of the alpha

receptor-mediated stimulation of phosphatidylinositol turnover in pineal gland. Life Sci 24:2179–2184.

73. Ebadi M (1984) Regulation of the synthesis of melatonin and its significance to neuroendocrinology. In: Reiter RJ (ed) The Pineal Gland. Raven, New York, pp 1–38.

74. King TS and Steinlechner S (1985) Pineal indoleamine synthesis and metabolism: Kinetic considerations. Pineal Res Rev 3:69–113.

75. Reiter RJ (1984) Pineal indoles: Production, secretion and actions. In: MacLeod RM, Müller EE (eds) Neuroendocrine Perspectives, Vol. 3 Elsevier, Amsterdam, pp 345–377.

76. Klein DC (1985) Photoneural regulation of the mammalian pineal gland. In: Photoperiodism, Melatonin and the Pineal. Pitman, London pp 38–51.

77. Namboodiri MAA, Sugden D, Klein DC et al (1983) 5-Hydroxytryptophan elevates serum melatonin. Science 221:659–661.

78. Quay WB (1963) Circadian rhythm in rat pineal serotonin and its modification by the estrous cycle and photoperiod. Gen Comp Endocrinol 3:473–479.

79. King TS, Steger RW, Richardson BA et al (1984) Day-night differences in the estimated rates of 5-hydroxytryptamine turnover in the rat pineal gland. Exp Brain Res 54:432–436.

80. Klein DC, Weller JL (1970) Indole metabolism in the pineal gland: A circadian rhythm in N-acetyltransferase. Science 169:1093–1095.

81. Deguchi T (1975) Ontogenesis of a biological clock for serotonin acetyl: co-enzyme A N-acetyltransferase in pineal gland of rat. Proc Nat Acad Sci USA 72:2814–2818.

82. Rudeen PK, Reiter RJ, Vaughan MK (1975) Pineal serotonin-N-acetyltransferase in four mammalian species. Neurosci Lett 1:225–229.

83. Pang SF, Brown GM, Grota LJ et al (1977) Determination of N-acetylserotonin and melatonin activities in the pineal gland, retina, Harderian gland, brain, and serum of rats and chickens. Neuroendocrinology 23:1–13.

84. Axelrod J, Weissbach H (1960) Enzymatic O-methylation of N-acetylserotonin to melatonin. Science 131:1312.

85. Wurtman RJ, Axelrod J, Phillips LS (1963) Melatonin synthesis in the pineal gland: Control by light. Science 142:1071–1073.

86. Axelrod J, Wurtman RJ, Snyder SH (1965) Control of hydroxyindole-O-methyltransferase activity in the rat pineal gland by environmental lighting. J Biol Chem 240:949–955.

87. Binkley S (1979) Pineal rhythms *in vitro*. Comp Biochem Physiol 64:201–206.

88. Binkley S, Brammer M (1981) Development of daily cycles in the pineal gland. In: Klein DC (ed) Melatonin Rhythm generating System. Karger, Basel. pp 124–131.

89. Reiter RJ, Hurlbut EC, Brainard GC et al (1983) Influence of light irradiance on hydroxyindole-O-methyltransferase, serotonin N-acetyltransferase activity, and radioimmunoassayable melatonin levels in the pineal gland of the diurnally active Richardson's ground squirrel. Brain Res 288:51–157.

90. Quay WB (1964) Circadian and estrous rhythm in pineal melatonin and 5-hydroxyindole-3-acetic acid. Proc Soc Exp Biol Med 115:710–713.

91. Rollag MD, Panke ES, Reiter RJ (1980) Pineal melatonin content in male hamsters throughout the seasonal reproductive cycle. Proc Soc Exp Biol Med 165:330–334.

92. Matthews SA, Evans KL, Morgan WW et al (1982) Pineal indoleamine metabolism in the cotton rat (*Sigmodon hispidus*): Studies on norepinephrine serotonin *N*-acetyltransferase and melatonin. In: Reiter RJ (ed) The Pineal and Its Hormones. Alan R. Liss, New York, pp 35–44.
93. Goldman BD (1983) The physiology of melatonin in mammals. Pineal Res Rev 1:145–182.
94. Smith I (1983) Indoles of pineal origin: Biochemical and physiological status. Psychoneuroendocrinology 8:41–60.
95. Brown, GM, Grota L et al (1981) Physiologic regulation of melatonin. In: Birau N, Schloot, W (eds). Pergamon, Oxford, pp 95–112.
96. Leyva H, Addiego L, Stabenfeldt G (1984) The effect of different photoperiods on plasma concentrations of melatonin, prolactin, and cortisol in the domestic cat. Endocrinology 115:1729–1736.
97. Arendt J (1985) Mammalian pineal rhythms. Pineal Res Rev 3:161–213.
98. Reiter RJ (1986) Normal patterns of melatonin levels in the pineal gland and body fluids of humans and experimental animals. J Neural Transm (Suppl) 2:35–54.
99. Wilson BW, Lynch HJ, Ozaki Y (1978) 5-Methoxytryptophol in rat serum and pineal: Detection, quantitation, and evidence of daily rhythmicity. Life Sci 23:1019–1024.
100. McIsaac WM, Tabarsky RG, Farrell G (1964) 5-Methoxytryptophol: Effect on estrus and ovarian weight. Science 145:63–64.
101. Talbot JA, Reiter RJ (1973/74) Influence of melatonin, 5-methoxytryptophol, and pinealectomy on pituitary and plasma gonadotropin and prolactin levels in castrated adult male rats. Neuroendocrinology 13:164–172.
102. Vaughan MK, Vaughan GM, Reiter RJ (1976) Inhibition of human chorionic gonadotrophin-induced hypertrophy of the ovaries and uterus in immature mice by some pineal indoles,6-hydroxymelatonin, and arginine vasotocin. J Endocrinol 68:397–400.
103. Balemans MGM, Van De Veerdonk FCG, Van DeKamer JC (1977) The influence of 5-methoxytryptophol, a pineal compound, on comb growth, ovarian weight, follicular growth and egg production of juvenile, maturing, and adult white leghorn hens (*Gallus domesticus L.*). J Neural Transm 41:37–46.
104. Mas M, Oaknin S (1978) Effects of pineal methoxyindoles on male sex behavior and spermatogenesis. J Neural Transm. (Suppl) 13:376–377.
105. Mullen PE, Leone RM, Hooper J et al (1979). Pineal 5-methoxytryptophol in man. Psychoneuroendocrinology 2:117–126.
106. Skene DJ, Smith I, Arendt J (1986) Radioimmunoassay of pineal 5-methoxytryptophol in different species: Comparison with pineal melatonin content. J Endocrinol 110:177–184.
107. Carter SJ, Laud CA, Smith I et al (1979) Concentration of 5-methoxytryptophol in pineal gland and plasma of rat. J Endocrinol 83:35–40.
108. Beck O, Pevet P (1984) Analysis of melatonin, 5-methoxytryptophol, and 5-methoxyindoleacetic acid in the pineal gland and retina of hamster by capillary gas chromatography-mass spectrometry. J Chromatogr Biomed Appl 311:1–8.
109. Beck O, Jonsson G, Lundman A (1981) 5-Methoxyindoles in pineal gland of cow, pig, sheep and rat. Naunyn Schniedbergs Arch Exp Pathol Pharmacol 318:49–55.

110. Kennaway DJ (1983) Radioimmunoassay of 5-methoxytryptophol in sheep plasma and pineal glands. Life Sci 32:2461–2469.
111. Reiter RJ, Vaughan MK (1977) Pineal antigonadotropic substances: Polypeptides and indoles. Life Sci 21:159–172.
112. Axelrod J, Weissbach H (1961) Purification and properties of hydroxyindole-O-methyltransferase. J Biol Chem 236:211–213.
113. Pevet P (1983) Is 5-methoxytryptamine a pineal hormone? Psychoeneuroendocrinology 8:61–73.
114. Pevet P, Haldar-Misra C, Öcal T (1981) Effect of 5-methoxytryptophan and 5-methoxytryptamine on the reproductive system of the male golden hamster. J Neural Transm 51:303–311.
115. Rollag MD, Stetson MH (1982) Melatonin injection into Syrian hamsters. In: Reiter RJ (ed) The Pineal and Its Hormones. Alan R. Liss, New York, pp 143–151.
116. Rollag MD (1983) Ability of tryptophan derivatives to mimic melatonin's action upon the Syrian hamster reproductive system. Life Sci 31:2699–2707.
117. Reiter RJ, Holtorf A, Champney TH et al (1984) Relative efficacy of melatonin and 5-methoxytryptamine in terms of their antigonadotrophic and counterantigonadotrophic actions in male Syrian hamsters. J Pineal Res 1:91–98.
118. Lang U, Aubert ML, Rivest RW et al (1985) 5-Methoxytryptamine, and 6-hydroxymelatonin on sexual maturation of male rats. Activity of 5-methoxytryptamine might be due to its conversion to melatonin: Biol Reprod 33:618–628.
119. Reiter RJ (1983) The role of light and age in determining melatonin production in the pineal gland. In: Axelrod J, Fraschini F, Velo GP (eds). The Pineal and Its Endocrine Role. Plenum, New York, pp 227–241.
120. Reiter RJ, Steinlechner S, Richardson BA et al (1983) Differential response of pineal melatonin levels to light at night in laboratory-raised and wild-captured 13-lined ground squirrels (Spermophilus tridecemlineatus). Life Sci 32:2625–2629.
121. Reiter RJ, Johnson LY, Steger RW et al (1980) Pineal biosynthetic activity and neuroendocrine physiology in the aging hamster and gerbil. Peptides 1 (Suppl 1):69–77.
122. King TS, Richardson BA, Reiter RJ (1981) Age-associated changes in pineal serotonin N-acetyltransferase activity and melatonin content in the male gerbil. Endocrinol Res Commun 8:253–262.
123. Ebihara S, Marks T, Hudson DJ et al (1986) Genetic control of melatonin synthesis in the pineal gland of the mouse. Science 231:491–493.
124. Goldman BD, Carter DS, Hall VD et al (1982) Physiology of melatonin in three species of hamsters. In: Klein DC (ed) Melatonin Rhythm Generating System. Karger, Basel. pp 210–231.
125. Johnson LY, Vaughan MK, Richardson BA et al (1982) Variation in pineal melatonin content during the estrous cycle of the rat. Proc Soc Exp Biol Med 169:416–419.
126. Reiter RJ, Peters JF (1984) Non-suppressibility by room light of pineal N-acetyltransferase activity and melatonin levels in two diurnally active rodents, the Mexican ground squirrel (Spermophilus mexicanus) and the eastern chipmunk (Tamias striatus). Endocrinol Res 10:113–121.
127. Vaughan GM (1984) Melatonin in humans. Pineal Res Rev 2:141–201.

128. Pelham RW, Vaughan GM, Sandock KL et al (1973) Twenty-four-hour cycle of a melatonin-like substance in the plasma of human males. J Clin Endocrinol Metab 37:341–344.
129. Arendt J, Wirz-Justice A, Bradtke J et al (1979) Long-term studies on immunoreactive human melatonin. Ann Clin Biochem 16:307–312.
130. Smith I, Mullen DE, Silman RE et al (1976) Absolute identification of melatonin in human plasma and cerebrospinal fluid. Nature 260:718–719.
131. Greiner AC, Chan SC (1978) Melatonin content of the human pineal gland. Science 199:83–84.
132. Anderson GM, Young JG, Cohn DJ et al (1982) Determination of indoles in human and rat pineal. J Chromatogr 228:155–163.
133. Beck O, Borg S, Lundman A (1982) Concentration of 5-methoxyindoles in the human pineal gland. J Neural Transm 54:111–116.
134. Otani T, Oyarkey F, Farrell G (1968) Enzymes of the human pineal body. J Clin Endocrinol Metab 28:345–354.
135. Smith JA, Padwick D, Mee TJ et al (1977) Synchronous nyctohemeral rhythms in human blood melatonin and in human post-mortem pineal enzymes. Clin Endocrinol (Oxford) 6:619–225.
136. Petterborg LJ, Richardson BA, Reiter RJ (1981) Effect of long or short photoperiod on pineal melatonin content in the white-footed mouse, *Peromyscus leucopus*. Life Sci 29:1623–1627.
137. Arendt J (1986) Role of the pineal gland and melatonin in seasonal reproduction function in mammals. Oxford Rev Reprod Biol 8:266–320.
138. Matthews SA, Evans KL, Morgan WW et al (1982) Pineal indoleamine metabolism in the cotton rat, *Sigmodon hispidus:* Studies on norepinephrine, serotonin, *N*-acetyltransferase and melatonin. In: Reiter RJ (ed) The Pineal and Its Hormones. Alan R. Liss, New York, pp 35–44.
139. Matthews SA (1982) Pineal indoleamine metabolism in the cotton rat, *Sigmodon hispidus:* Photoperiodic and adrenergic control. Master's thesis, University of Texas Health Science Center at San Antonio.
140. Wilkinson M, Arendt J, Bradtke J et al (1977) Determination of a dark-induced increase of pineal *N*-acetyltransferase activity and simultaneous radioimmunoassay of melatonin in pineal, serum, and pituitary tissue of the male rat. J Endocrinol 72:243–244.
141. Rollag MD, Panke ES, Trakulrungsi Wk et al (1980) Quantification of daily melatonin synthesis in the hamster pineal gland. Endocrinology 106:231–236.
142. Lewy AJ, Wehr TA, Goodwin RK et al (1980) Light suppresses melatonin secretion in humans. Science 210:1267–1269.
143. Thiele G, Holtorf A, Steinlechner S et al (1983) The influence of different light irradiances on pineal *N*-acetyltransferase activity and melatonin in the cotton rat, *Sigmodon hispidus*. Life Sci 33:1543–1547.
144. Reiter RJ (1985) Action spectra, dose-response relationships, and temporal aspects of light's effects on the pineal gland. Ann NY Acad Sci 453:215–229.
145. Vaughan GM, Pelham RW, Pang SF et al (1976) Nocturnal elevation of plasma melatonin and urinary 5-hydroxyindoleacetic acid in young men: Attempts at modification by brief changes in environmental lighting and sleep and by autonomic drugs. J Clin Endocrinol Metab 42:752–764.
146. Reiter RJ, Hurlbut EC, Esquifino AI et al (1984) Changes in serotonin levels, *N*-acetyltransferase activity, hydroxyindole-*O*-methyltransferase activity,

and melatonin levels in the pineal gland of the Richardson's ground squirrel in relation to light-dark cycle. Neuroendocrinology 39:356–360.

147. Lewy AJ (1983) Human melatonin secretion, its endogenous circadian pacemaker, and the effects of light. In Axelrod J, Fraschini F, Velo GP (eds). The Pineal and Its Endocrine Role. Plenum, New York, pp 535–550.

148. Webb SM, Champney TH, Lewinski AK et al (1985) Photoreceptor damage and eye pigmentation: Influence on the sensitivity of rat pineal N-acetyltransferase activity and melatonin to light at night. Neuroendocrinology 40:205–209.

149. Brainard GC, Richardson BA, King TS et al (1983) The suppression of pineal melatonin content and N-acetyltransferase activity by different light irradiances in the Syrian hamster: A dose-response study. Endocrinology 113:293–296.

150. Brainard GC, Richardson BA, Petterborg LJ et al (1982) The effect of different light intensities on pineal melatonin content. Brain Res 233:75–84.

151. Illnerova H, Vanecek J (1979) Effect of one-minute exposure to light at night on rat pineal serotonin N-acetyltransferase. Progr Brain Res 52:241–243.

152. Illnerova H, Vanecek J (1979) Response of rat pineal serotonin N-acetyltransferase to one minute light pulse at different night times. Brain Res 167:431–434.

153. Nürnberger F, Joshi BN, Heinzeller T et al (1985) Responsiveness of pineal N-acetyltransferase and melatonin in the cotton rat exposed to either artificial or natural light at night. J Pineal Res 2:375–386.

154. Reiter RJ, Joshi BN, Heinzeller T et al (1986) A single 1 or 5 second light pulse at night inhibits hamster pineal melatonin. Endocrinology 118:1906–1909.

155. Hoffman RA (1982) Seasonal reproductive cycles in golden hamsters: Speculation on data and dogma. In: Reiter RJ (ed) The Pineal and Its Hormones. Alan R. Liss, New York, pp 153–166.

156. Brainard GC, Richardson BA, Hurlbut EC et al (1984) The influence of various irradiances of artificial light twilight and moonlight on the suppression of pineal melatonin content in the Syrian hamster. J Pineal Res 1:105–119.

157. Cardinali DP, Larin F, Wurtman RJ (1972) Action spectra for effects of light on hydroxyindole-O-methyltransferases in rat pineal, retina, and Harderian gland. Endocrinology 91:877–886.

158. Brainard GC, Richardson BA, King TS et al (1984) The influence of different light spectra on the suppression of pineal melatonin content in the Syrian hamster. Brain Res 294:333–339.

159. Brainard GC, Podolin PL, Leivy SW et al (1986) Near-ultraviolet radiation suppresses pineal melatonin content. Endocrinology 119:2201–2205.

160. Brainard GC, Lewy AJ, Manaker M et al (1985) Effect of light wavelength on the suppression of nocturnal plasma melatonin in normal volunteers. Ann NY Acad Sci 453:376–378.

161. Wilson BW, Anderson LE, Hilton DI et al (1981) Chronic exposure to 60-Hz electric fields: Effects on pineal function in the rat. Bioelectromagnetics 2:371–380.

162. Wilson BW, Chess EK, Anderson LE (1986) 60-Hz electric-field effects on

pineal melatonin rhythms: Time course for onset and recovery. Bioelectromagnetics 7:239–242.

163. Reiter RJ, Richardson BA, Johnson LY et al (1980) Pineal melatonin rhythm: Reduction in aging Syrian hamsters. Science 210:1372–1373.

164. Weiss B, Greenberg L, Cantor E (1979) Age-related alterations in the development of adrenergic denervation supersensitivity. Fed Proc 38:1915–1921.

165. Reiter RJ, Vriend J, Brainard GC et al (1982) Reduced pineal and plasma melatonin levels and gonadal atrophy in old hamsters kept under winter photoperiods. Exp Aging Res 8:27–30.

166. Reiter RJ, Craft CM, Johnson LY et al (1981) Age-associated reduction in nocturnal pineal melatonin levels in female rats. Endocrinology 109:1295–1297.

167. Reiter RJ, Johnson LY, Steger RW et al (1980) Pineal biosynthetic activity and neuroendocrine physiology in the aging hamster and gerbil. Peptides 1 (Suppl):69–77.

168. King TS, Richardson BA, Reiter RJ (1981) Age-associated changes in pineal serotonin N-acetyltransferase activity and melatonin content in the male gerbil. Endocrinol Res Commun 8:253–262.

169. Pang SF, Tang F, Tang PL (1983) Decreased serum and pineal concentrations of melatonin and N-acetylserotonin in aged male hamsters. Horm Res 17:228–234.

170. Pang SF, Tang F, Tang PL (1984) Negative correlation of age and the levels of pineal melatonin, pineal N-acetylserotonin, and serum melatonin in male rats. J Exp Zool 229:41–47.

171. Arendt J, Wirz-Justice A, Bradtke J et al (1979) Long-term studies on immunoreactive human melatonin. Ann Clin Biochem 16:307–312.

172. Wetterberg L, Halberg F, Haus E et al (1981) Circadian rhythmic urinary melatonin excretion in four seasons by clinically healthy Japanese subjects in Kyushu. Chronobiologia 8:88–89.

173. Beck-Friis J, Von Rosen D, Kjellman BF et al (1984) Melatonin in relation to body measures, sex, age, season, and use of drugs in patients with major affective disorders and healthy subjects. Psychoneuroendocrinology 9:261–277.

174. Illnerova H, Zvokky P, Vanecek J (1985) The circadian rhythm in plasma melatonin concentration of the urbanized man: The effect of summer and winter time. Brain Res 328:186–1898.

175. Iguchi H, Kato K, Ibayashi H (1982) Age-dependent reduction in serum melatonin concentrations in healthy human subjects. J. Clin Endocrinol Metab 55:27–29.

176. Toutou Y, Fevre M, Bogdan A (1984) Patterns of plasma melatonin with aging and mental condition: Stability of nyctohemeral rhythms and differences in season variation. Acta Endocrinol 106:145–151.

177. Nair NPV, Hariharasubramanian N, Pilapil C et al (1986) Plasma melatonin: an index of brain aging in humans? Biol Psychiatry 21:141–150.

178. Sack RL, Lewy AJ, Erb DL et al (1986) Human melatonin production decreases with age. J Pineal Res 3:379–388.

179. Arendt J, Bajkowski C, Franey C et al (1985) Immunoassay of 6-hydroxymelatonin sulfate in human plasma and urine: Abolition of the

urinary 24-hour rhythm with atenolol. J Clin Endocrinol Metab 60:1166–1173.
180. Quay WB (1974) Pineal Chemistry. CC Thomas, Springfield.
181. Vollrath L (1981) The Pineal Organ. Springer, Berlin.
182. Welsh MG (1985) Pineal calcification: Structural and functional aspects. Pineal Res Rev 3:41–48.
183. Löfgren FO (1958) Vertebral angiography in the diagnosis of tumors in the pineal region. Acta Radiol (Stockholm) 50:108–114.
184. Tapp E, Blumfield M (1970) The weight of the pineal gland in malignancy. B J Cancer 24:67–75.
185. Pende V (1953) La calcificazione della glandola pineale: Correlazioni con le endocrinopatie. Studio clinico radiologico su 1121 casi. Folia Endocrinol (Roma) 6:191–197.
186. Kitay JI, Altschule MD (1954) The Pineal Gland. Harvard, Cambridge.
187. Krstic R (1986) Pineal calcification: Its mechanism and significance. J Neural Transm (Suppl) 21:415–432.
188. Angervall L, Buger S, Röcket H (1958) A microradiographic and X-ray crystallographic study of calcium in the pineal body in intracranial tumors. Acta Pathol Microbiol Scand 44:113–119.
189. Earle KM (1965) X-ray diffraction and other studies of the calcareous deposits in human pineal glands. J Neuropathol Exp Neurol 24:108–118.
190. Krstic R (1976) A combined scanning and transmission electron microscopic study and electron probe microanalysis of human pineal acervuli. Cell Tiss Res 174:129–137.
191. Michotte Y, Lowenthal A, Knaepen L et al (1977) A morphological and chemical study of calcification of the pineal gland. J Neural Transm 215:209–219.
192. Japha JL, Eder TJ, Goldsmith ED (1976) Calcified inclusions in the superficial pineal gland of the Mongolian gerbil, *Meriones unguiculatus*. Acta Anat 94:533–544.
193. Welsh MG, Reiter RJ (1978) The pineal gland of the gerbil, *Meriones unguiculatus*. Cell Tissue Res 193:323–336.
194. Erdinc F (1977) Concrement formation encountered in the rat pineal gland. Experientia 33:514.
195. Diehl BJM (1978) Occurrence and regional distribution of calcareous concretions in the rat pineal gland. Cell Tissue Res 195:359–366.
196. Jung D, Vollrath L (1982) Structural dissimilarities in different regions of the pineal gland of Pribright white guinea pigs. J Neural Transm 54:117–128.
197. Japha JL, Eder TJ, Goldsmith ED (1974) Morphological and histochemical features of the gerbil system. Anat Rec 178:381 (Abstract).
198. Reiter RJ, Welsh MG, Vaughan MK (1976) Age-related changes in the intact and sympathetically denervated gerbil pineal gland. Am J Anat 146:427–432.
199. Lewinski A, Vaughan MK, Champney TH et al (1983) Dark-exposure increases the number of pineal concretions in male gerbils (*Meriones unguiculatus*). IRCS Med Sci 11:977–78.
200. Vaughan MK, Joshi BN, Reiter et al (1986) Daily propranolol administration reduces pineal concretion formation in the Mongolian gerbil. Proc Soc Exp Biol Med 182:372–374.
201. Champney TH, Joshi BN, Vaughan MK et al (1985) Superior cervical

ganglionectomy results in the loss of pineal concretions in the adult male gerbil, *Meriones unguiculatus*. Anat Rec 211:465–468.

202. Lukaszyk A, Reiter RJ (1975) Histophysiological evidence for the secretion of polypeptides by the pineal gland. Am J Anat 143:451–464.

203. Hinterberger H, Pickering J (1976) Catecholamine, indolealkylamine, and calcium levels of human pineal glands in various clinical conditions. Pathology 8:221–229.

204. Miline R (1980) The role of the pineal gland in stress. J Neural Transm 47:191–220.

205. Vaughan MK, Spanel-Borowski K, Karasek M et al (1983) Action of subcutaneous implants or injections of melatonin on reproductive and metabolic variables and pineal concretions in male gerbils (*Meriones unguiculatus*). Biomed Res 4:329–336.

206. Maestroni GJM, Conti A, Pierpoali W (1986) Role of the pineal gland in immunity. Circadian synthesis and release of melatonin modulates the antibody response and antagonizes the immunosuppressive effect of corticosterone. J Neuroimmunol 13:19–30.

207. Pang SF, Yu Hs, Suen HC et al (1980) Melatonin in the retina of rats: A diurnal rhythm. J Endocrinol 87:89–93.

208. Pang SF, Allen AE (1986) Extra-pineal melatonin in the retina. Its regulation and physiological function. Pineal Res Rev 4:55–96.

209. Bubenik GA, Purtill RA, Brown GM et al (1978) Melatonin in the retina and the Harderian gland; Ontogeny, diurnal variations, and melatonin treatment. Exp Eye Res 27:323–333.

210. Hoffman RA, Johnson LB, Reiter RJ (1985) Harderian glands of golden hamsters: Temporal and sexual differences in immunoreactive melatonin. J Pineal Res 2:161–168.

211. Reiter RJ, Richardson BA, Matthews SJ et al (1983) Rhythms in immunoreactive melatonin in the retina and Harderian glands of rats: Persistence after pinealectomy. Life Sci 32:1229–1236.

212. Khoory R, Dubbek R, Schloot W (1986) Melatonin content in the pineal gland of different rat trains and stocks. Z Versuchstierkd 28:141–148.

213. Wirz-Justice A, Arendt J, Marston A (1980) Antidepressant drugs elevate rat pineal and plasma melatonin. Experientia 36:442–444.

214. Ozaki Y, Lynch HJ, Wurtman RJ (1976) Melatonin in rat pineal, plasma, and urine: 24-hour rhythmicity and effect of chlorpromazine. Endocrinology 98:1418–1424.

215. Withyachumnarnkul B, Knigge KM (1980) Melatonin concentration in cerebrospinal fluid, peripheral plasma, and plasma of the confluence sinuum of the rat. Neuroendocrinology 30:382–388.

216. Wetterberg L (1978) Melatonin in humans. Physiological and clinical studies. J Neural Transm (Suppl 13):289–310.

217. Birau N (1981) Melatonin in human serum: Progress in screening investigation and clinic. Adv Biosci 29:297–326.

218. Reiter RJ (1987) Pineal response to stress: Implications for reproductive physiology. In: Pancheri P. Zichella L (eds) Biorhythms and Stress in Pathophysiology of Reproduction. Hemisphere, New York, in press.

219. Parfitt AG, Klein DC (1976) Sympathetic nerve endings in the pineal gland

protect against acute stress-induced increase in N-acetyltransferase activity. Endocrinology 99:840–851.

220. Illnerova H (1976) The effect of immobilization on the activity of serotonin N-acetyltransferase in the rat epiphysis. In: Usdin E, Kvetnansky R, Kopin IJ (eds). Pergamon, New York, pp 129–136.

221. Champney TH, Steger RW, Christie DS et al (1985) Alterations in components of the pineal melatonin synthetic pathway by acute insulin stress in the rat and Syrian hamster. Brain Res 338:25–32.

222. Seggie J, Campbell L, Brown GM et al (1985) Melatonin and N-acetylserotonin stress responses: Effects of type of stimulation and housing conditions. J Pineal Res 2:39–49.

223. Oxenkrug GF, McIntyre IM (1985) Stress-induced synthesis of melatonin: Possible involvement of the endogenous monoamine oxidase inhibitor (Tribulin). Life Sci 37:1743–1746.

224. Welker HA, Vollrath L (1984) The effects of a number of short-term exogenous stimuli on pineal serotonin N-acetyltransferase activity in rats. J Neural Transm 59:69–80.

225. Reiter RJ, Hurlbut EC, Tannenbaum MG et al (1987) Melatonin synthesis in the pineal gland of the Richardon's ground squirrel (*Spermophilus richardsonni*): Influence of age and insulin-induced hypoglycemia. J Neural Transm (in press).

226. Joshi BN, Troiani ME, Milin J et al (1986) Adrenal-mediated depression of N-acetyltransferase activity and melatonin levels in the rat pineal gland. Life Sci 38:1573–1580.

227. Troiani ME, Reiter RJ, Vaughan MK et al (1987) Swimming depresses nighttime melatonin content without changing N-acetyltransferase activity in the rat pineal gland. Neuroendocrinology (in press).

228. Rollag MD, O'Callaghan DL, Niswender GD (1978) Dynamics of photo-induced alterations in pineal blood flow. J Endocrinol 113:547–548.

229. Vaughan GM, McDonald SD, Jordan RM et al (1978) Melatonin concentration in human blood and cerebrospinal fluid: Relationship to stress. J Clin Endocrinol Metab 47: 220–223.

230. Vaughan GM, McDonald SD, Jordan RM et al (1979) Melatonin, pituitary function, and stress in humans. Psychoneuroendocrinology 4:351–355.

231. Vaughan GM, Taylor TJ, Pruitt BA Jr et al (1985) Pineal function in burns: Melatonin is not a marker for general sympathetic activity. J Pineal Res 2:1–12.

232. Carr DG, Reppert SM, Bullen B et al (1981) Plasma melatonin increases during exercise in women. J Clin Endocrinol Metab 53:224–225.

233. Reiter RJ (1980) The pineal and its hormones in the control of reproduction in mammals. Endocrinol Rev 1:109–131.

234. Reiter RJ (1986) Pineal function in the human: Implications for reproductive physiology. J Obstet Gynaecol (Suppl 2) 6:577–581.

235. Dole E, Gerlach DH, Wilwhite AL (1979) Menstrual dysfunction in distance runners. Obstet Gynecol 54:47–53.

236. Reiter RJ, Trakulrungsi WK, Trakulrungsi C et al (1982) Pineal melatonin production: Endocrine and age effects.In: Klein DC (ed) Melatonin Rhythm Generating System. Karger, Basel, pp 143–154.

237. Quay WB (1964) Circadian and estrous rhythms in pineal melatonin and 5-hydroxyindole-3-acetic acid. Proc Soc Exp Biol Med 115:710–716.
238. Johnson LY, Vaughan MK, Richardson BA et al (1982) Variation in pineal melatonin content during the estrous cycle of the rat. Proc Soc Exp Biol Med 169:416–419.
239. Wurtman RJ, Axelrod J, Snyder SH (1965) Changes in the enzymatic synthesis of melatonin in the pineal during the estrous cycle. Endocrinology 76:798–800.
240. Cardinali DP, Nagle CA, Rosner JM (1974) Effects of estradiol and protein synthesis in the rat pineal organ. Horm Res 5:304–310.
241. Wallen EP, Yochim JM (1975) An analysis of pineal hydroxyindole-O-methyltransferase rhythm during the estrous cycle in the rat. Adv Exp Med Biol 54:79–84.
242. Cardinali DP, Vacas MI (1978) Feedback control of pineal function by reproductive hormones: A neuroendocrine paradigm. J Neural Transm (Suppl 13):175–201.
243. Reiter RJ, Sorrentino S Jr (1971) Inhibition of luteinizing hormone release and ovulation in PMS-treated rats by peripherally administered melatonin. Contraception 4:385–392.
244. Rollag MD, Chen HJ, Ferguson BN et al (1979) Pineal melatonin content throughout the hamster estrous cycle. Proc Soc Exp Biol Med 162:211–213.
245. Wetterberg L, Arendt J, Paunier L et al (1976) Human serum melatonin changes during the menstrual cycle. J Clin Endocrinol Metab 42:185–188.
246. Arendt J (1978) Melatonin assay in body fluids. J Neural Transm (Suppl 13) 13:265–278.
247. Hariharasubramanian N, Nair NPV, Pilapl C et al (1985) Plasma melatonin levels during menstrual cycle: Changes with age. In: Gupta D, Reiter RJ (eds), The Pineal Gland during Development. Croom-Helm, London, pp 166–173.
248. Champney TH, Craft CM, Webb SM et al (1985) Hormonal modulation of pineal melatonin synthesis in rats and Syrian hamsters: Effect of adrenalectomy and corticosteroid implants. J Neural Transm 64:67–79.
249. Champney TH, Webb SM, Richardson BA et al (1986) Hormonal modulation of cyclic melatonin production in the pineal gland of rats and Syrian hamsters: Effects of thyroidectomy or thyroxine administration. Chronobiol Int 2:177–183.
250. Cardinali DP (1981) Melatonin. A mammalian pineal hormone. Endocrinol Rev 2:327–354.
251. Vaughan GM, Allen JP, Tullis W et al (1978) Overnight plasma profiles of melatonin and certain adenohypophyseal hormones in men. J Clin Endocrinol Metab 47:566–571.
252. Blum K, Merritt JH, Reiter RJ et al (1973) A possible relationship between the pineal gland and ethanol preference in the rat. Curr Ther Res 15:25–30.
253. Reiter RJ, Blum K, Wallace JE et al (1973) Effect of the pineal gland on alcohol consumption in congenitally blind male rats. Q J Stud Alcohol 34:937–940.
254. Reiter RJ, Blum K, Wallace JE et al (1974) Pineal gland: Evidence for an influence on ethanol preference in male Syrian hamsters. Comp Biochem Physiol 47A:11–16.

255. Moss HB, Tamarkin L, Majchrowicz E et al (1986) Pineal function during ethanol intoxication, dependence, and withdrawal. Life Sci 39:2209–2214.
256. Hiatt RA, Bawol RD (1984) Alcoholic beverage consumption and breast cancer incidence. Am J Epidemiol 120:676–683.
257. Tamarkin L, Cohen M, Roselle D et al (1981) Melatonin inhibition and pinealectomy enhancement of 7,12-dimethyl-benz (a) anthracene-induced mammary tumors in the rat. Cancer Res 41:4432–4436.
258. Blask DE (1984) The pineal: An oncostatic gland? In: Reiter RJ (ed) The Pineal Gland. Raven, New York, pp 253–284.
259. Tamarkin L, Danforth D, Lichter A et al (1982) Decreased nocturnal plasma melatonin peak in patients with estrogen positive breast cancer. Science 216:1003–1005.
260. Reiter RJ, Vaughan MK, Blask DE (1975) Possible role of the cerebrospinal fluid in the transport of pineal hormones in mammals. In: Knigge KM, Scott DE, Kobayashi H, Ishii S (eds) Brain-Neuroendocrine Interaction III. Karger, Basel, pp 337–354.
261. Mess B, Trentini GP, Kovacs L et al (1975) Melatonin, cerebrospinal fluid, pineal gland interrelationships. In: Knigge KM, Scott DE, Kobayashi H, Ishii S (eds) Brain-Endocrine Interaction II. Karger, Basel, pp 355–364.
262. Pavel S (1973) Arginine vasotocin release into cerebrospinal fluid of cats induced by melatonin. Nature 246:183–184.
263. Hedlund L, Leschko MM, Rollag MD et al (1977) Melatonin: Daily cycle in plasma and cerebrospinal fluid in calves. Science 195:686–687.
264. Rollag MD, Morgan RJ, Niswender GD (1978) Route of melatonin secretion in sheep. Endocrinology 102:1–8.
265. Reppert SM, Perlow MJ, Klein DC (1980) Cerebrospinal fluid melatonin production. In: Wood JH (ed) Neurobiology of Cerebrospinal Fluid. Plenum, New York, pp 579–589.
266. Reppert SM, Perlow MJ, Tamarkin L et al (1979) A diurnal melatonin rhythm in primate cerebrospinal fluid. Endocrinology 104:295–301.
267. Perlow MJ, Reppert SM, Tamarkin L et al (1980) Photic regulation of the melatonin rhythm: Monkey and man are not the same. Brain Res 182:211–216.
268. Reppert SM, Perlow MJ, Tamarkin L et al (1981) The effects of environmental lighting on the daily melatonin rhythm in primate cerebrospinal fluid. Brain Res 223:313–323.
269. Wilson BW, Snedden W, Silman RE et al (1977) A gas chromatography-mass spectrometry method for the quantitative analysis of melatonin in plasma and cerebrospinal fluid. Anal Biochem 81:283–291.
270. Smith JA, Mee TJX, Barnes ND et al (1976) Melatonin in serum and cerebrospinal fluid. Lancet 2:425.
271. Smith JA, Barnes JLC, Mee TJX (1979) The effect of neuroleptic drugs on serum and cerebrospinal fluid melatonin concentrations in psychiatric subjects. J Pharm Pharmacol 31:246–248.
272. Tan CH, Khoo JC (1981) Melatonin concentrations in human serum, ventricular, and lumbar cerebrospinal fluids as an index of the secretory pathway of the pineal gland. Horm Res 14:224–233.
273. Brown GM, Young SN, Gauthier S et al (1979) Melatonin in human

cerebrospinal fluid in the daytime: Its origin and variation with age. Life Sci 25:929–936.

274. Vakkuri O, Leppäluoto J, Vuolteenako O (1984) Development and validation of a melatonin radioimmunoassay using radioiodinated melatonin as tracer. Acta Endocrinol 106:152–157.

275. Miles A, Philbrick DRS, Tidmarsh SF et al (1985) Direct radioimmunoassay of melatonin in saliva. Clin Chem 31:1412–1413.

276. Miles A, Philbrick DRS, Shaw DM et al (1985) Salivary melatonin estimation in clinical research. Clin Chem 31:2041–2042.

277. Vakkuri O (1985) Diurnal rhythm of melatonin in human saliva. Acta Physiol Scand 124:409–412.

278. Vakkuri O, Leppäluoto J, Kauppila A (1985) Oral administration and distribution of melatonin in human serum, saliva and urine. Life Sci 37:489–495.

279. Lynch HJ, Wurtman RJ, Moskowitz MA et al (1979) Daily rhythm in human urinary melatonin. Science 187:169–171.

280. Waldhauser F, Lynch HJ, Wurtman RJ (1984) Melatonin in human body fuilds: Clinical significance. In: Reiter RJ (ed) The Pineal Gland. Raven, New York, pp 345–370.

281. Lang U, Kornemark M, Aubert ML et al (1981) Radioimmunological determination of urinary melatonin in humans. Correlation with plasma levels and typical 24-hour rhythmicity. J Clin Endocrinol Metab 53:645–650.

282. Wetterberg L, Eriksson O, Friberg V et al (1978) A simplified radioimmunoassay for melatonin and its application to biological fluids. Preliminary observations on the half-life of plasma melatonin in man. Clin Chim Acta 86:169–177.

283. Fellenberg AJ, Phillipou G, Seamark RF (1981) Urinary 6-hydroxymelatonin excretion and melatonin production rate: Studies in sheep and man. In: Matthews CD, Seamark RJ (eds), Pineal Function. Elsevier/North Holland, Amsterdam, pp 143–150.

284. Matthews CD, Kennaway DJ, Fellenberg AJG et al (1981) Melatonin in man. Adv Biosci 29:371–384.

285. Tetsuo M, Markey SP, Kopin IJ (1980) Measurement of 6-hydroxymelatonin in urine and its diurnal variation. Life Sci 27:105–109.

286. Tetsuo M, Markey SP, Colburn RW et al (1981) Quantitative analysis of 6-hydroxymelatonin in human urine by gas chromatography negative chemical ionization mass spectrometry. Anal Biochem 110:208–215.

287. Fellenberg AJ, Phillipou G, Seamark RF (1980) Specific quantitation of urinary 6-hydroxymelatonin sulfate by gas chromatography-mass spectrometry. Biomed Mass Spectrom 7:84–87.

288. Reiter RJ (1973) Comparative physiology: Pineal gland. Annu Rev Physiol 35:305–323.

289. Hoffmann K (1981) Pineal involvement in the photoperiodic control of reproduction and other functions in the Djungarian hamster. *Phodopus sungorus*. In: Reiter RJ (ed) The Pineal Gland. Vol. II, Reproductive Effects. CRC Press, Boca Raton, pp 83–102.

290. Rusak B (1982) Circadian organization in mammals and birds: Role of the pineal gland. In: Reiter RJ (ed), The Pineal Gland, Vol. III, Extra-Reproductive Effects. CRC Press, Boca Raton, pp 27–52.

291. Johnson LY (1982) The pineal gland as a modulator of the adrenal and thyroid axes. In: Reiter RJ (ed) The Pineal Gland, Vol. II, Reproductive Effects. CRC Press, Boca Raton, pp 107–152.
292. Wainwright S (1982) Role of the pineal gland in the vertebrate master biological clock. In: Reiter RJ (ed) The Pineal Gland, Vol. II, Reproductive Effects. CRC Press, Boca Raton, pp 53–80.
293. Ralph CL (1984) Pineal bodies and thermoregulation. In: Reiter RJ (ed), The Pineal Gland. Raven, New York, pp 193–220.
294. Vriend J (1983) Pineal-thyroid interactions. Pineal Res Rev 1:183–207.
295. Reiter RJ (1984) Pineal function in mammals including man. In: Neuwelt EA (ed), Diagnosis and Treatment of Pineal Region Tumors. Williams and Wilkins, Baltimore, pp 86–107.
296. Erlich SS, Apuzzo MLJ (1985) The pineal gland: Anatomy, physiology and clinical significance. J Neurosurg 63:321–341.

5

Neuropeptides and Glucose Metabolism

HIDEO SASAKI, SEIJIRO MARUBASHI, YOSHIKAZU
YAWATA, KEIICHI YAMATANI, MAKOTO TOMINAGA,
AND TADASHI KATAGIRI

Introduction

In recent years, numerous new biologically active peptides have been discovered in the gastrointestinal tract. While immunohistochemical evidence suggests that at least 20 of them are present in the brain (1–3), their functions remain at issue since the effects of their direct intracerebral administration vary according to dose and experimental condition, and may include psychologic, neurologic, or autonomic functions, such as the emotions, consciousness and behavior, sensory and motor activity, libido, appetite, gut motility, blood pressure, respiration, body temperature, blood glucose, endocrine, and other functions. This chapter focuses on the role of some of these peptides in the central regulation of glucose metabolism.

History

Claude Bernard's discovery that hyperglycemia could be produced by "piqûre" of the base of the 4th ventricle led to an intensive investigation of the brain as a regulator of glucose metabolism, along three major lines. The first one concerns the hypothalamus as the control center of pituitary function. This role was discovered by Harris et al. (4) and is exemplified by the control of the adrenocorticotropic hormone (ACTH)–adrenocortical system by the supraoptic nucleus, by stress-induced hyperglycemia, and by the secretion of somatostatin (5), growth hormone releasing factor (6), and of thyrotropin releasing hormone (TRH).

The second line of investigation deals with the ventro-medial (VMH) and the lateral hypothalamic areas (LHA) of the hypothalamus as autonomic nervous system centers for the control of blood glucose concentration and hepatic glucose production (7–10). In this system, VMH and LHA have opposite actions. Thus, the stimulation of the VMH leads to an increase in hepatic phosphorylase activity, to a decrease in hepatic glycogen, and to an increase in blood glucose concentration, phenomena that can be observed also following stimulation of the splanchnic nerve or the injection of norepinephrine into the VMH and that can be blocked by local pretreatment with propranolol or hexamethonium. On the other hand, the stimulation of the lateral hypothalamic area or of the vagus nerve and the injection of acetylcholine or carbachol into the LHA lead to an increased activity of hepatic glycogen synthetase and hepatic glycogen content, effects that can be inhibited by the intracerebral injection of atropine or scopolamine or by the intraperitoneal administration of methylatropine. These observations led to the hypothesis that the VMH exerts its activities through central and peripheral β-adrenergic mechanisms while the lateral hypothalamus acts through central and peripheral cholinergic mechanisms. In addition, it has been demonstrated that stimulation of the LHA decreases the activity of phosphoenolpyruvate carboxykinase (PEPCK), a key enzyme of gluconeogenesis, while stimulation of the VMH increases the activity of PEPCK and decreases that of pyruvate kinase, a key enzyme for glycolysis. Thus, stimulation of the VMH may result in hyperglycemia due to inhibition of glycolysis and stimulation of glycogenolysis and of gluconeogenesis, while stimulation of the LHA may lead to hypoglycemia due to acceleration of glycogen synthesis and inhibition of gluconeogenesis. In addition, stimulation of the VMH increases the blood levels of epinephrine and pancreatic glucagon and decreases that of insulin, leading to the suggestion that the stimulus for glucose mobilization follows the route: VMH → sympathetic nerves → adrenal medulla → epinephrine → changes in pancreatic hormones → liver.

The importance of the hypothalamic VMH and LHA as centers for the regulation of glucose has been confirmed by evidence that the increase in plasma levels of insulin in the VMH–lesioned rat is vagus mediated and that the increase in blood glucose levels following stimulation of the VMH requires an intact sympathetic system (11,12).

The third line of investigation deals with the control of appetite. Oomura et al. (13,14) have demonstrated the presence of glucose-sensitive neurons that inhibit the electric discharge activity of the LHA and the presence of glucoreceptor neurons that accelerate such activity in the VMH. Thus, glucose would inhibit feeding activity, and stimulate insulin secretion, on the one hand, but cause a sense of satiety while decreasing the secretion of insulin on the other hand. The occurrence of obesity, increased insulin secretion, and increased appetite after destruction of the VMH are well known and are believed to be associated with excitation of the vagus nerve and inhibition of the β-adrenergic system (11,12).

From the evidence summarized above, one could conclude that the hypothalamus participates in the regulation of glucose metabolism primarily through a neural modulation of pancreatic hormone secretion.

TABLE 5,1 Brain-gut peptides capable of inducing hyperglycemia when administered centrally or peripherally

Central Administration (intracerebroventricular, intracisternal, intracerebral)	Peripheral Administration (intravenous, intraperitoneal, intramuscular)
Bombesin, GRP	Bombesin, GRP
Glucagon	Glucagon
TRH	Neurotensin
Neurotensin	VIP
β-Endorphin	Substance P
Substance P	Angiotensin II
CCK	Secretin
Angiotensin II	Vasopressin
CRF	Galanin
GRF	

However, since the secretion of insulin by pancreatic islets may be stimulated by hypothalamic homogenates (15), and since hypothalamic extracts, especially extracts of VMH, increase immunoreactive glucagon (IRG) levels and decrease immunoreactive insulin (IRI) levels (16,17), the possibility of humoral mechanisms can not be ruled out. In this sense, the neuropeptides may represent a third, or neurohumoral system for the transmission of information.

Neuropeptides and Changes in Blood Glucose Levels

Many neuropeptides induce hyperglycemia when they are injected parenterally or directly into the lateral ventricle, although the effective dose varies (Table 5.1). Glucagon is the most effective peptide, as it causes a significant increase in plasma glucose levels at doses as low as 3×10^{-12} moles and achieves maximum effects at a dose of 10^{-11} moles. Bombesin (BBS), gastrin releasing peptide (GRP), β-endorphin (β-E), and porcine GRP are effective at doses greater than 5×10^{-11} moles and have a half-maximal effect at 1 to 2×10^{-10} moles. Angiotensin II (Angio II), substance P (SP), and cholecystokinin octapeptide (CCK-8) are effective at approximately 10^{-8} moles, TRH at 10^{-7} moles, and neurotensin (NT) at higher doses. Somatostatin (SRIF) had no significant effect on the level of blood glucose at doses up to 10^{-8} moles, but caused death in all cases at a dose of 10^{-7} moles.

There are three types of hyperglycemic response to neuropeptide injection (Fig. 5.1): one that increases progressively for at least 90 to 120 minutes, one that reaches its peak in 10 to 15 minutes and decreases gradually thereafter, and one that increases immediately after the injec-

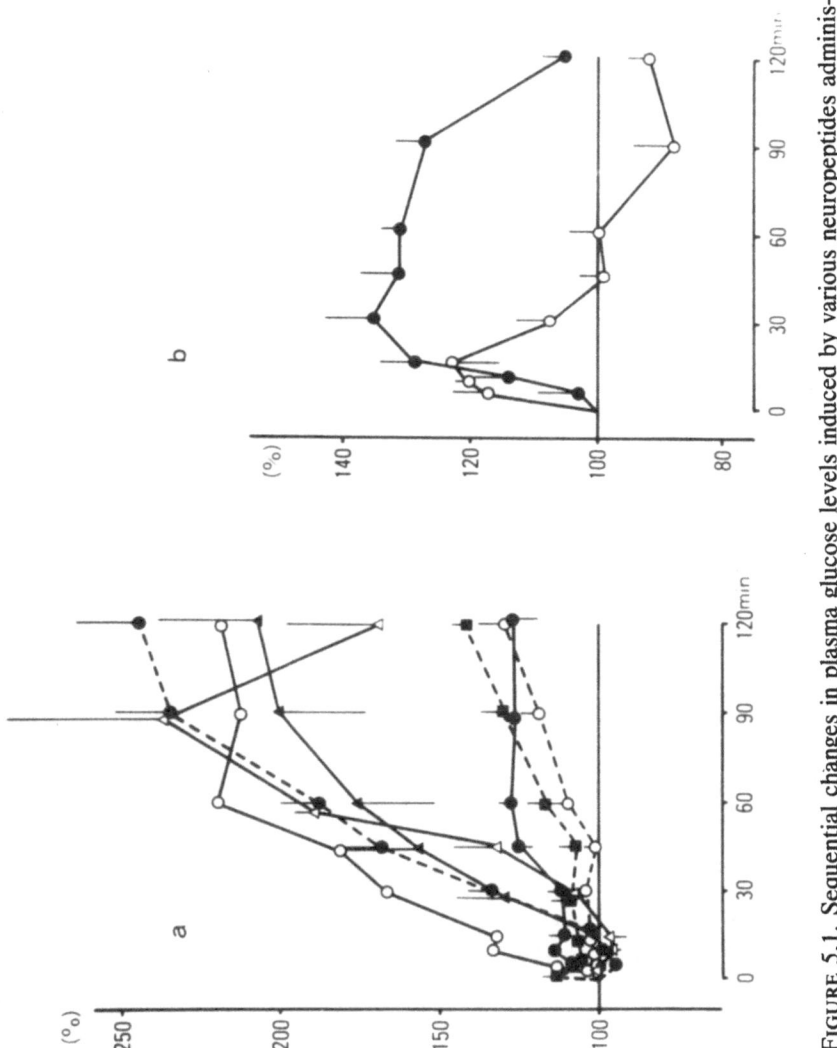

FIGURE 5.1. Sequential changes in plasma glucose levels induced by various neuropeptides administered icv in the rat. **a**: sustained type. BBS (●––●); pGRP, porcine gastrin releasing peptide (○——○); pGRP 1–27, ▲——▲ pGRP 14–27); β-E (△---△); ang. II, angiotensin II (■--■); CRF, corticotropin releasing factor (●——●); substance P (○——○). **b**: Transient type. TRH (○--○); glucagon (●--●). Each value represents the mean ± SEM.

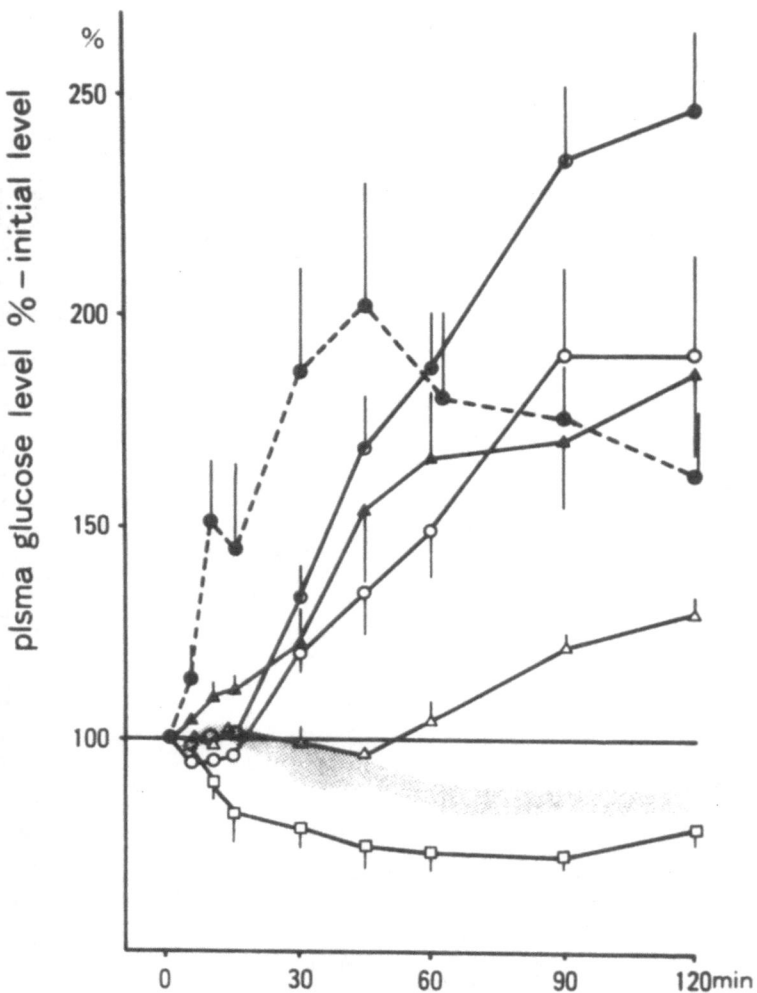

FIGURE 5.2. Modulation of plasma glucose levels by the icv or iv administration of BBS. BBS 10^{-11} moles icv (□——□), 5×10^{-11} moles (△---△), 10^{-10} moles (▲——▲), 10^{-9} moles (●——●), and 10^{-8} moles (○——○), and iv BBS 10^{-9} moles (●---●) were tested. Each value represents the mean ± SEM.

tion and then decreases rapidly. The first response is typical of BBS, GRP, β-E, Angio II, SP, and corticotropin-releasing hormone (CRF), the second of glucagon, and the third of TRH. These differences in response may be due to different mechanisms of action, although different rates of peptide absorption, and rates of metabolism cannot be ignored. The action of representative peptides is described below.

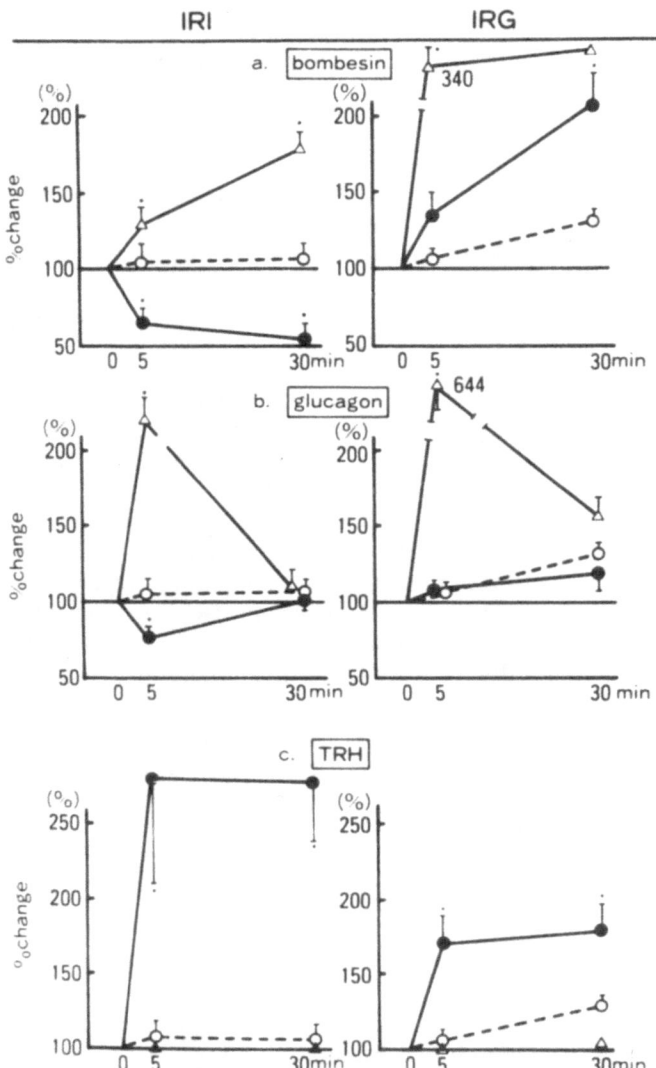

FIGURE 5.3. Percent changes in plasma levels of IRI and IRG following the icv (●——●) or iv (△——△) administration of BBS (a), glucagon (b), and TRH (c) in the rat. Buffer solution was injected into the rat lateral ventricle as a control (○---○). Each value represents the mean ± SEM.

Bombesin

Bomsesin, a peptide originally isolated from the skin of the frog Bombina bombina, is abundant in the brain, as shown by radioimmunoassay (18,19,20) or immunocytochemical methods (21) and has specific intracerebral receptors (22). Its structure is similar to that of the mammalian GRP (23,24). Indeed, BBS and GRP have identical C-terminal amino acid active sites. Thus, it is believed that the biologic effects of BBS in mammalian brain reflect those of GRP (25).

Modulation of Plasma Levels of Glucose, Insulin, and Glucagon by BBS Injected icv

The injection of BBS into a rat's lateral ventricle under pentobarbital anesthesia (Fig. 5.2) causes a rise in plasma glucose, except the lowest amount tested (10^{-11} moles) causes a decrease in plasma glucose similar to that obtained with control solution. In these experiments, no consistent dose–response relationship was observed. Although intravenous BBS also causes hyperglycemia, the pattern of response is different (Fig. 5.2). An iv injection of 10^{-9} moles of BBS has maximal hyperglycemic effect, and causes a significant increase in plasma IRI levels in contrast to icv injections which cause a significant decrease (Fig. 5.3). While plasma levels of IRG markedly increased after the iv administration of BBS, they increased only slightly after icv administration. Thus, iv BBS caused a marked hyperglucagonemia associated with significant hyperinsulinemia in contrast with the mild hyperglucagonemia associated with hypoinsulinemia caused by icv BBS. Marked hyperglycemia was also observed after the injection of BBS into the VMH (26), while icv injections cause hyperglycemia associated with marked hyperglucagonemia and hypoinsulinemia (27).

Effect of Autonomic Blockade on icv BBS-Induced Hyperglycemia

The involvement of the autonomic nervous system in BBS-induced hyperglycemia was confirmed by studies using various blocking agents. The results of these experiments (Fig. 5.4) show that the phenomenon is strongly inhibited by pretreatment with hexamethonium, a nicotinic cholinergic blocker, and with phentolamine and, to a lesser extent, by pretreatment with propranolol, or atropine, a muscarinic cholinergic blocker. Clearly, the hyperglycemic response to icv BBS is mediated, at least in part by α-adrenergic, β-adrenergic, and cholinergic mechanisms.

Effect of Adrenalectomy on icv BBS-Induced Hyperglycemia

The hyperglycemia induced by icv BBS can be prevented markedly by bilateral adrenalectomy (Fig. 5.4), suggesting involvement of the adrenals. This finding is in agreement with previous observations (28), which

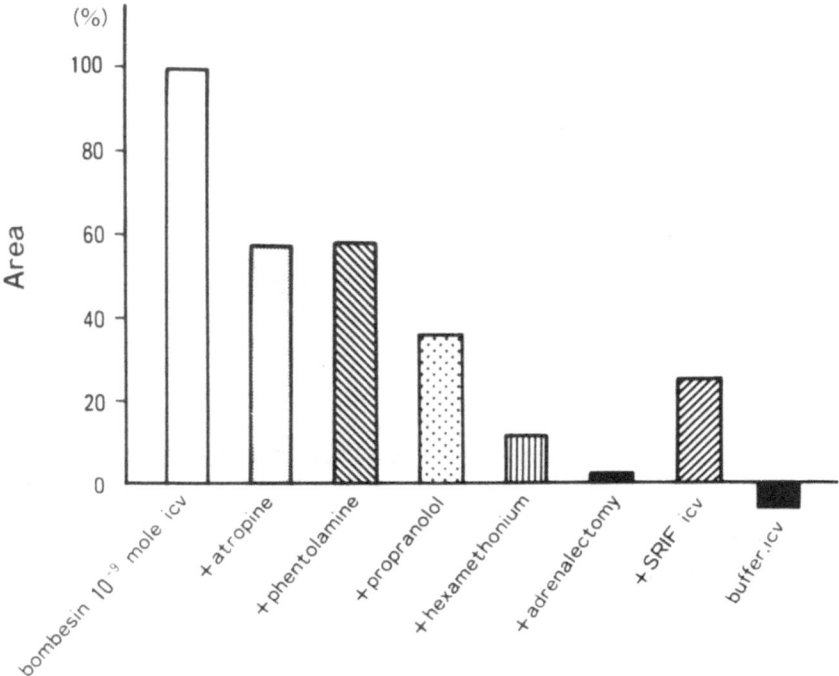

FIGURE 5.4. Effects of the ip administration of various autonomic blocking agents, bilateral adrenalectomy, and of icv administration of SRIF on hyperglycemia induced by the icv administration of BBS. Changes are depicted as the total increment of plasma glucose levels in 120 minutes, as indicated by percent change of the area under the curve.

indicate that this type of hyperglycemia is associated with a marked rise in plasma epinephrine levels, and is also abolished by adrenalectomy.

EFFECT OF ICV SRIF ON HYPERGLYCEMIA INDUCED BY ICV BBS

The icv injection of SRIF alone causes a slight decrease in plasma glucose comparable to that observed after the injection of control buffer solution, whereas pretreatment with SRIF markedly inhibits the glycemic response to icv BBS, confirming previously reported observations (26–28). In addition, pretreatment with SRIF inhibits the rise in plasma levels of glucose and IRG induced by iv epinephrine (28–30), confirming the notion that epinephrine-induced hyperglycemia is in part the result of an action of the hormone on the secretory activity of the pancreatic islets.

SUMMARY

The findings described above suggest that centrally administered BBS acts through a mechanism involving α- and β-adrenergic pathways, and the adrenal medulla (30). Atropine, a drug which readily crosses the

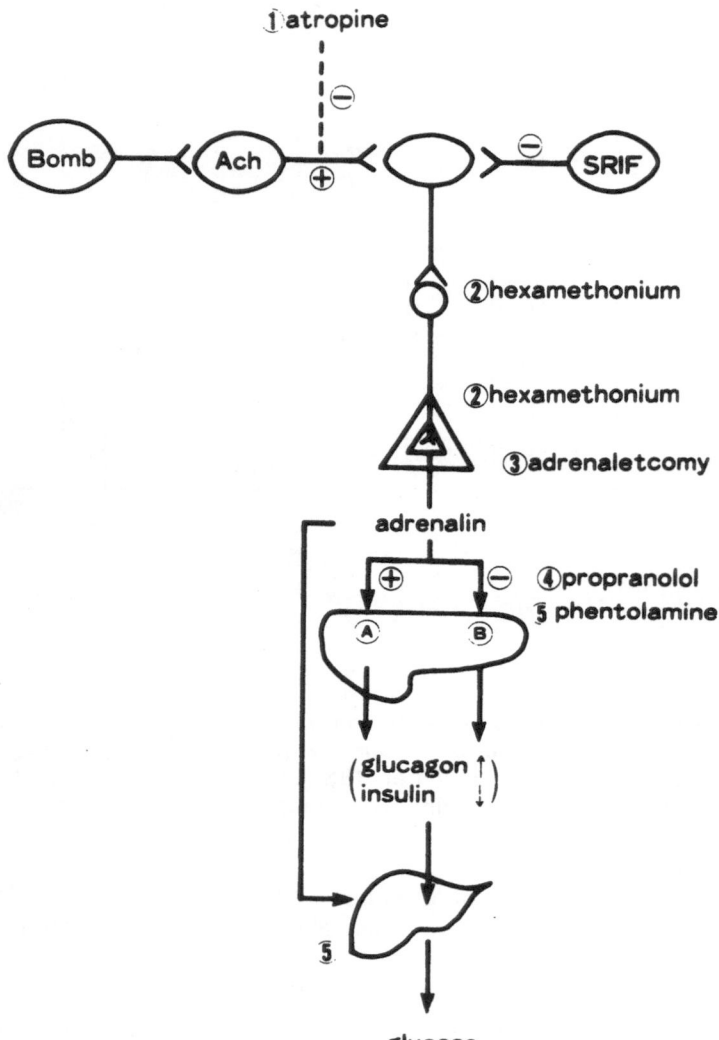

FIGURE 5.5. Schematic representation of the hyperglycemic effect of bombesin administered icv in the rat.

blood–brain barrier, may exert its action centrally, while hexamethonium may act on the sympathetic and parasympathetic ganglia, and on the adrenal medulla, which can be considered a modified ganglion. Epinephrine, in turn, could bring about hyperglycemia by stimulating the secretion of glucagon, and inhibiting that of insulin through a β-adrenergic mechanism, or through an effect on the hepatic α-adrenergic receptors (31). Thus, the effect of epinephrine could be inhibited by α- or

β-blockers. SRIF is believed to act centrally through an, as yet, poorly defined cholinergic pathway (Fig. 5.5).

Glucagon

Glocagon is one of the few gastrointestinal hormones which has not been positively identified in the brain, although glucagon-like substances have been found in mammalian, including human, brains by radioimmunoassay or immunohistochemical methods (32–39). That these substances may have a physiologic role is suggested by the presence of glucagon receptors in rat brain (40) and by the observation that these glucagon-like substances compete with ^{125}I-glucagon for binding to rat hepatocyte membrane and activate adenyl cyclase (33). In addition, it has been reported that the content of glucagon-like substances in the canine hypothalamus is increased by starvation, alloxan diabetes (41), and insulin hypoglycemia (42), and is decreased by the combined infusion of glucose and insulin (42). Thus, it appears to be related to changes in glucose metabolism. The highest concentrations of active material are found in the hypothalamus, followed at a distance by the brain stem including the midbrain, the pons, and the medulla oblongata, and the spinal cord. None has been found in the cerebral cortex, basal ganglia, cerebellum, or pituitary of the dog, man, pig and cattle, and only a small amount in the cerebral cortex and cerebellum of the rat. Almost identical data have been obtained using C-terminal or N-terminal-specific antibodies.

Modulation of Plasma Glucose, Insulin, and Glucagon Levels by icv Glucagon

As stated above (Fig. 5.2), the injection of control buffer solution into the lateral ventricle of the rat resulted in mild hypoglycemia lasting at least 120 minutes. On the other hand, the administration of 10 ng glucagon (3×10^{-12} moles) was followed by a rapid and significant increase in plasma glucose lasting about 45 minutes (Fig. 5.6). The administration of a larger dose of glucagon (100 ng or 3×10^{-11} moles) had a greater and longer effect (43). In contrast, iv injections of 100 ng glucagon caused greater, more rapid, but more transient responses, while the effect of intramuscular injection was altogether less pronounced. Thus, the hyperglycemia induced by icv glucagon cannot be considered the result of transfer of the injected hormone into the blood. The changes in plasma levels of IRG and IRI following icv and iv glucagon are shown in Fig. 5.3. It will be noted that the intravenous administration caused a marked rise in plasma IRG and IRI levels, while the icv administration caused only a statistically insignificant rise in plasma IRG and a decrease in plasma IRI, which also did not reach statistical significance. Thus, it is probable that the increase in the secretion of insulin produced by the iv administration of glucagon

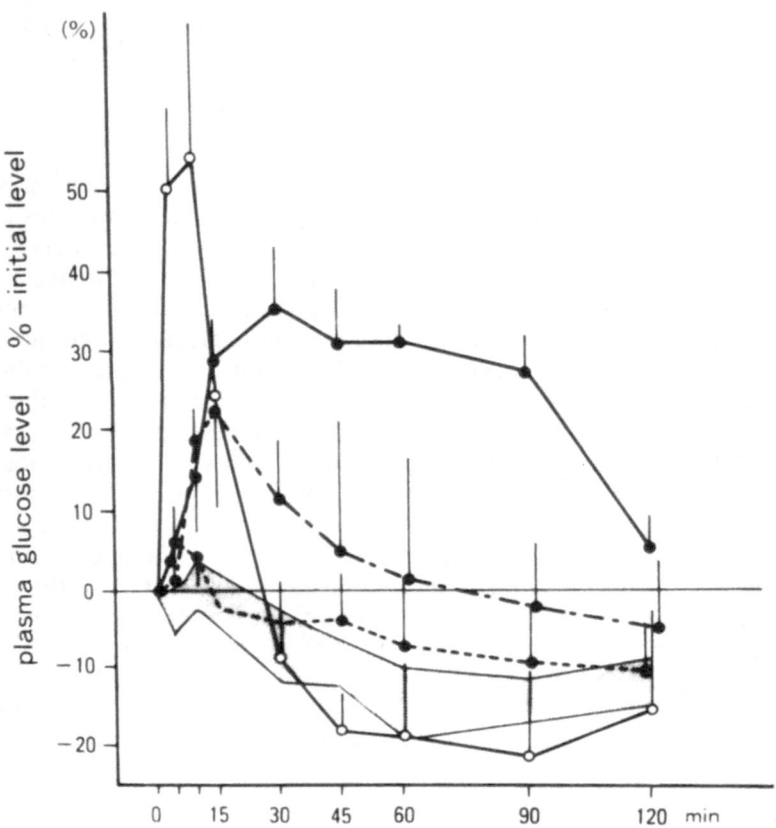

FIGURE 5.6. Modulation of plasma glucose levels by 100 ng of glucagon adminis-tered icv (●——●), iv (○——○), and im (●---●), and 10 ng glucagon icv (●--●). Bars represent the standard error of the mean. (Reprinted with permission of the Biomed Res 34.)

contributed to the drastic secondary decline of plasma glucose. On the other hand, since the icv administration of glucagon caused no increase in plasma levels of IRG, it appears that none of it found its way into the blood and that its effect on blood glucose was the result of a centrally mediated mechanism. In order to examine this mechanism, we studied the effect of various autonomic blocking agents.

EFFECT OF AUTONOMIC BLOCKADE ON HYPERGLYCEMIA INDUCED BY ICV GLUCAGON

Figure 5.7 represents the sum of the increments in plasma glucose levels for 120 minutes after the icv injection of 100 ng glucagon under different experimental conditions. It can be seen that the effect was totally

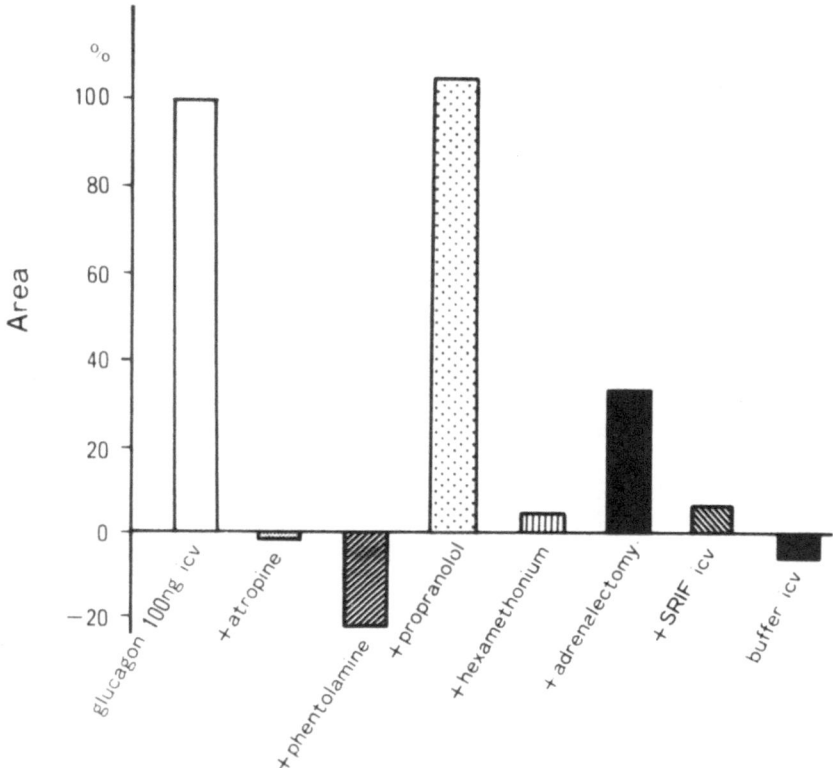

FIGURE 5.7. Effects of the ip administration of various autonomic blockers, bilateral adrenalectomy, and icv administration of SRIF on hyperglycemia induced by 100 ng glucagon given icv. Percent changes of the glucose area above the initial level for 120 minutes. (Reprinted with permission of the Biomed Res 34.)

inhibited by atropine or phentolamine, while the prior administration of hexamethonium inhibited the glycemic response only partially, and that of propranolol had no significant effect. Thus, it appears likely that the hyperglycemic effect of icv glucagon is mediated by an α-adrenergic and a cholinergic mechanism. The partial inhibition by hexamethonium could be explained by the short half-life of this agent.

EFFECT OF ADRENALECTOMY ON HYPERGLYCEMIA
INDUCED BY ICV GLUCAGON

The hyperglycemic effect of icv glucagon was partially inhibited by bilateral adrenalectomy (Fig. 5.7), in contrast to the total inhibition of the effect of BBS.

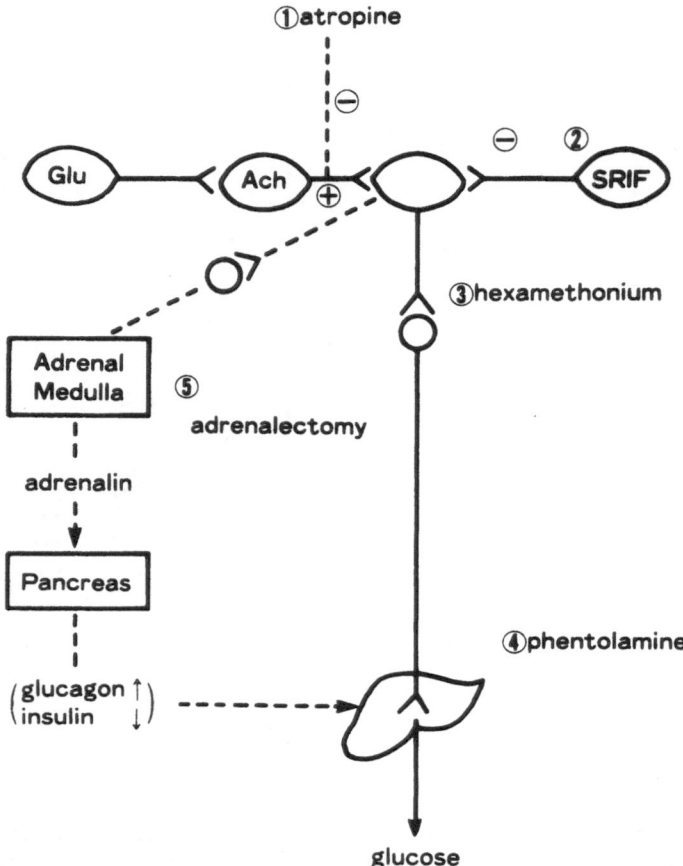

FIGURE 5.8. Schematic representation of the hyperglycemic effect of glucagon administered icv in the rat.

EFFECT OF ICV SRIF ON ICV GLUCAGON-INDUCED HYPERGLYCEMIA

The effect of icv glucagon was markedly reduced when it was given five minutes after the icv administration of SRIF (Fig. 5.7), suggesting a central control by somatostatin.

SUMMARY

The above results, summarized in Fig. 5.8, suggest that the hyperglycemic effect of icv glucagon is mediated by autonomic nervous system pathways that can be blocked completely by atropine and by phentolamine, partially by hexamethonium and adrenalectomy, and not at all by propranolol. It appears that hexamethonium acts at the presynaptic level, whereas the ineffectiveness of propranolol confirms the relative lack of adrenomedullary involvement, in contrast with the mechanism of action of BBS. The effectiveness of phentolamine supports the notion that, in the rat, the release of glucose from the liver is exerted through the α-adrenergic receptor and suggests that icv glucagon may act on the liver. Finally, icv SRIF is presumed to suppress the central mechanism of icv glucagon-induced hyperglycemia. It has been well recognized that the transitory nature of the hyperglycemic effect of glucagon when injected intravenously or when secreted under physiologic conditions is due to the counterregulatory action of insulin. On the other hand, glucagon injected directly into the brain can produce sustained hyperglycemia and appears to act as a neurotransmitter since it characteristically is not accompanied by an increase in plasma levels of either IRI or IRG. Thus, glucagon appears to regulate glucose homeostasis through central and peripheral mechanisms, just as central and peripheral actions appear involved in its regulation of feeding behavior (44–46).

Thyrotropin Releasing Hormone (TRH)

Thyrotropin releasing hormone was first purified and identified in porcine hypothalamic extracts (47). Since then TRH has been found widely distributed throughout the brain, and therefore is considered a neuroactive substance (48,49). Centrally administered TRH has a variety of biologic effects, including hyperthermia (50), tachycardia (51), tachypnea (52), enhancement of gut motility (53) and of gastric secretion (54), hyperglycemia (55), and hypersecretion of pancreatic juice (56). Although the intravenous administration of TRH induces secretion of TSH and prolactin from the pituitary, it does not cause any changes in plasma glucose levels (57). Thus, hyperglycemia induced by icv TRH must be the result of a central mechanism, possibly the activation of a sympathetic pathway leading to epinephrine-mediated hyperglucagonemia (55). On the other hand, the effects of TRH on the digestive tract are thought to be exerted through a cholinergic pathway (53,54,56).

Modulation of Plasma Levels of Glucose by icv TRH

The icv injection of TRH in doses of 10^{-7} moles or greater caused transient hyperglycemia, followed by a rapid decline to control levels (Fig. 5.9). In agreement with previous observations (55), this effect was not altered significantly by the prior icv administration of SRIF (Fig. 5.9), providing no support for the contention that SRIF participates in the hyperglycemia induced by the TRH analog DN-1417 (58).

Effect of Adrenalectomy on icv TRH-Induced Modulation of Plasma Glucose Levels

Icv TRH-induced hyperglycemia was prevented by bilateral adrenalectomy. Indeed, sustained hypoglycemia was observed (Fig. 5.9). Brown et al. (55) reported that the intracisternal administration of TRH induced

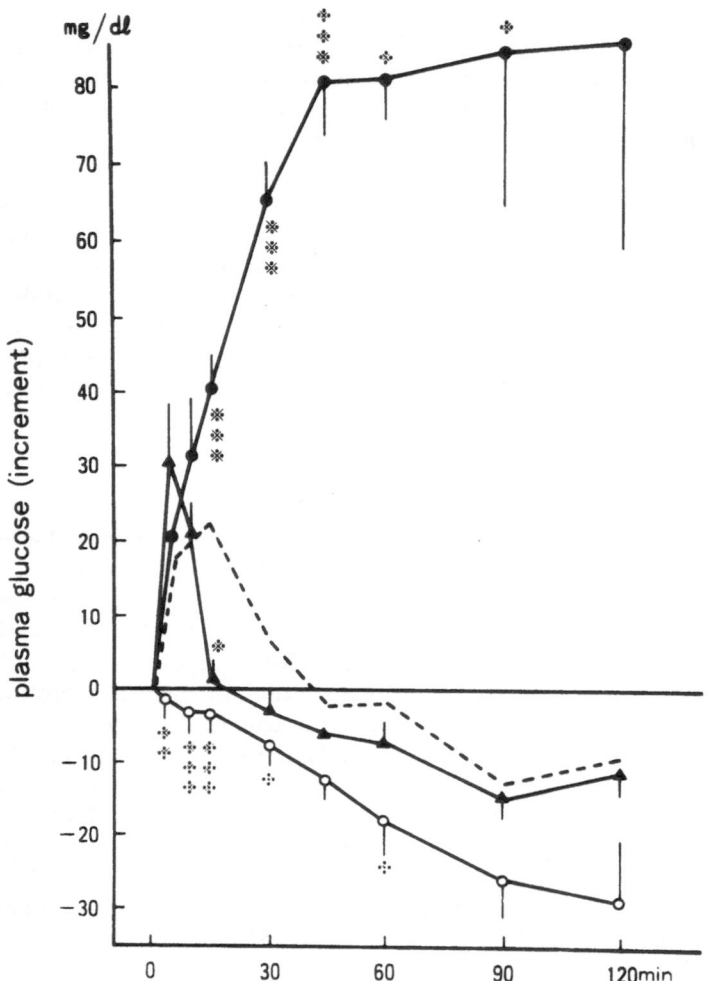

FIGURE 5.9. Effect of bilateral adrenalectomy (O———O), icv administration of somatostatin (▲———▲), and streptozotocin-induced diabetes (●———●) on the increment of plasma glucose levels induced by icv administration of 10^{-7} moles of TRH (----). Bars represent SEM. *$p < 0.05$; **$p < 0.01$; ***$p < 0.005$ compared with TRH alone.

concomitant increases of plasma glucose, catecholamines, and IRG. Since these changes were abolished by adrenalectomy, they concluded that TRH-induced hyperglycemia is mediated by the sympathico-adrenal system. Kabayama et al. (58) also reported that icv TRH induced hyperglycemia associated with hypersecretion of catecholamines. They further reported that icv DN-1417 induced more pronounced hyperglycemia and catecholamine responses than those obtained with TRH.

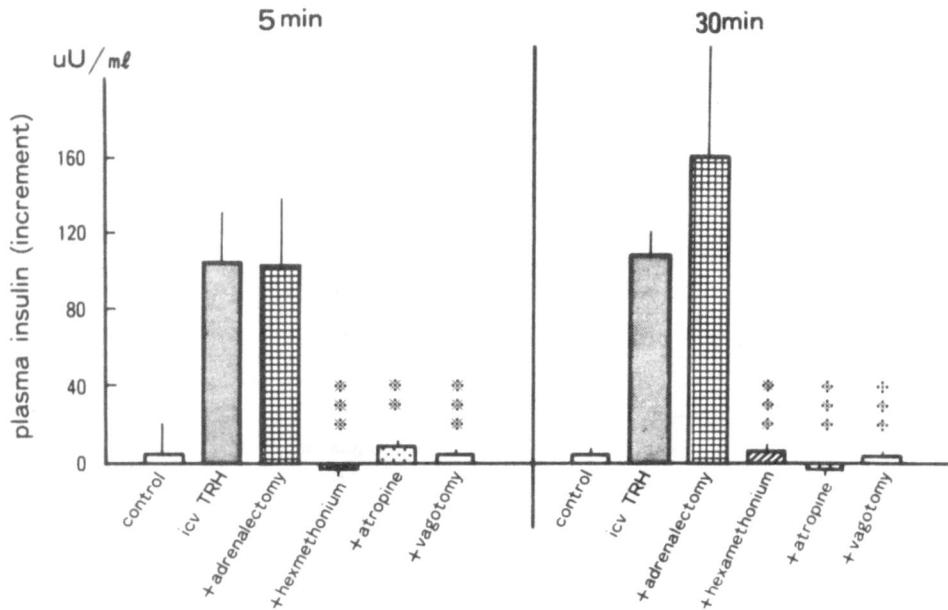

FIGURE 5.10. Modulation of plasma levels of IRI caused by the icv administration of TRH by bilateral adrenalectomy, cholinergic blockade, and subdiaphragmatic vagotomy in the rat. **$p < 0.01$ and ***$p < 0.005$

MODULATION OF PLASMA LEVELS OF GLUCOSE, INSULIN, AND GLUCAGON BY ICV TRH

The icv injection of 10^{-7} moles of TRH resulted in an increase in the plasma levels of IRI and of IRG in normal (Fig. 5.3) as well as adrenalectomized (Fig. 5.10) rats. We believe that the insulinogenic response was the result of a mechanism other than the hyperglycemia induced by icv TRH and that it explains the transient nature of this hyperglycemia. Indeed, when the experiments were repeated in strep-tozotocin-diabetic rats, in whom the plasma IRI levels did not respond to stimulation by iv arginine or oral glucose and in whom the levels of plasma IRG were normal, the icv TRH-induced hyperglycemia was more marked and sustained than that observed in nondiabetic rats (Fig. 5.9) and was prevented by the administration of SRIF (57).

EFFECT OF AUTONOMIC BLOCKADE AND OF VAGOTOMY ON ICV TRH-INDUCED HYPERINSULINEMIA

To examine the participation of the autonomic nervous system in icv TRH-induced hyperinsulinemia, we studied the effects of vagotomy and autonomic nerve blockade and found that pretreatment with either hexamethonium bromide or atropine sulfate, or subdiaphragmatic vagot-

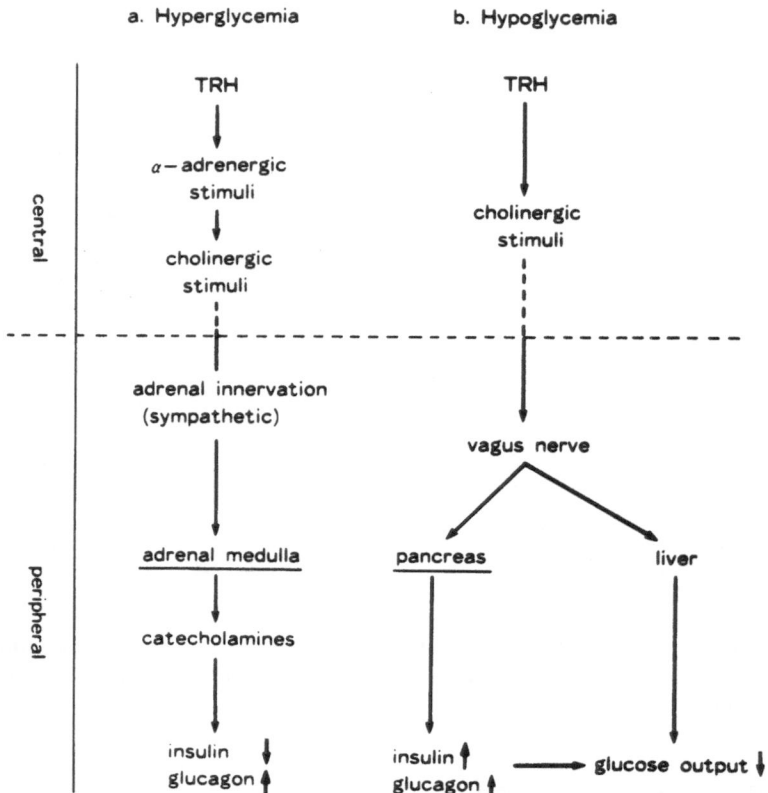

FIGURE 5.11. Schematic representation of icv TRH-induced modulation of the plasma levels of glucose, IRI, and IRG.

omy prevented icv TRH-induced hyperinsulinemia (Fig. 5.10). On the other hand, hyperglucagonemia was inhibited by pretreatment with hexamethonium and with atropine, but not by vagotomy. Thus, it seems reasonable to suggest that cholinergic mechanisms contribute to the stimulation of insulin and to glucagon secretion, and that the latter may be induced primarily by an increase in adrenomedullary function.

MECHANISM OF ICV TRH-INDUCED HYPERGLYCEMIA

As mentioned above, it is known that icv TRH increases plasma catecholamine levels and accelerates noradrenalin turnover in the brain (59), while DN-1417 stimulates norepinephrine and dopamine release from perifused rat hypothalamic fragments (60). It is also known that norepinephrine stimulates TRH release from the rat hypothalamus (61). In either case, norepinephrine could cause hyperglycemia. Indeed, the icv administration of norepinephrine (5×10^{-8} moles) results in a spike of hyperglycemia similar to that obtained with icv TRH (57). This response, like TRH-induced hyperglycemia, is suppressed significantly by the prior icv administration of atropine but not, in contrast to TRH-induced hyperglycemia, by pretreatment with 6-hydroxydopamine. Shen et al. (62) reported that hyperglycemia was the highest when TRH was injected into the lateral hypothalamus of the rat and was inhibited by pretreatment with atropine. These results suggest the following pathway of stimulation: TRH→ α-adrenergic system→ intracerebral cholinergic transmission→ cholinergic outflow to the adrenal medulla→ hyperglycemia, caused partly by catecolamine-mediated changes in pacreatic hormone secretion and partly by a direct effect of the catecholamines on the liver.

Summary

In summary (Fig. 5.11), we believe that icv TRH stimulates both the sympathetic nervous system to induce hyperglycemia and the vagus nerve to induce hypoglycemia. The combined effect may result in transient hyperglycemia followed by sustained hypoglycemia. This dynamic balance may shift either way depending on the strength of the stimulus. In fact, centrally administered TRH in the mouse results in hypoglycemia (62).

Neurotensin (NT)

Neurotensin is a peptide first isolated form the bovine hypothalamus (64) and later found in other areas of the brain and in the gut of many animals (65). Its biologic effects include hypotension, hypothermia, inhibition of gut mobility and secretion, and hyperglycemia (65).

Modulation of Plasma Glucose, Insulin, and Glucagon Levels by Neurotensin Infusion

In the rat, icv NT at a dose of 10^{-6} moles caused only slight hyperglycemia and, although an iv dose of 2×10^{-9} moles had a greater effect, the animals became cyanotic, blood flow decreased, and sampling became so difficult that all subsequent experiments were performed in dogs. In this species, as shown in Fig. 5.12, the iv administration of NT was followed by a significant increase in plasma glucose which was dose related between 0.1 and 1 μg/kg $\cdot$ min^{-1}. On the other hand, doses as low as 0.01 μg/kg $\cdot$ min^{-1} are known to cause hypotensin (66). The plasma insulin level also rose, but returned to basal values before the rise in plasma glucose had become significant. On the other hand, the plasma level of IRG rose to maximum values before the hyperglycemia had reached its peak. Thus, we thought that the sharp rise in plasma glucagon had contributed to the observed increase in plasma glucose levels. However,

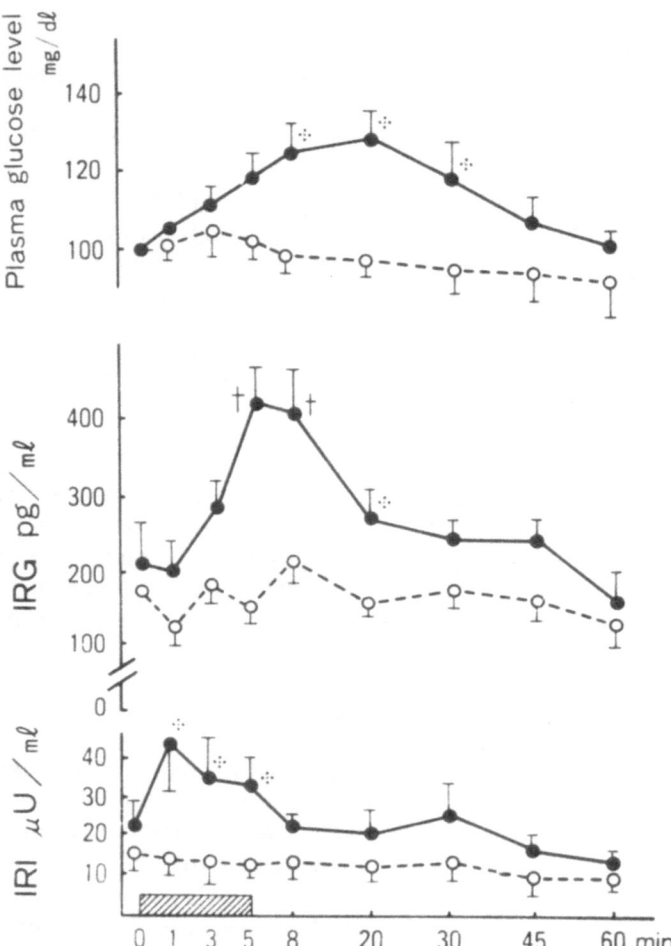

FIGURE 5.12. Modulation of the plasma levels of glucose, IRG, and IRI by the iv infusion of (1 μg/kg/min) NT (●——●) and control solution (○---○) in the dog. Bars represent SEM. (Reprinted with permission of the Tohoku J Exp Med, Ref. 66.)

this explanation could not be sustained when it was found that iv SRIF, a hormone that in itself causes a gradual decline of the plasma glucose level (67), did not prevent NT-induced hyperglycemia (Fig. 5.13), even though it blocked the glucagon response entirely. Therefore, other factors had to be considered, among them, a possible direct effect of neurotensin on the liver. To examine this possibility we injected the peptide directly into the portal vein. The resulting hyperglycemia was not significantly different from that observed when NT was injected into the femoral vein (66). Pretreatment with SRIF did not inhibit the response, suggesting that NT

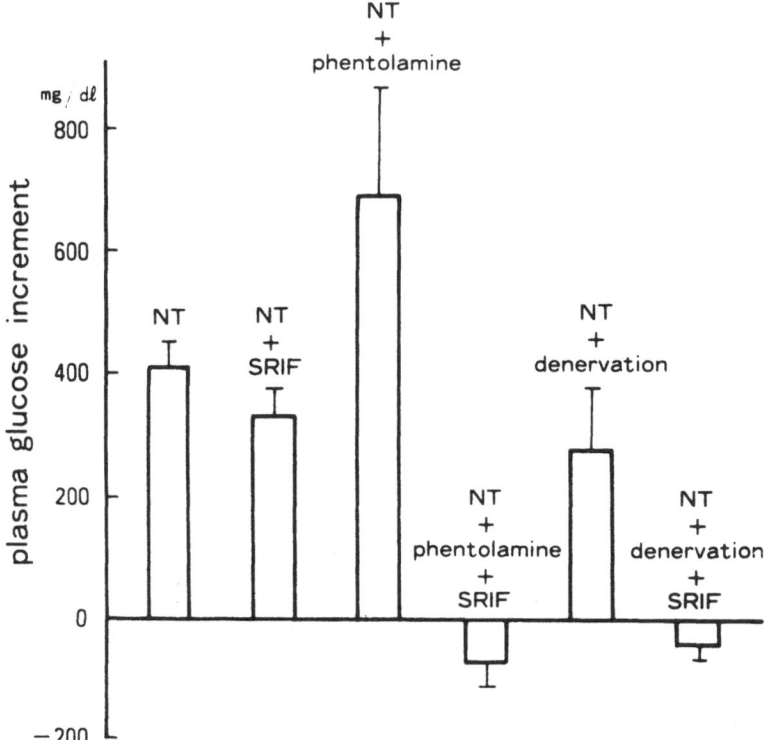

FIGURE 5.13. Modulation of plasma glucose levels caused by the iv infusion of NT by the iv infusion of adrenergic blocking agents and SRIF, and by baroreceptor denervation in the dog. Changes are shown as the total increment of plasma glucose levels in 60 minutes. Bars represent SEM.

does not have a direct effect on the liver but that it exerts an indirect effect, possibly through the intervention of glucagon. The problem was investigated further by experiments in vitro. In a first set of experiments, using rat hepatic cell membranes prepared according to Emellot (68) and a radioreceptor techique (69), we found evidence of glucagon but not of neurotensin receptors. Accordingly, NT at concentrations between 10^{-9} and 10^{-5} M did not increase the production of cyclic AMP by these membranes (69), although a marked and concentration-dependent increase was noted after the addition of glucagon (10^{-9} to 10^{-7} M). Nor did the addition of NT (10^{-5} M) have a significantly greater effect than glucagon alone. The possibility that NT receptors may have been damaged by the isolation procedure was ruled out by additional studies using isolated perfused rat livers. As shown in Fig. 5.14, glucose output from the liver was markedly increased with glucagon 10^{-10} M or epinephrine 10^{-6} M, but not with NT 10^{-6} M. We could only conclude that NT has no direct effect on the liver.

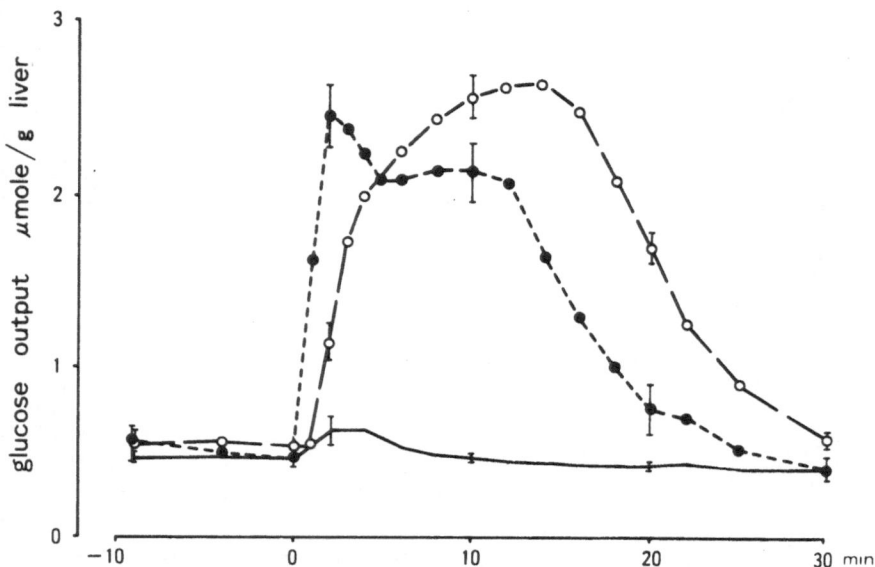

FIGURE 5.14. Changes in glucose output from the isolated rat liver induced by the administration of glucagon (10^{-10} M) (O---O), epinephrine (10^{-6} M) (●---●), and NT (10^{-6} M) (——). Bars represent SEM.

MODULATION OF PITUITARY HORMONES BY NT

In dogs, the iv injection of NT does not appear to modify the blood levels of growth hormone (GH), adrenocorticotropic hormone (ACTH), and cortisol (66,70). Thus it is unlikely that these hormones participate in NT-induced hyperglycemia.

EFFECT OF ADRENERGIC BLODCADE ON NT-INDUCED HYPERGLYCEMIA

The probability that NT-induced hyperglycemia may be exerted through the autonomic nervous system was confirmed by the observation that the hyperglycemic effect of NT was enhanced if administered after the start of a phentolamine infusion at the rate of 0.1 μg/min (Fig. 5.13). The increases in plasma IRG and IRI were also enhanced by the blocking agent (66). We interpret these results as indicating that the increase in plasma glucose levels resulted from an increase in glucagon secretion brought about by a predominant activity of the β-adrenergic system in the face of the α-adrenergic block. In fact, the combined administration of α-adrenergic blocker and SRIF completely inhibited NT hyperglycemia. On the other hand, the hyperglycemic effect was decreased by the administration of propranolol at the rate of 0.1 μg/min and completely inhibited by the combined action of SRIF and propranolol. The combination of SRIF and propranolol also inhibited the rise in the plasma levels of IRG and IRI (66). Since neither the α- nor the β-adrenergic block alone could inhibit the increase in plasma glucose levels completely, in the absence of SRIF, we conclude that NT-induced hyperglycemia needs the participation of both α- and β-agonistic effects and of glucagon.

PARTICIPATION OF BAROCEPTORS IN NT-INDUCED HYPERGLYCEMIA

Since NT has a kinin-like hypotensive effect, it is reasonable to suggest that a hypotension-mediated catecholamine secretion participates in the glycemic response. Indeed, in the rabbit, blood glucose levels are increased by phlebotomy, an effect that can be prevented by the resection of the depressor nerve and that is associated with an increased discharge activity of the hepatic efferent nerve (71). Hyperglycemia and hyperglucagonemia have also been observed in patients with severe traumatic shock (72) and in the exsanguinated dog (73) and both can be blocked by the administration of propranolol. The hypothesis of baroreceptor participation was confirmed by the observation that under SRIF infusion NT does not produce significant hyperglycemia in dogs whose vagosympathetic trunk has been crushed and whose carotid sinus has been denervated (Fig. 5.13), nor does the plasma IRG level increase significantly in these animals. (66).

EFFECT OF HEMOCONCENTRATION ON NT-INDUCED HYPERGLYCEMIA

The kinin-like effects of NT include vasodilatation, cyanosis, increased vascular permeability, and hemoconcentration. Indeed, it has been observed that NT-induced hyperglycemia parallels the increase in hematocrit (Ht) (74). As shown in Fig. 5.15, in the rat, hyperglycemia caused by an iv injection of NT alone was accompanied by a significant increase in Ht. On the other hand, the administration of NT in animals pretreated with phentolamine, propranolol, and SRIF, resulted in a slight decrease in blood glucose even though there was a comparable increase in Ht. Thus it is unlikely that hyperglycemia caused by NT is due merely to hemoconcentration.

SUMMARY

The results described above, summarized in Fig. 5.16, suggest that the hyperglycemic effect of NT is due primarily to an increase in the secretion of pancreatic glucagon with the participation of the sympathetic nervous system or the catecholamines. Indeed, the hyperglycemia is completely inhibited by a combination of SRIF and α-adrenergic blockade and is enhanced when β-adrenergic activity prevails following the blockade of the α-adrenergic system (Fig. 5.16). Conversely, the moderation of the hyperglycemic response following β-adrenergic block may be attributable

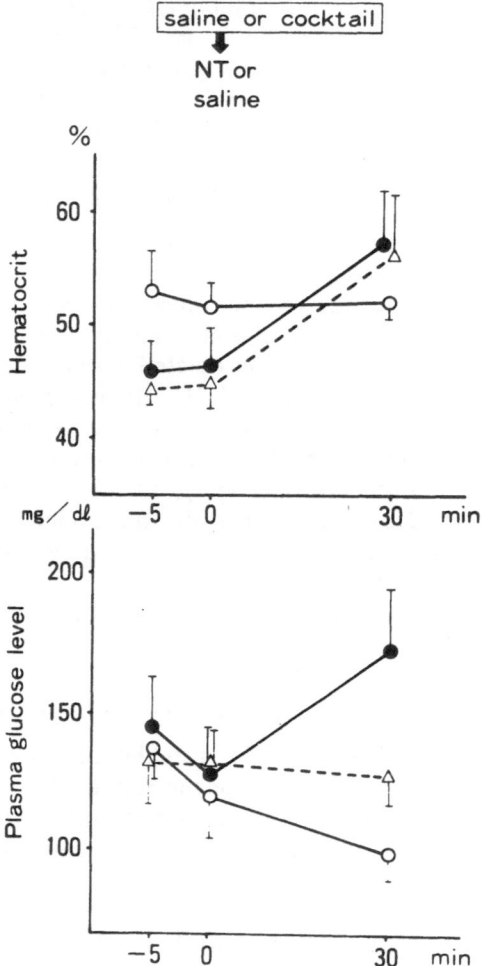

FIGURE 5.15. Changes in hematocrit and blood glucose brought about by the iv administration of NT (2 μmoles/kg) in the absence (●——●) or presence (△---△) of a pretreatment cocktail consisting of phentolamine (0.1 mg/kg), propranolol (0.1 mg/kg), and SRIF (0.5 μg/kg). Saline solution was injected intravenously as a control (○——○). Bars indicate SEM.

to α-adrenergic dominance during which glucagon secretion is not stimulated, although an increased hepatic glucose production through the α-adrenergic receptor in the rat may occur. In either event, a contribution of catecholamine secretion from the sympathetic ganglia or the adrenal medulla is postulated. This adrenergic mechanism may be activated also through the baroceptors, responding to NT-induced hypotension, possi-

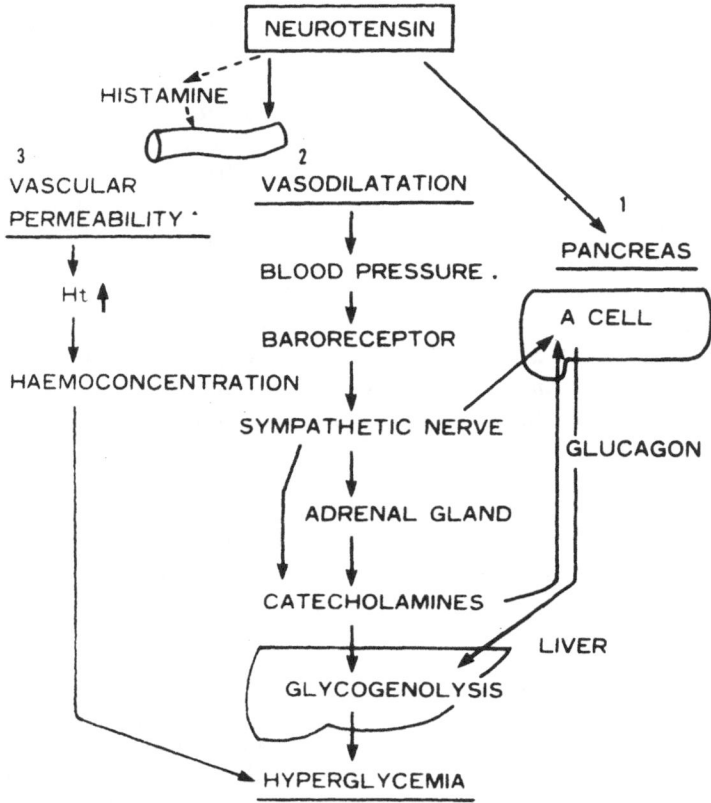

FIGURE 5.16. Schematic representation of NT-induced hyperglycemia.

bly with the participation of histamine (75). On the other hand, there is some question regarding the participation of hemoconcentration, as there is for that of insulin, in view of the uncertain effect of NT upon its secretion (76–78,80). Recently it has been reported that the pancreas contains NT (79) and that NT stimulated glucagon secretion by the isolated dog pancreas and the isolated pancreatic islets of the rat (77). Thus, neurotensin may induce hyperglucagonemia through direct action on the pancreas. Neurotensin has no hyperglycemic effect when injected into the cerebral ventricles (26) and is believed to exert its action mainly through a peripheral mechanism. This conclusion must be tempered by the observation that centrally administered NT inhibits the increase in appetite induced by starvation and by the icv administration of norepinephrine or dynorphin, although not that induced by the administration of GABA agonist, muscinol, or by insulin hypoglycemia (81). In addition, there is a report that icv NT increases SRIF concentration in pituitary portal blood (82).

Other Neuropeptides Inducing Hyperglycemia

β-ENDORPHIN

Intracerebroventricular β-E, a peptide abundant in brain tissue, causes an increase in blood glucose similar to that observed after the injection of BBS (83). This response is inhibited by pretreatment with naloxone, suggesting mediation by the opioid receptor (84,85). Van Loon and his collaborators demonstrated that hyperglycemia induced by the administration of β-E into the cerebral cistern of the rat was associated with increased levels of dopamine, norepinephrine, and especially epinephrine, and was abolished after bilateral denervation of the adrenal medulla or by pretreatment with the ganglionic blocking agent chlorisondamine (86). The hyperglycemic response to intracisternal β-E was also inhibited by pretreatment with hemicholinium-3, a compound that decreases acetylcholine (Ach) concentration by inhibiting the intracerebral uptake of choline, but was not inhibited by pretreatment with apomorphine. Thus, β-E may act mainly through cholinergic mechanisms and less so through the dopamine receptors. β-Endorphin also causes hyperglycemia when injected intravenously, possibly through an increase in glucagon secretion, although a reduction in insulin secretion cannot be excluded (87,88).

CHOLECYSTOKININ (CCK)

Cholecystokinin is widely distributed in the brain, where it has a variety of functions (89–92). In rats anesthetized with pentobarbital, icv CCK-8 caused hyperglycemia in a dose dependent manner (93). This was accompanied by a significant increase in the plasma levels of IRG but only when the maximal dose of CCK-8 was used. No changes in plasma IRI were noted at any dose. Thus, it appears that pancreatic hormones are not involved in CCK-induced hyperglycemia. On the other hand, it has been reported that icv CCK-induced hyperglycemia is prevented by adrenalectomy (95). Thus, the hyperglycemic effect of icv CCK seems to be mediated by the adrenal medulla and does not involve the secretion of pancreatic hormones. On the other hand, after an intravenous injection of CCK, there is a marked secretion of IRI (96), which cannot be inhibited by vagotomy, atropine, or propranolol, and which therefore is presumed to be a manifestation of a direct effect of CCK. When given centrally CCK inhibits appetite (90,91,94), while intravenously it stimulates the secretion of pancreatic polypeptide (PP), which in turn has an anorexic effect (97). It is known that long-term administration of CCK, or a chronic increase in the secretion of CCK induced by the administration of trypsin inhibitor, results in hypertrophy of the pancreatic islets and of the exocrine pancreatic tissue (98). Thus, CCK appears to influence glucose metabolism by promoting glucose assimilation through hyperinsulinemia, while stimulating glucose production through its action on the brain.

PANCREATIC POLYPEPTIDE

Pancreatic Polypeptide, a peptide produced by the pancreas and the upper intestine, has recently been discovered in the brain (99) and is considered to be a new neuropeptide, though its relation with the structurally similar Neuropeptide Y, recently isolated from porcine brain and known to be widely distributed throughout the central and peripheral nervous systems (180–182), is uncertain. The plasma concentration of PP increases under a variety of conditions, suggesting that it has an endocrine role, although its actual functions are unclear. The secretion of PP is stimulated by food intake, an effect inhibited by atropine (100). PP injected either iv or icv inhibits glucose-stimulated insulin secretion (101). Thus, the hyperglycemic effect of PP is believed to be the result of a centrally mediated inhibition of the pancreatic B cells.

VASOACTIVE INTESTINAL POLYPEPTIDE (VIP), PEPTIDE HISTIDINE ISOLEUCINE (PHI), AND PEPTIDE HISTIDINE METHIONINE (PHM)

Vasoactive intestinal polypeptide a peptide discovered by Said et al (102) in the porcine small intestine and widely distributed in the brain, spinal cord, peripheral nerve, and intestinal nerve plexus, is now considered a typical neuropeptide. Another porcine intestinal peptide widely distributed in the nervous tissue is PHI (103), which differs by only two amino acid from PHM, a peptide derived from the NB-1 strain of human neuroblastoma (104). PHI and VIP have a common precursor and are considered homologous peptides. PHI has a wide intracerebral distribution (105,106). VIP activates hepatic adenylate cyclase, stimulates hepatic glycogenolysis and gluconeogenesis (107,108), and increases blood glucose levels (109) when given intravenously. These effects are inhibited by insulin (108). In addition, VIP causes a transient increase in IRI and IRG secretion from the perfused pancreas (110). A similar transient release of IRI has been observed after single doses of PHI (111). It has been shown recently that VIP, PHI, and PHM at doses which have no effect on the secretion of pancreatic hormones have been found to markedly enhance the insulinogenic action of glucose (112). Thus, VIP, possibly released from nerve terminals (113), may have an incretin-like potentiating role on insulin secretion similar to that of gastric inhibitory polypeptide (GIP). This may be a peripheral effect, since icv VIP has no effect on plasma glucose unless given in high doses (26).

GALANIN

Galanin, recently isolated from porcine intestine, has an amino acid sequence different from that of other known gut peptide, and is named from its N-terminal glycine and C-terminal alanine residues (183). Galanin immunoreactivity was found to be widely distributed throughout the central and peripheral nervous systems and in the pancreas (183–186) and thus it is considered a neuropeptide. When infused, galamin induces smooth muscle contraction in the gastrointestinal and genitourinary tracts (183), mild sustained hyperglycemia associated with fall in IRI and elevation of IRG, both in fasting and glucose-infused dogs (186,188).

In addition, galanin caused a reduction of IRI in high glucose- or arginine-stimulated state (187), and an increase in IRG during arginine infusion from the perfused rat pancreas (189). However, galanin caused no change in plasma glucose levels when injected icv, nor glucose production from the cultured rat hepatocytes (187,188).

Thus, galanin-induced hyperglycemia seems to be due to a direct effect on the pancreas.

GROWTH HORMONE RELEASING FACTOR (GRF)

Another peptide widely distributed in the brain is GRF (114). When injected iv or icv, GRF causes an increase in blood glucose similar to that caused by BBS, in which the participation of the sympathetic nervous system-adrenal medulla is assumed.

CORTICOTROPIN-RELEASING FACTOR (CRF)

CRF, a well known hypothalamic peptide (115), also has a hyperglycemic effect when injected icv (116). In addition, icv CRF increases the plasma levels of epinephrine, norepinephrine, and IRG, but not that of IRI. The blood glucose response is not affected by hypophysectomy and, therefore, is not mediated by ACTH. On the other hand, the response appears to be mediated by the nervous system because, although not affected by adrenalectomy, the increases in plasma levels of epinephrine and glucose, but not that of norepinephrine, are inhibited by pretreatment with des-AA^{1,2,4,5,12,13}-[D-Trp^8]somatostatin (ODT-SS), a derivative of SRIF (116). The increase in catecholamine secretion by CRF is consistent with the increase in blood pressure and heart rate caused by this peptide (117).

ANGIOTENSIN II (ANGIO II)

Large amounts of Angio II and Angio II receptors are found in the brain (118). Angiotensin II is a potent hypertensive agent when injected iv, while its icv administration causes significant hyperglycemia. This response appears mediated by the adrenal medulla because it does not occur in adrenalectomized animals (57).

SUBSTANCE P

Substance P was first found in porcine intestine and then in bovine brain (119,120). Its presence and release from the nerve terminals, including the nerves of the pancreas, are well documented (120–123). The predominant distribution of substance P in the posterior horns and its action on nerve cells suggest that it functions as an excitatory neurotransmitter (122,124). Substance P produces hypotension, increased vascular permeability and gut contraction (125), hyperglycemia, hyperglucagonemia, and hypoinsulinemia (76). It is believed that substance P may modulate insulin release through the adrenal medulla (76) and through a direct effect on the pancreas (126). Considering the strong hypotensive effect of this peptide, a baroceptor-mediated catecholamine response is possible, as in the case of neurotensin. The microinjection of substance P into VMH or LHA appears to have a negligible effect on the concentration of glucose in the hepatic vein (26). Substance P icv in doses greater than 10^{-8} moles induces sustained mild hyperglycemia, which can be suppressed by either icv SRIF or adrenalectomy and thus is thought to involve a mechanism similar to that of bombesin (57).

SOMATOSTATIN (SRIF)

SRIF, a peptide secreted by the neurons of the hypothalamus and by the D cell of the pancreatic islets, is a potent inhibitor of GH, insulin, and glucagon secretion (5,127). In addition, SRIF has been found in the exocrine pancreas, in the endocrine cells of the mucosa, and in the myenteric nerve plexus of the gastrointestinal tract, where it may function as a hormone and/or a neurotransmitter (5,128). In the gastrointestinal tract, SRIF inhibits the secretion of hormones such as gastrin, secretin, CCK, GIP, VIP, and motilin, and of gastric acid, pepsin, gastric intrinsic factor, pancreatic and intestinal juices, inhibits peristalsis, decreases blood flow, and inhibits the absorption of water, calcium, glucose, fat, and other substances. Thus, SRIF is a general inhibitor of the digestion, absorption, and assimilation of nutrients (130). In the central nervous system, SRIF is most abundant in the hypothalamus, followed by the cerebral cortex, limbic system, brain stem, and spinal cord (5,129). The major functions of intracerebral SRIF are to modulate the secretion of pituitary and hypothalamic hormones such as GH, TSH, PRL, and TRH (5,131), and the release of dopamine, serotonin, norepinephrine, and other neurotransmitters (132). The iv administration of SRIF causes little change or only a gradual decrease in plasma glucose levels, in spite of a complete inhibition of IRG and IRI secretion. However, SRIF enhances the hyperglycemic or hypoglycemic effect of small doses of glucagon or insulin, which by themselves would not be effective (133). Upon intracerebral administration of SRIF, doses of 10^{-10} to 10^{-8} moles cause a gradual decrease in plasma levels of glucose, but this effect does not differ significantly from that seen in animals receiving control solutions (57), while higher doses are lethal for most rats. As mentioned above, pretreatment with a 10^{-8} mole dose of SRIF inhibits the hyperglycemic effect of icv BBS, glucagon, CCK, β-E, and GRF. Such inhibitory effects may result from the inhibition of pancreatic hormone secretion or from a direct inhibitory effect of SRIF on the sympathico-adrenal system exerted through the glucoregulatory center (29,134). This would explain why SRIF alone has little effect on plasma glucose but inhibits hyperglycemia induced by the above mentioned neuropeptides. In accordance with this hypothesis, the hyperglycemia induced by the icv injection of DN-1417, a TRH analog (58), or of GRF is enhanced by pretreatment with cysteamine, which is believed to reduce the hypothalamic content of SRIF. Thus, it is believed that many of the hyperglycemic effects of neuropeptides are influenced by changes in intracerebral SRIF or by somatostatinergic neuron activity (135).

INSULIN

It has long been held that insulin does not play an important role in the uptake of glucose by the brain. However, recently it has been reported that insulin increases the binding of gold thioglucose to the ventromedial nucleus of the hypothalamus of the mouse (136) and also enhances the activity of glucose-sensitive neurons of the lateral nucleus of the rat hypothalamus (13). Additionally, it has also been reported that insulin, indistinguishable from pancreatic insulin by RIA, radioreceptor assay, bioassay or gel filtration profile, can be extracted from the rat brain and, in particular, from the hypothalamus and olfactory bulb (137–140). Data indicating the presence of insulin receptors in the brain have also been reported (140–142), although the distribution of these receptors was slightly different from that of insulin itself (143,144). The amount and distribution of insulin and of its receptors in the brain is not modified by obesity, diabetes, or starvation (137,140) suggesting that brain insulin acts as a neurotransmitter or as a modulator of neurotransmitter activity. Detailed investigations of the distribution of the receptor, its functions, and of the abundance of insulin in areas surrounding the ventricles, coupled with the fact that ^{125}I-insulin is transferred easily from the blood into the brain and that proinsulin-like peptides are absent in the brain, indicate the CNS insulin derives from the blood (141,142). On the other hand, the marked hypoglycemia which follows injection of small amounts of insulin into the carotid artery of the rat led Szabo et al. (145) to postulate the existence of an insulin-sensitive glucoregulatory center in the brain. These investigators also suggested that the response required an intact parasympathetic outflow, since it was inhibited by pretreatment with atropine and cervical vagotomy. Other evidence for a regulatory role of insulin in the CNS is that the intracerebral administration of insulin suppresses appetite in the rat (146), and that the injection of insulin into rat lateral, as well as ventromedial, hypothalamic nuclei causes hypoglycemia (147). In addition, Oomura et al. (13) have demonstrated that the activity of glucoreceptor neurons in the rat ventromedial nucleus is inhibited by insulin, but stimulated by insulin and glucose, while the activity of glucose-sensitive neurons in the lateral hypothalamic area is stimulated, and appetite is increased by insulin in a dose-dependent manner. These results suggest that insulin acts on the brain to produce hypoglycemia, that this action is mediated by the parasympathetic nervous system, and that an inhibition of the sympathetic nervous system may be involved.

Clinical Implications of Hyperglycemic Neuropeptides

Glucose modulating peptides appear to be concerned with neurometabolic regulation in two major ways: chronic and mild changes in their concentration in the brain may relate to feeding behavior, while acute and relatively profound changes may be involved in rapid changes of plasma glucose levels.

One of the most interesting aspects of the problem in clinical setting is the possible role of neuropeptides and biogenic amines in the regulation of food intake. This matter has been reviewed by many investigators (9,11,12,14,148–150). A large number of neuropeptides inhibit feeding activity when administered into the brain in animals, whereas at least three kinds of neuropeptides: the endogenous opioids, the peptides of the PP family [NPY and Polypeptide YY (PPY)], and insulin enhance it when administered either centrally or peripherally.

Hyperphagia leading to obesity results from VMH lesions and is associated with hyperinsulinemia resulting from autonomic nervous system derangement (12). This experimental model finds its clinical counterpart in the syndrome of hypothalamic obesity caused by a variety of hypothalamic lesions including neoplasms, inflammatory or cerebrovascular lesions, and leukemic cell infiltration (151). Simple obesity and obesity seen in patients with Fröhlich, Prader-Willi, and other obesity-hypogonadism syndromes are also thought to be associated with disorders of the hypothalamus and with relative or absolute hyperinsulinemia suggestive of a decreased sympathetic outflow (12,152).

Obese women have elevated plasma levels of β-E (153), and a hyperendorphinemia was observed in an obese patient (154). Additionally, in genetically obese animals such as fa/fa rats and ob/ob mice, increased amounts of a pituitary β-E, enkephalin, and dynorphin have been reported (155,156), while overeating was suppressed by the administration of naloxone or naltrexone (156–159). The peripheral administration of naloxone to patients with Prader–Willi syndrome (160) or simple obesity (161) caused reduction of their food intake resulting in body weight loss, while naltrexone decreased food intake in narcotic addicts (162). These results indicate a role of the opioid peptides in food intake.

Centrally administered PYY or NPY also strongly stimulate feeding, while inhibiting sexual behavior (163), and there is some experimental evidence that these two functions may be inversely regulated (164). For example, opioid peptides enhanced feeding in doses that inhibit sexual behavior (165), while the frequent association of obesity and hypogonadism suggests the participation of these peptides in the pathogenesis of obesity.

Anorexia is less commonly found in patients with brain tumors involving the hypothalamus, such as ectopic germinoma and craniopharyngioma (166–168). In general, destruction of the hypothalamus is more prominent in anorexia than in obesity syndromes, and anorexia is frequently associated with diabetes insipidus (151,168). In turn, the development of diabetes insipidus in anorectic patients requires the destruction of the lateral hypothalamic nuclei and the preservation of the ventromedial nuclei. Anorexia nervosa is believed to be a functional disorder, similar to that observed in rats with lesions of the lateral hypothalamus, which frequently suffer multiple neuroendocrine deficiencies. Thus, lack of β-E or cortisol associated with anterior pituitary hypofunction could cause anorexia.

On the other hand, increased plasma or CSF concentration of β-E (169), increased levels of CRF (170), and high urinary excretion of a potent anorexigenic peptide (171) have been found in patients with anorexia nervosa. Increase of β-E may be the consequence of mental stress or may represent a compensatory process to restore appetite (169). On the other hand, it has been reported that naloxone causes weight gain and has an antilipolytic effect in patients with anorexia nervosa (172). These contrasting results may be due to differences in the acting sites of β-E.

The central administration of hyperglycemic peptides, except for β-E, reduces feeding (148,149). While most patients under stress have a decreased appetite and show increased levels of β-E, CRF, β-lipotropin, and ACTH in plasma and CSF (148,170), feeding induced by tail-pinching stress in rats is attenuated by naloxone (148). Stress-induced CRF production increases brain β-E, leading to bulimia, whereas CRF itself, and the elevated plasma levels of β-E or FFA, resulting from CRF-induced pituitary β-lipotropin secretion, suppress feeding. Similar dissociation between central and peripheral effects on feeding was reported in the case of norepinephrine (148). These apparent paradoxic phenomena may be explained by the balance between inhibition of feeding by plasma β-E and its stimulation by brain endorphin or by differences in sensitivity to β-E between men and animals.

Insulin injected peripherally increases appetite mainly through hypoglycemia but sometimes irrespective of plasma glucose levels (148,173), whereas centrally administered insulin inhibits feeding (146,174). Plasma levels of insulin tend to decrease under stress because of inhibition by epinephrine secretion. These hormonal changes tend to inhibit feeding. The effect of peptides of the PP family also vary of with the route of administration. Thus, the central administration of PYY, or NPY, or human PP results in a strong stimulation of appetite leading to bulimia in animals, while the peripheral administration of NPY or PYY has no effect and that of PP inhibits feeding (149,163,175). Postprandial increase of PP may be related to satiety, but anorexia has not been noticed in patients with PP-producing tumors.

The second role of hyperglycemic neuropeptides is the counterregulation of neuroglycopenia. Thus, hyperglycemia induced by the central administration of neuropeptides (except TRH) is associated with reduction or no change in plasma insulin levels, and with increased levels of counterregulatory hormones, such as epinephrine or glucagon (11,28,29,43,57,58), while central glycopenia induced by icv 2 deoxyglucose is blocked by naloxone (84). The prolonged hyperglycemia obtained by icv glucagon and alterations of brain glucagon concentration in response to changes in intracellular glucose levels (41–43) support the notion that glucagon is involved in the defense mechanism against neuroglycopenia.

Hypoglycemia in insulin-treated, especially long-standing, diabetic patients could be the result of excess amounts of insulin and deficient counterregulatory hormone response, which in turn, may be due to a primary secretory defect or to autonomic neuropathy (176). Diabetic autonomic neuropathy may also decrease the production of neuropeptides or interrupt an essential neural pathway such as the sympathetic outflow.

The most prominent neuroendocrinologic feature of stress is an enhancement of sympathetic nervous system response leading to hyperglycemia, hyperglucagonemia, and hypoinsulinemia (72,150,170), and is similar to the enhancement induced by BBS acting mainly through stimulation of the adrenal nerves (28–30). Through this mechanism, persistent stress could cause glycogenolysis, and gluconeogenesis resulting in marked hyperglycemia, and a cachectic state (177). A sustained increase of β-E or CRF may also occur in chronic stress, further contributing to sustained hyperglycemia (178,179). This "stress diabetes" may occur easily in patients with latent disease and may be aggravated further by an increased secretion of GH and cortisol.

Severe hyperglycemia causing nonketotic hyperosmolar diabetic coma has been described in patients with head injury, cerebral infarction, encephalitis, brain neoplasms, or heart attack (152). Although, in some cases, the hyperglycemia may be explained by hemoconcentration, it is possible that stimulation of hyperglycemic neuropeptides may be a contributing factor and may represent a counterpart of the marked hyperglycemia secondary to stimulation of VMH (11).

Other vasoactive peptides, such as neurotensin, substance P, and VIP induce hyperglycemia when given peripherally. This type of hyperglycemia is associated with hypotension, and may be the result of changes in the secretion of catecholamines, and of pancreatic hormones similar to those observed in shock due to the loss of blood. Thus, vasoactive peptides may play a role in the prevention of neuroglycopenia due to ischemia of the brain.

Since clinical data concerning the secretion of neuropeptides are fragmentary and their physiologic and pathologic implications are poorly understood, further intensive studies are mandatory.

Summary and Conclusions

This chapter outlines the role of various neuropeptides in the regulation of glucose metabolism, with particular reference to their central effects. Many neuropeptides promote glucose release from the liver through the autonomic nervous system. The adrenal medulla and the islets of Langerhans hold important intermediate positions in the transduction of signals from the nervous system to the liver. The hyperglycemic neuropeptides, when administered centrally, act directly on, or transmit information to a putative regulatory center of glucose metabolism through cholinergic or adrenergic pathways, from whence the information is transmitted to the adrenal medulla or to the pancreatic islets via the autonomic nervous system, mainly as sympathetic outflow. Epinephrine released from the adrenal medulla stimulates the secretion of pancreatic glucagon and inhibits the secretion of insulin, thus causing an increase in hepatic glucose production. Additional information is transmitted directly from the glucoregulatory center to the pancreas and to the liver through adrenergic pathways. Somatostatin may inhibit central cholinergic transmission and, possibly, adrenergic stimuli to the glucoregulatory center, and thus hold back sympathetic outflow to the periphery. Insulin-induced hypoglycemia may also play a role in glucose metabolism. Moreover, TRH appears to stimulate the secretion of insulin through neural pathways originating in the lateral nucleus and reaching the pancreas via the vagus nerve.

The above summary does not take into account species differences, dose and method of administration, conditions of anesthesia, circadian cycles, and other variables. Some of the observed phenomena may be pharmacologic and may not relate directly to physiologic functions. However, the unexpected fact that the central administration of some neuropeptides, such as BBS and glucagon, causes far more marked and more sustained hyperglycemia than their peripheral administration, suggests that there is much more to be learned about the physiopathologic roles of these substances.

Acknowledgements. We wish to thank Dr. Piero P. Foa for his kind help in drafting this paper and Ms. N. Kureyama for typing the manuscript.

References

1. Polak JM, Bloom SR (1980) Neural and cellular origin of gastrointestinal hormonal peptides in health and disease. In: Glass GBJ (ed) Gastrointestinal Hormones. Raven Press. New York, pp 19–51.
2. Said SI (1980) Peptides common to the nervous system and the gastrointestinal tract. In: Martini L, Ganong WF (eds) Frontiers in Neuroendocrinol. vol. 6. Raven Press, New York, pp 293–331.
3. Roberts GW, Woodhams PL, Polak JM et al (1984) Distribution of neuropeptides in the limbic system. The hippocampus. Neuroscience 11: 35–77.
4. Harris GW (1955) Neural control of the pituitary gland. Edward Arnold Ltd, London.
5. Guillemin R, Gerich JE (1976) Somatostatin: physiological and clinical significance. Annu Rev Med 27:379–388.
6. Block B, Brazeau P, Ling N et al (1983) Immunohistochemical detection of growth hormone releasing factor in brain. Nature 301:607–608.
7. Ban T (1975) Fiber connections in the hypothalamus and some autonomic functions. Pharmacol Physiol Behav 3(Suppl 1):3–13.
8. Shimazu T, Ogasawara S (1975) Effect of hypothalamic stimulation on gluconeogenesis and glycolysis in rat liver. Am J Physiol 228:1787–1793.
9. Shimazu T (1981) Central nervous system regulation of liver and adipose tissue metabolism. Diabetologia 20:343–356.
10. Shimazu T (1983) Reciprocal innervation of the liver: Its significance in metabolic control. In: Szabo AJ (ed) CNS Regulation of Carbohydrate Metabolism. Academic Press, New York, pp 355–384.
11. Frohman LA and Bernardis LL (1971) Effect of hypothalamic stimulation on plasma glucose, insulin, and glucagon levels. Am J Physiol 221:1596–1603.
12. Bray GA, Inoue S, Nishizawa Y (1981) Hypothalamic obesity. The autonomic hypothesis and the lateral hypothalamus. Diabetologia 20:366–377.
13. Oomura Y, Kita H (1981) Insulin acting as a modulator of feeding through the hypothalamus. Diabetologia 20 Suppl. 20:290–298.
14. Oomura Y (1983) Glucose as a regulator of neuronal activity. In: Szabo AJ (ed) CNS Regulation of Carbohydrate Metabolism. Academic Press, New York, pp 32–67.
15. Idahl LA, Martin JM (1971) Stimulation of insulin release by a ventrolateral hypothalamic factor. J Endocrinol 51:601–602.
16. Moltz JH, Dobbs RE, McCann SM et al. CP (1977) Effect of hypothalamic factors on insulin and glucagon release from the islets of Langerhans. Endocrinology 101:196–202.
17. Molz JH, Dobbs RE, McCann SM, et al. (1979) Preparation and properties of hypothalamic factors capable of altering pancreatic hormone release in vitro. Endocrinology 105:1262–1268.
18. Brown MR, Allen R, Villarreal J et al (1978) Bombesin-like activity: radioimmunologic assessment in biological tissues. Life Sci 23:2721–2728.
19. Moody TR, Pert CB (1979) Bombesin-like peptides in rat brain: quantitation and biochemical characterization. Biochem Biophys Res Commun 90:1–4.
20. Soveny C and Hansky J (1982) Bombesin-like immunoreactivity in rat brain

and gastrointestinal tract: influence of diet and starvation. Regul Pept 3:325–331.

21. Panula P, Yang H-YT, Costa E (1982) Neuronal location of the bombesin-like immunoreactivity in the central nervous system of the rat. Regul Pept 4:275–283.
22. Moody TW, Pert CB, Rivier J et al (1979) Bombesin: specific binding to rat brain membranes. Proc Natl Acad Sci USA 75:5372–5376.
23. McDonald TJ, Jornvall H, Nilsson G et al (1979) Characterization of a gastrin-releasing peptide from porcine non-antral gastric tissue. Biochem Biophys Res Commun 90:72–233.
24. Brown MR, Marki W, Rivier J (1980) Is gastrin releasing peptide mammalian bombesin? Life Sci 27:25–128.
25. Yanaihara N, Yanaihara C, Mochizuki T et al (1981) Immunoreactive GRP. Peptides 2 Suppl:185–191.
26. Iguchi A, Sakamoto N, Burleson PD (1983) The effect of neuropeptides on glucoregulation. In: Szabo AJ (ed) CNS regulation of carbohydrate metabolism. Academic Press, New York, pp 421–434.
27. Brown MR, Rivier J, Vale W (1977) Bombesin affects the central nervous system to produce hyperglycemia in rats. Life Sci 21:1729–1734.
28. Brown MR, Tache Y, Fisher D (1979) Central nervous system action of bombesin: Mechanism to induce hyperglycemia. Endocrinology 105: 660–665.
29. Brown M (1981) Neuropeptides: central nervous system effects on nutrient metabolism. Diabetologia 20:299–304.
30. Brown M, Allen R, Fisher D (1987) Bombesin alters the sympathetic nervous system response to cold exposure. Brain Res 400:35–39.
31. Huston NJ, Brumley FT, Assimacopoulos FD et al (1976) Studies on the α-adrenergic activation of hepatic glucose output. I. Studies on the α-adrenergic activation of phosphorylase and gluconeogenesis and in activation of glycogen synthase in isolated rat parenchymal cells. J Biol Chem 251:5200–5208.
32. Conlon JM, Samson WR, Dobbs RE et al (1979) Glucagon-like polypeptide in canine brain. Diabetes 28:700–702.
33. Sasaki H, Ebitani I, Tominaga M et al (1980) Glucagon-like substances in the canine brain. Endocrinol Jap SRI, 135–140.
34. Sasaki H, Tominaga M, Marubashi S et al (1985) Glucagon as a neuro-transmitter. Biomed Res 6 Suppl:91–99.
35. Tominaga M, Ebitani I, Marubashi S et al (1981) Species difference of glucagon-like materials in the brain. Life Sci 29:1577–1581.
36. Loren I, Alumets J, Hakanson R et al (1979) Gut-type glucagon immunoreactivity in nerves of the rat brain. Histochem 61:335–341.
37. Tager H, Hohenboken M, Markese J et al (1979) Identification and localization of glucagon related peptides in rat brain. Proc Natl Acad Sci 77:6229–6233.
38. Dorn A, Bernstein HG, Hahn HJ et al (1980) Regional distribution of glucagon-like immunoreactivity material in the brain of rats and sand rats. Acta Histochem 66:269–272.
39. Dorn A, Rinne A, Bernstein HG et al (1981) Immunoreactive glucagon in neurons of various parts of the human brain. Demonstration by immunofluorescence technique. Acta Histochem 69:234–247.

40. Hoosein NM, Gurd RS (1984) Identification of glucagon receptors in rat brain. Proc Natl Acad Sci USA 81:4368–4372.
41. Tominaga M, Marubashi S, Kamimura T et al (1984) The effects of starvation and alloxan treatment on glucagon-like materials in the dog brain. Bull Yamagata Univ (Med Sci) 2:185–193.
42. Kaneda H, Tominaga M, Marubashi S et al (1984) Effect of insulin-induced hypoglycemia and glucose insulin infusion on brain glucagon-like materials in the dog. Biomed Res 5:61–66.
43. Marubashi S, Tominaga M, Katagiri T et al (1985) Hyperglycemic effect of glucagon administered intracerebroventricularly in the rat. Acta Endocrinol 108:6–10.
44. Geary N, Smith GP, Gibbs J (1982) Pancreatic glucagon and postprandial satiety in the rat. Physiol Behav 28:313–322.
45. Geary N, Smith GP, Gibbs J (1986) Pancreatic glucagon and bombesin inhibit meal size in ventromedial hypothalamus-lesioned rats. Regul Pept 15:261–268.
46. Inokuchi A, Oomura Y, Nishimura H (1984) Effect of intracerebroventricularly infused glucagon on feeding behavior. Physiol Behav 33:397–400.
47. Schally AV, Bowers CY, Redding TW et al (1966) Isolation of thyrotropin releasing factor (TRH) from porcine hypothalamus. Biochem Biophys Res Commun 25:165–169.
48. Ishikawa K, Inoue K, Tosaka H et al (1984) Imunohistochemical characterization of thyrotropin-releasing hormone-containing neurons in rat septum. Neuroendocrinology 39:448–452.
49. Lechan RM, Jackson IMD (1982) Immunohistochemical localization of thyrotropin releasing hormone in the rat hypothalamus and pituitary. Endocrinology 111:55–65.
50. Cario MA, Smith JR, Weich BG et al (1976) Effects of thyrotropin releasing hormone (TRH) microinjected into various brain areas of conscious and pentobarbital-pretreated rabbits. Life Sci 19:1687–1692.
51. Tsay BC, Lin MT (1982) Effects of intracerebroventricular administration of thyrotropin-releasing hormone on cardiovascular function in the rat. Neuroendocrinology 35:173–177.
52. Koranyi L, Tamasy V, Lissak K et al (1976) Effect of thyrotropin releasing hormone (TRH) and antidepressant agents on brain stem and hypothalamic multiple unit activity in the cat. Psychopharmacology 49:197–200.
53. Lahann TR, Horita A (1982) Thyrotropin releasing hormone: Centrally mediated effects on gastrointestinal motor activity. J Pharmacol Exp Ther 222:66–70.
54. Tache Y, Vale W, Brown M (1980) Thyrotropin-releasing hormone-CNS action to stimulate gastric acid secretion. Nature 287:149–151.
55. Brown MR (1981) Thyrotropin releasing factor: a putative CNS regulator of the autonomic nervous system. Life Sci 28:1789–1795.
56. Kato Y, Kanno T (1983) Thyrotropin releasing hormone injected intracerebroventricularly in the rat stimulates exocrine pancreatic secretion via the vagus nerve. Pegul Pept 7:347–356.
57. Marubashi S, Kunii Y, Sasaki H (1986) Effect of intracerebroventricular administration of TRH on central glucoregulation in the rat. Prog Med 6:860–867.
58. Kabayama Y, Kato Y, Tojo K et al (1985) Central effect of DN 1417, a novel

TRH analog, on plasma glucose and catecholamines in conscious rat. Life Sci 36:1287–1294.

59. Keller HH, Bartholini G, Pletscher A (1974) Enhancement of cerebral noradrenaline turnover by thyrotropin-releasing hormone. Nature 248: 528–529.

60. Kabayama Y, Kato Y, Shimatsu A et al (1986) Stimulation by gastrin-releasing peptide, neurotensin and DN 1417, a novel TRH analog, of dopamine and norepinephrine release from perifused rat hypothalamic fragments in vitro. Brain Res 372:394–399.

61. Hirooka Y, Hollander CS, Suzuki S et al (1978) Somatostatin inhibits release of thyrotropin releasing factor from organ cultures of rat hypothalamus. Proc Natl Acad Sci USA 75:4509–4513.

62. Shen DC, Lin MT, Shian LR (1985) Thyrotropin releasing hormone-induced hyperglycemia: possible involvement of cholinergic receptors in the lateral hypothalamus. Neuroendocrinology 41:499–503.

63. Amir S, Rirkind IA, Harel M (1985) Central thyrotropin-releasing hormone elicits systemic hypoglycemia in mice. Brain Res 344:387–391.

64. Carraway R, Leeman SE (1973) The isolation of a new hypotensive peptide, neurotensin, from bovine hypothalami. J Biol chem 248:6854–6861.

65. Fernström MH, Carraway RE, Leeman LE (1980) Neurotensin. In: Martini L, Ganong WF (ed) Frontiers in Neuroendocrinology 6, Raven Press, New York, pp 103–127.

66. Yawata Y, Yamatani K, Tominaga M et al (1984) Hyperglycemic effect of neurotensin. Tohoku J Exp Med 143:185–196.

67. Chideckel EW, Palmer J, Koerker DJ et al (1975) Somatostatin blockade of acute and chronic stimuli of the endocrine pancreas and the consequence of this blockade on glucose homeostasis. J Clin Invest 55:754–762.

68. Emellot PC, Bos J, Van Hoeven RP et al (1974) Isolation of plasma membranes from rat and mouse livers and hepatomas. Method Enzymol 31, Biomembrane Part A 75–90.

69. Yamatani K, Yawata Y, Tominaga M et al (1980) Does neurotensin act on the liver directly? Biomed Res 1 Suppl:165–168.

70. Vijayan E, McCann SM (1978) Effect of intraventricular injection of substance P (SP), neurotensin (NT) and gastrin (G) on pituitary hormone release in conscious, ovariectomized rats. Endocrinology 102:271 A.

71. Nijima A (1976) Baroceptor effects on renal and adrenal nerve activity. Am J Physiol 230:1733–1736.

72. Lindsey A, Santeusanito F, Braaten J et al (1974) Pancreatic alpha-cell function in trauma. JAMA 227:757–761.

73. Lindsey A, Falooa GR, Unger RH (1975) Plasma glucagon levels during rapid exsanguination with and without adrenergic blockade. Diabetes 24:313–316.

74. Carraway RE, Demers LM, Leeman SE (1976) Hyperglycemic effect of neurotensin, a hypothalmic peptide. Endocrinology 99:1452–1462.

75. Nagai K, Frohman LA (1978) Neurotensin hyperglycemia: evidence for hitamine mediation and the assessment of a possible physiologic role. Diabetes 27:577–582.

76. Brown M, Vale W (1976) Effects of neurotensin and substance P on plasma insulin, glucagon and glucose levels. Endocrinology 98:819–922.

77. Dolais-Kitabgi J, Kitabgi P, Brazeau P et al (1979) Effect of neurotensin on insulin, glucagon and somatostatin release from isolated pancreatic islets. Endocrinology 105:256–260.

78. Kaneto A, Kaneko T, Kajinuma H et al (1978) Effect of substance P and neurotensin infused intrapancreatically on glucagon and insulin secretion. Endocrinology 102:393–401.

79. Feurle G, Helmstaedter V, Tischbirek K et al (1981) A multiple tumor of the pancreas producing neurotensin. Digest Dis Sci 26:1125–1133.

80. Ishida T (1976) Stimulatory effect of neurotensin on insulin and gastrin secretion in dogs. Endocrinol Jap 24:335–342.

81. Levine AS, Keip J, Grace M et al (1983) Effect of centrally administered neurotensin on multiple feeding paradigms. Pharmacol Biochem Behav 18:19–23.

82. Abe H, Chihara K, Chiba T et al (1981) Effect of intraventricular injection of neurotensin and other various bioactive peptides on plasma immunoreactive somatostatin levels in rat hypophyseal portal blood. Endocrinology 108:1939–1943.

83. Ipp E, Dobbs R, Unger RH (1978) Morphine and β-endorphin influence the secretion of the endocrine pancreas. Nature 276:190–191.

84. Ipp E, Garberogio C, Richter H et al (1984) Naloxone decreases centrally induced hyperglycemia in dogs. Evidence for an opioid role in glucose homeostasis. Diabetes 33:619–621.

85. Van Loon GR and Appel NM (1981) β-Endorphin-induced hyperglycemia is mediated by increased central sympathetic outflow to adrenal medulla. Brain Res 204:236–241.

86. Van Loon GR, Appel NM, Ho D (1981) β-Endorphin-induced stimulation of central sympathetic outflow: β-endorphin increases plasma concentrations of epinephrine, norepinephrine, and dopamine in rats. Endocrinology 109: 46–53.

87. Green IC, Ray K, Perrin D (1983) Opioid peptide effect on insulin release and c-AMP in islets of Langerhans. Horm Metab Res 15:124–128.

88. Rudman D, Berry CJ, Riedebrug CH et al (1983) Effects of opioid peptides and opiate alkaloids on insulin secretion in the rabbit. Endocrinology 112:1702–1710.

89. Beinfeld MC (1985) An overview of the biochemistry and anatomy of cholecystokinin (CCK) peptides in the brain. In: Endocoids. Liss, New York, pp 83–94.

90. Della-Fera MA, Baile CA (1979) Cholecystokinin octapeptide: continuous picomole injections into the lateral ventricle of sheep suppress feeding. Science 206:471–473.

91. Morley JE (1980) The neuroendocrine control of appetite: the role of the endogenous opiates, cholecystokinin, TRH, gamma-aminobutyric acid and diazepam receptor. Life Sci 27:355–368.

92. Goldman SA, Monahan J, Schneider BS (1985) The regional and subcellular development of cholecytokinin immunoreactivity in vertebrate brain. Dev Brain Res 22:237–246.

93. Morley JE, Levine AS (1981) Intraventricular cholecystokinin octapeptide produces hyperglycemia in rats. Life Sci 28:2187–2190.

94. Schick RR, Yaksh TL, Go VLW (1986) Intracerebroventricular injections of cholecystokinin octapeptide suppress feeding in rats—pharmacological characterization of this action. Regul Pept 14:277–291.
95. Levine AS, Morley JE (1981) Cholecystokinin-octapeptide suppresses stress-induced eating by inducing hyperglycemia. Regul Pept 2:353–357.
96. Sakamoto C, Otsuki M, Ohki A et al (1982) Glucose-dependent insulinotropic action of cholecystokinin and caerulein in the isolated perfused rat pancreas. Endocrinology 110:398–402.
97. Lu QH, Greeley G, Zhu XG et al (1984) Intracerebroventricular administration of cholecystokinin-8 elevates plasma pancreatic polypeptide levels in awake dog. Endocrinology 114:2415–2417.
98. Fujita T, Matsunari Y, Sato K et al (1979) Effect of oral administration of trypsin inhibitor and repeated injections of pancreozymin on the insulin and glucagon contents of rat pancreas. Endocrinol Jap 26:35–39.
99. Fujii S, Baba S, Fujita T (1982) Pancreatic polypeptide immunoreactive cells and nerves in the canine pituitary. Biomed Res 3:525–533.
100. Adrian TE, Besterman HS, Bloom SR (1979) The importance of cholinergic tone in the release of pancreatic polypeptide by gut hormones in man. Life Sci 24:1989–1994.
101. Lundquist I, Sundler F, Ahren B et al (1979) Somatostatin, pancreatic polypeptide, substance P, and neurotensin: distribution and effects on stimulated insulin secretion in the mouse. Endocrinology 104:832–838.
102. Said SI (1982) Vasoactive Intestinal Polypeptide. Raven Press, New York.
103. Tatemoto K, Mutt V (1981) Isolation and characterization of the intestinal peptide porcine PHI (PHI-27), a new member of the glucagon secretin family. Proc Natl Acad Sci 78:6603–6607.
104. Itoh N, Obata K, Yanaihara N et al (1983) Human prevasoactive intestinal polypeptide contains a novel PHI-27-like peptide, PHM-27. Nature 304:547–549.
105. Inoue T, Kato Y, Tatsuoka Y et al (1985) Subcellular distribution of the immunoreactive peptide histidine isoleucine. Neurosci Lett 57:165–168.
106. Christofides ND, Yiangou Y, McGregor GP et al (1982) Distribution of PHI in the rat brain. Biomed Res 3:573–574.
107. Kerins C, Said SI (1973) Hyperglycemic and glycogenolytic effects of vasoactive intestinal polypeptides. Proc Soc Exp Biol Med 142:1014–1017.
108. Felin JE, Mojena M, Silvestre RA et al (1983) Stimulatory effect of vasoactive intestinal peptide on glycogenolysis and gluconeogenesis in isolated rat hepatocyte: antagonism by insulin. Endocrinology 112:2120–2127.
109. Makhlouf GM, Yau WM, Zfass AM et al (1978) Comparative effects of synthetic and natural vasoactive intestinal peptide on pancreatic and biliary secretion and on glucose and insulin blood levels in the dog. Scand J Gastroent 13:759–765.
110. Szecowka J, Tendler D, Efendic S (1980) The interaction of vasoactive intestinal polypeptide (VIP), glucose, and arginine on the secretion of insulin, glucagon and somatostatin in the perfused rat pancreas. Diabetologia 19:137–142.
111. Szecowka J, Tendler D, Efendic S (1983) Effects of PHI on hormonal secretion from perfused rat pancreas. Am J Physiol 245:E313–E317.

112. Yanaihara C, Hashimoto Y, Takeda Y et al (1986) PHI structural requirements for potentiation of glucose-induced insulin release. Peptides 7 Suppl1:83–88.
113. Fujita T and Kobayashi S (1979) Proposal of a neurosecretory system in the pancreas. An electron microscope study in the dog. Arch Histol Jap 42:259–277.
114. Merchenthaler I, Vigh S, Schally AV et al (1984) Immunocytochemical localization of growth hormone-releasing factor in the rat hypothalamus. Endocrinology 114:1082–1085.
115. Swachenko PE, Swanson LW (1985) Localization, colocalization, and plasticity of corticotropin-releasing factor immunoreactivity in rat brain. Fed Proc 44:221–227.
116. Brown MR, Fisher DA, Spiess J et al (1982) Corticotropin-releasing factor: actions on the sympathetic nervous system and metabolism. Endocrinology 111:928–931.
117. Fisher DA, Jessen G, Brown MR (1983) Corticotropin-releasing factor (CRF): mechanism to elevate mean arterial pressure and heart rate. Regu Pept 5:153–161.
118. Phillips MI (1978) Angiotensin in the brain. Neuroendocrinology 25:354–377.
119. Chang MM, Leeman SE (1970) Isolation of sialogogic peptide from bovine hypothalamus. J Biol Chem 245:4784–4790.
120. Hökfelt T, Johansson O, Kellerth JD et al (1977) Immunohistochemical distribution of substance P. In: von Euler US, Pernow B (eds) Substance P. Raven Press, New York, pp 117–145.
121. Otsuka M, Konishi S (1977) Electrophysiological and neurochemical evidence for substance P as a transmitter of primary sensory neurons. In: von Euler US, Pernow B. (eds) Substance P. Raven Press, New York, pp 207–214.
122. Otuska M, Konishi S, Takahashi T (1972) The presence of a motoneuron-depolarizing peptide in bovine dorsal roots of spinal nerves. Proc Jap Acad 48:342–346.
123. Larsson LI (1979) Innervation of the pancreas by substance P, enkephalin, vasoactive intestinal polypeptide, and gastrin/CCK immunoreactive nerves. J Histochem Cytochem 27:1283–1284.
124. Hökfelt T, Lungdahl A, Ternius L et al (1977) Immunochemical analysis of peptide pathways possibly related to pain and analgesia: enkephalin and substance P. Proc Natl Acad Sci USA 74:3081–3085.
125. Rossel S, Bjorkroth U, Chang D et al (1977) Effects of substance P and analogs on isolated guinea pig ileum. In: von Euler US, Pernow B (eds) Substance P. Raven Press, New York, pp 83–88.
126. Hermansen K (1980) Effects of substance P and other peptides on the release of somatostatin. Endocrinology 107:256–261.
127. Unger RH and Orci L (1977) Possible role of the pancreatic D-cell in the normal and diabetic states. Diabetes 26:241–244.
128. Arimura A, Sato H, Dupont A et al (1975) Somatostatin: Abundance of immunoreactive hormone in rat stomach and pancreas. Science 189:1007–1009.
129. Brownstein MJ, Arimura A, Sato H et al (1975) The regional distribution of somatostatin in the rat brain. Endocrinology 96:1456–1461.

130. Schusdziarra V (1980) Somatostatin-A regulatory modulator connecting nutrient entry and metabolism. Horm Metab Res 12:563–577.
131. Lumpkin MD, Negro-Vilar A, McCann SM (1981) Paradoxical elevation of growth hormone by intraventricular somatostatin: possible ultrashort feed back. Science 211:1072–1074.
132. Garcia-Sevila JA, Magunusson T, Carlson A (1978) Effect of intracerebroventricularly administered somatostatin on brain monoamine turnover. Brain Res 155:159–164.
133. Sakurai H, Dobbs RE, Unger RH (1975) The role of glucagon in the pathogenesis of the endogenous hyperglycemia of diabetes mellitus. Metabol 24:1287–1297.
134. Brown MR, Rivier J, Vale W (1979) Somatostatin: central nervous system actions on glucoregulation. Endocrinology 104:1709–1714.
135. Fisher DA, Brown MR (1980) Somatostatin analog: plasma catecholamine suppression mediated by the central nervous system. Endocrinology 107:714–719.
136. Debons AF, Krimsky I, From A (1970) A direct action of insulin on the hypothalamic satiety center. Am J Physiol 219:938–943.
137. Sakamoto Y, Oomura Y, Kita H et al (1980) Insulin content and insulin receptor in the rat brain. Effect of fasting and streptozotocin treatment. Biomed Res 1:334–340.
138. Baskin DG, Porte D, Guest K, et al. (1983) Regional concentration of insulin in the rat brain. Endocrinology 112:898–903.
139. Havrankova J, Schmechel D, Kath J et al (1978) Identification of insulin in rat brain. Proc Natl Acad Sci 75:5737–5741.
140. Havrankova J, Brownstein M, Roth J (1981) Insulin and insulin receptors in rodent brain. Diabetologia 20:268–273.
141. Van Houten M, Posner BI, Kopriwa BM et al (1979) Insulin binding sites in the rat brain: in vivo localization to the circumventricular organs by quantitative radioautography. Endocrinology 105:666–673.
142. Van Houten M, Posner BI (1981) Cellular basis of direct insulin action in the central nervous system. Diabetologia 20:255–267.
143. Pansky B, Hatfield JS (1978) Cerebral localization of insulin by immunoflurescence. Am J Anat 153:459–467.
144. Dorn A, Bernstein HG, Rinne A et al (1982) Insulin-like immunoreactivity in the human brain. Histochem 74:293–300.
145. Szabo AJ, Szabo O (1975) Influence of the insulin-sensitive central nervous system glucoregulator receptor on hepatic glucose metabolism. J Physiol 253:121–133.
146. Woods SC, Lotter EC, McKay LD et al (1979) Chronic intracerebroventricular infusion of insulin reduces food intake and body weight of baboons. Nature 282:503–505.
147. Iguchi A, Burleson PD, Szabo AJ (1981) Decrease in plasma glucose concentration after microinjection of insulin into VMN. Am J Physiol 240 (Endocrinol Metab 3):E95–E100.
148. Morley JE, Bartness TJ, Gosnell BA et al (1985) Peptidergic regulation of feeding. Int Rev Neurobiol 27:207–298.
149. Morley JE, Levine AS, Gosnell BA et al (1985) Peptides as central regulators of feeding. Brain Res Bull 14:511–519.

150. Woods SC, West DB, Stein DJ et al (1981) Peptides and the control of meal size. Diabetologia 20:305–313.
151. Bray GA, Gallagher PFJr (1975) Manifestations of hypothalamic obesity in man: a comprehensive investigation of eight patients and review of literature. Medicine 54:301–330.
152. Frohman LA (1980) Hypothalamic control of metabolism. In: Morgane PJ, Paksepp, (eds) Handbook of the Hypothalamus Vol 2, Physiology of the Hypothalamus. Marcel Decker Inc, New York. pp 519–555.
153. Givens JR, Wiedemann E, Anderson RN et al (1980) β-Endorphin and β-lipotropin plasma levels in hirsute women: correlation with body weight. J Clin Endocrinol Metab 50:975–976.
154. Dunger DB, Wolff OH, Leonard JV et al (1980) Effect of naloxone in a previously undescribed hypothalamic syndrome: a disorder of the endogenous opioid peptide system? Lancet 1:1277–1281.
155. Marguless DL, Moisett B, Lewis MJ et al (1978) β-Endorphin is associated with overeating in genetically obese mice (ob/ob) and rats (fa/fa). Science 202:988–991.
156. Ferguson-Segall M, Flynn JJ, Walker J et al (1981) Increased immunoreactive dynorphin and Leu-enkephalin in posterior pituitary of obese mice (ob/ob) and supersensitivity to drugs that act kappa receptors. Life Sci 31:2233–2236.
157. Brands B, Thornhill JA, Hirst M et al (1979) Suppression of food intake and body weight gain by naloxone in rats. Life Sci 24:1773–1778.
158. Wallace M, Fraser CD, Clements JA et al (1981) Naloxone, adrenalectomy, and steroid replacement: evidence against a role for circulating β-endorphin in food intake. Endocrinology 108:189–192.
159. Lowy MT, Yim GKW (1981) The anorexigenic effect of naltrexone is dependent of its suppressant effect on water intake. Neurophamacology 20:883–886.
160. Kyriakides M, Silverstone T, Jeffcoat W et al (1980) Effect of naloxone on hyperphagia in Prader-Willi's Syndrome. Lancet 1:876–877.
161. Atkinson RL (1982) Naloxone decreases food intake in obese humans. J Clin Endocrinol Metab 55:196–198.
162. Sternbach HA, Annitto W, Pottash ALC et al (1982) Anorexic effect of naltrexone in man. Lancet 1:388–389.
163. Clark JT, Kalra PS, Kalra SP (1985) Neuropeptide Y stimulates feeding but inhibits sexual behavior in rats. Endocrinology 117:2435–2442.
164. Edmonds ES, Withyachumnarnkul B (1980) Sexual behavior of the obese male Zucker rat. Physiol Behav 24:1139–1141.
165. Morley JE, Levine AS, Yim GK et al (1983) Opioid modulation of appetite. Neurosci Biobehav Rev 7:281–305.
166. Kagan H (1985) Anorexia and severe inanition associated with a tumor involving the hypothalamus. Arch Dis Child 33:257–260.
167. White LE, Hain RF (1959) Anorexia in association with a destructive lesion of the hypothalamus. Arch Pathol 68:275–281.
168. Imura H (1982) Clinical aspects of neuroendocrinology. J Jap Soc Int Med 71:901–917.
169. Kaye WH, Pickar D, Naber D et al (1982) Cerebrospinal fluid opioid activity in anorexia nervosa. Am J Psychiatry 139:643–645.

170. Rose RM, Sacher E (1981) Psychoendocrinology. In: Williams RH (ed) Textbook of Endocrinology, pp 646–671.
171. Trygstad O, Foss I, Edminson PD et al (1978) Humoral control of appetite: a urinary anorexigenic peptide. Chromatographic patterns of urinary peptides in anorexia nervosa. Acta Endocrinol 89:196–208.
172. Moore R, Mills IH, Forster A (1981) Naloxone in the treatment of anorexia nervosa: effect on weight gain and lipolysis. J Royal Soc Med 74:129–131.
173. Baile CA, Della-Ferra MA, McLaughlin CL (1983) Hormones and food intake. Proc Nutr Soc 42:113–127.
174. Brief DJ, Davis JD (1984) Reduction of food intake and body weight by chronic intraventricular insulin infusion. Brain Res Bull 12:571–575.
175. Malaisse-Lagae F, Carpenter JL, Patel YC et al (1977) Pancreatic polypeptide: a possible role in the regulation of food intake in the mouse. Experientia 33:915–917.
176. Cryer PE (1985) Does central nervous system adaptation to antecedent glycemia occur in patients with insulin-dependent diabetes mellitus? Ann Intern Med 103:284–286.
177. Long CL, Kinney JM, Geiger JW (1976) Nonsuppressibility of gluconeogenesis by glucose in septic patients. Metabolism 25:193–201.
178. Ipp E, Rubinstein AH (1980) Opioide peptides and stress hyperglycemia. Lancet 2:1082–1091.
179. Ipp E, Dhorajiwala J, Pugh W et al (1982) Effect of an enkephalin analog on pancreatic endocrine function and glucose homeostasis in normal and diabetic dogs. Endocrinology 111:2110–2116.
180. Tatemoto K, Carlquist M, Mutt V (1982) A novel brain peptide with structural similarities to peptides YY and pancreatic polypeptide. Nature 296:659–660.
181. O'Donohue TL, Chronwall BM, Pruss BM et al (1985) Neuropeptide Y and peptide YY neuronal and endocrine systems. Peptides 6:755–768.
182. Everitt BJ, Hokfelt T, Terenius L et al (1984) Differential co-existence of neuropeptide Y (NPY)-like immunoreactivity with catecholamines in the central nervous system of the rat. Neuroscience 11:443–462.
183. Tatemoto K, Rokaeus A, Jornvall H et al (1983) Galanin-a novel biologically active peptide from porcine intestine. FEBS Lett 164:124–128.
184. Rokaeus A, Melander T, Hokfelt T et al (1984) A galanin-like peptide in the central nervous system and intestine of the rat. Neurosci Lett 47:161–166.
185. Melander T, Hokfelt T, Rokaeus A et al (1985) Distribution of galanin-like immunoreactivity in the gastrointestinal tract of several mammalian species. Cell Tissue Res 239:253–270.
186. Dunning BE, Ahren B, Veith RC et al (1986) Galanin-a novel pancreatic neuropeptide. Am J Physiol 251:E127–133.
187. McDonald TJ, Dupre J, Greenberg GR et al (1986) The effect of galanin on canine plasma glucose and gastroentero-pancreatic hormone response to oral nutrients and intravenous arginine. Endocrinology 119:2340–2345.
188. Silvestre RA, Miralles P, Monge L et al (1987) Effects of galanin on hormone secretion from the in situ perfused rat pancreas and on glucose production in rat hepatocyte in vitro. Endocrinology 121:378–383.
189. Manabe T, Yoshimura T, Kii E et al (1986) Galanin-induced hyperglycemia: effect on insulin and glucagon. Endocrinol Res 12:93–98.

6

Feedback Regulation of Growth Hormone Secretion

Shlomo Melmed

Introduction

Growth hormone (GH) is synthesized, stored, and secreted by the somatotroph cells of the anterior pituitary. These cells are densely granulated, with secretory granules measuring 250–500 nm on electron microscopy. Somatotrophs comprise about 50% of the anterior pituitary cells; their number does not change with age or systemic illness, and is similar in both sexes. GH is the predominant secretory product of the anterior pituitary and is a 22,000-dalton polypeptide consisting of 191 amino acids (1). The human GH gene consists of at least five exons and is located on the long arm of chromosome 17 at position q 22–q 24 (2,3). The somatotroph expression of GH is under selective and specific hormonal control, emanating from both central and peripheral sources, that regulates the secretion of GH by a complex set of cellular signals. This chapter discusses hormonal modulation of GH secretion, and focuses on the feedback regulation of GH expression by insulin-like growth factor-I (IGF-I).

Regulation of GH Secretion

The secretion of GH is under dual stimulatory and inhibitory hypothalamic influences (Table 6.1). Growth hormone releasing hormone (GHRH) is secreted by the hypothalamus into the hypophyseal portal system, and acts on the somatotroph cell to stimulate GH gene expression and secretion (4–7). Somatostatin (SRIF), also secreted by the hypothalamus, inhibits the secretion of GH (8). Other hormones, including triiodothyronine, hydrocortisone, and insulin also influence the secretion of GH and the transcription of the rat GH gene (9–11). These hormones act directly on the pituitary somatotroph. Interestingly, hydrocortisone enhances GHRH stimulation of the somatotroph (12). The peripheral growth promoting actions of GH are, to a large extent, mediated by soma-

TABLE 6.1 Hormonal regulation of rat GH gene expression and secretion

		Effects	
		In vitro	In vivo
Hypothalamic	Growth hormone releasing hormone	↑	↑
	Somatostatin	↓	↓
Pituitary	Growth hormone	↓	↓
	Insulin-like growth factor I	↓	?
Peripheral	Insulin-like growth factor I	↓	↓
	Triiodothyronine	↑	↑
	Hydrocortisone	↑	?
	Gonadal steroids	±	±
	Insulin	↓	↓

↑ = stimulated; ↓ = suppressed; ? = unknown; ± = conflicting data.

tomedin-C (IGF-I), a growth factor which stimulates sulfate incorporation by cartilage (13). IGF-I has a high homology with pro-insulin, exhibits a 50% homology with insulin (14), and has a molecular weight of 7650 (15). The IGF-I and IGF-II molecules both contain the C-peptide sequence, resulting in single-chain polypeptides. IGF-I is secreted predominantly by the liver, but the IGF-I gene is also expressed in multiple tissues of the rat including kidney, brain, lung, muscle, heart, and the pituitary itself (16,17). The hepatic secretion of IGF-I appears to be dependent on GH.

Regulation of GH secretion by IGF-I may represent a classic negative feedback loop, analogous to that for thyroid hormones and adrenal steroids on their respective pituitary trophic hormones.

Receptors for Insulin, IGF-I, and IGF-II in the Pituitary

Specific binding sites for insulin, IGF-I, and IGF-II have been well characterized in normal rat anterior pituitary cells and membranes (18,19), cloned rat pituitary cell lines (20), and human pituitary adenoma tissue (21).

Normal Rat Pituitary Cells

IGF-I was found to be more potent than IGF-II in displacing the ^{125}I-labeled IGF-I ligand from intact rat anterior pituitary cells grown in primary culture. Fifty percent displacement of the radioligand occurred at an IGF-I concentration of 4–20 ng/ml. Specific binding per 100,000 pituitary cells averaged about 10% for ^{125}I-IGF-II, 0.83% for ^{125}I-IGF-I, and only 0.11% for ^{125}I-insulin. IGF-II was twice as potent as IGF in displacing ^{125}I-IGF-II binding. IGF-I was fivefold more potent than

IGF-II in displacing ^{125}I-IGF-I binding and 1,000-fold more potent than insulin (18).

Specific IGF-I and IGF-II binding sites have also been demonstrated on rat anterior pituitary, hypothalamic, and brain membrane preparations (19). IGF-II was significantly more potent than IGF-I in displacing its tracer from pituitary membrane binding sites. Binding of ^{125}I-IGF-I to pituitary membranes was specific and, interestingly, binding of IGF-II was three to five times that of IGF-I. This is curious in view of the lack of correlation between IGF-I and IGF-II specific binding and their respective biologic potencies in inhibiting GH secretion at the level of the pituitary (42). The function of the relatively abundant pituitary IGF-II receptors is unclear, especially in view of the lack of any obvious biologic effect of IGF-II on anterior pituitary hormone secretion. Nevertheless, the existence of abundant high-affinity IGF receptors on rat anterior pituitary cells is consistent with a biologic role for the somatomedin peptides in the regulation of GH secretion.

Cloned Rat Pituitary Cells

Interestingly, although the specific binding of ^{125}I-insulin was extremely low in normal rat pituitary cells, it was relatively high in cloned rat pituitary cell lines (GH$_3$, GH1, and GC) (20). In these lines, the K_D for the high-affinity insulin receptor ranged from 0.1 to 5 nM. The presence of specific high-affinity receptors for insulin on these transformed GH secreting cell lines is also consistent with experimental observations on the feedback regulation of GH in these cells by insulin (22).

Human Pituitary Adenomas

Specific receptors for insulin, IGF-I, and IGF-II have been characterized in membranes derived from human GH-secreting adenomas. Specific binding (per 100 μg protein) was similar (ranging from 3 to 10%) for all three peptides (21).

Feedback Regulation of GH by IGF-I

In Vivo Studies

Autoregulation of GH secretion has been demonstrated in various models. Early observations that rat pituitary GH content was suppressed by exogenous administration of GH (23,24) were followed by further evidence for the autoregulation of GH secretion in rats (25–30), rhesus monkey (31), and man (32–34). It has been postulated that the stimulation of hypothalamic SRIF secretion by high levels of GH in vitro (35) and in vivo (36–40) is responsible for the negative feedback regulation of GH secretion by GH. In any event, GH appears to regulate its own secretion, either by acting on the hypothalamus to increase hypothalamic SRIF release, or possibly by a direct inhibitory effect on pituitary cells. The peripheral effects of pituitary GH include stimulation of IGF-I generation, and this peptide also participates in the feedback regulation of GH secretion both at the level of the hypothalamus, as well as by direct effects on the somatotroph (41–46).

Pituitary GH mRNA levels are suppressed in animals bearing GH-secreting tumors, in which the circulating concentrations of GH and IGF-I are elevated (47). The suppression of GH mRNA in these animals appears selective since levels of pituitary gamma-actin mRNA, a constitutive cellular RNA, and of total pituitary RNA were not different in control or tumor-bearing rats (Fig. 6.1). This suppression of pituitary GH mRNA sequences could result from autoregulation by GH either at the hypothalamic or the pituitary level. Although hypothalamic stimulation of SRIF could be implicated in the inhibition of GH secretion, as yet there is no evidence that SRIF directly suppresses GH mRNA levels in vitro. Alternatively, a recent preliminary report describing hypothalamic GHRH depletion in GH-secreting tumor-bearing rats suggests another hypothalamic effect of excessive GH secretion (48). GH itself has not consistently been shown to directly inhibit pituitary GH secretion.

Intraventricular injection of somatomedin-C to conscious, freely moving, chronically catheterized rats markedly suppressed the episodic pulsatile secretion of pituitary GH (49,50) (Fig. 6.2). The action of the administered peptide may therefore occur at the level of the hypothalamus by stimulating SRIF production, or decreasing GHRH secretion, as well as by direct action on the pituitary gland.

Thus, the high circulating levels of IGF-I in rats bearing GH-secreting tumors appear to participate in a long loop negative feedback of pituitary GH mRNA. High circulating levels of both GH and IGF-I contribute to the profound suppression of pituitary GH mRNA levels. Additional evidence from in vitro studies, presented below, suggests that inhibition occurs directly at the level of the pituitary.

In Vitro Studies

The recent availability of recombinant human IGF-I analogues (51,52) and of pure native peptide have facilitated the in vitro study of the effects of IGF-I on rat and human GH gene expression.

RAT

Effects of IGF-I on Basal GH Secretion and mRNA Levels

A purified somatomedin-C preparation was shown to inhibit dibutyryl cyclic AMP-stimulated GH secretion by rat pituitary cells (41). In these

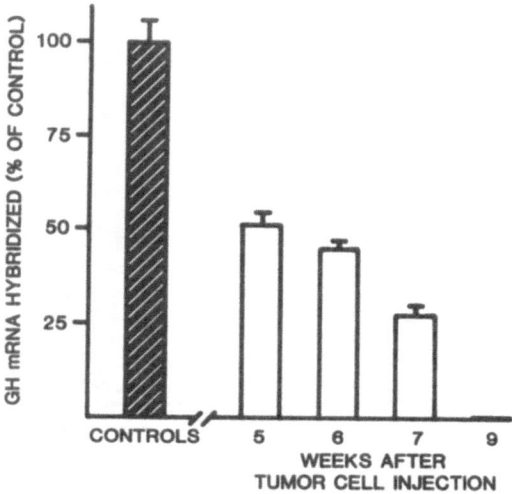

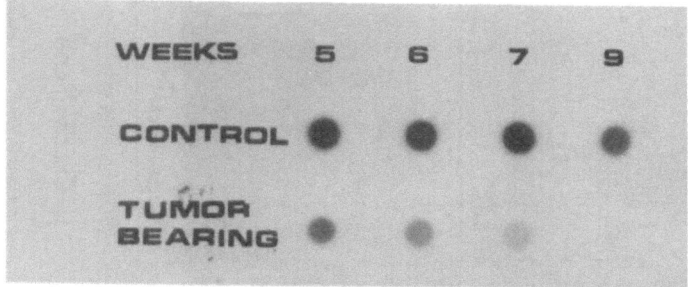

FIGURE 6.1. Autoregulation of pituitary GH mRNA levels. Dot blot hybridization of total pituitary RNA. Triplicate blots (3 μg total RNA each) from each pituitary gland were immobilized and hybridized with rGH [^{32}P]cDNA. Representative autoradiograph of rGH mRNA hybridization after 9 weeks in control (C) and tumor-bearing (T) rats. (Reprinted with permission from ref. 47, © The Endocrine Society, 1986.)

experiments, somatomedin-C also stimulated the in vitro secretion of hypothalamic SRIF, further delineating the participation of this growth factor in feedback regulation of GH secretion (41). A recombinant IGF-I analog clearly suppressed GH secretion by rat pituitary cells grown in a serum-free defined medium (Fig. 6.3). Basal and GHRH-stimulated GH secretion in vitro were both similarly suppressed by other IGF-I prepara-

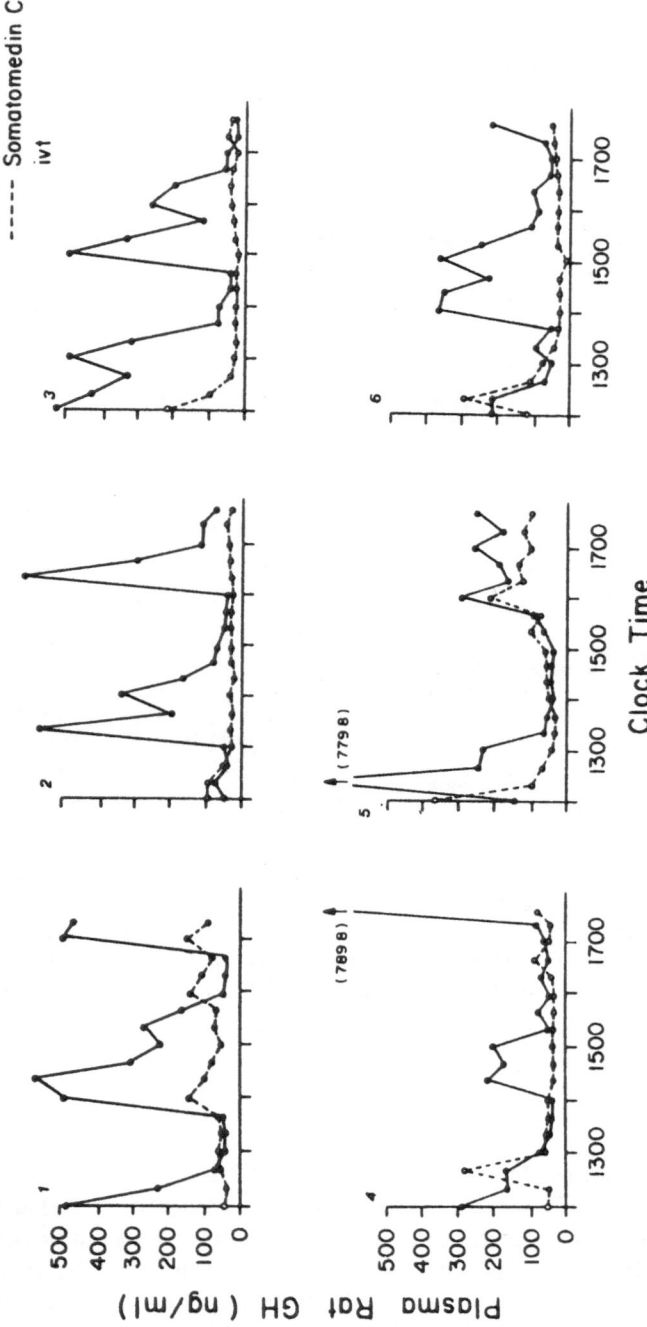

FIGURE 6.2. Effects of intraventricular injection of somatomedin C on spontaneous GH release in individual rats. (Reprinted with permission from ref. 49, © The Endocrine Society, 1983.)

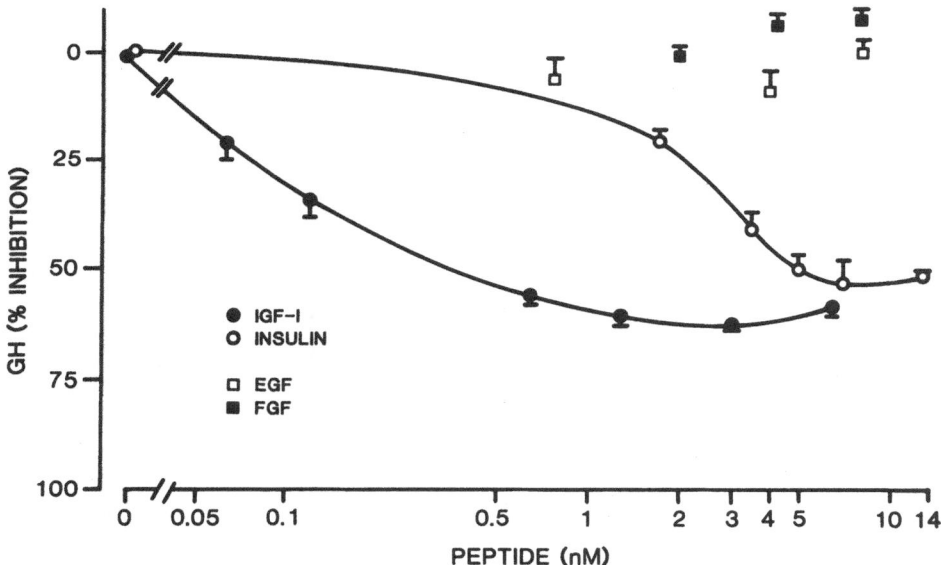

FIGURE 6.3. Dose-response effect of IGF-I on GH secretion. Pituitary cells were treated for 72 hours with the indicated doses of peptide added to serum-free medium. GH secretion by control cells (no added peptide) was $3.47 \pm 0.42\ \mu g/10^4$ cells/72 hours. On a weight basis, the highest concentrations of FGF and EGF were 100 and 50 ng/ml, respectively. Data from nine separate experiments (three to six wells per dose point) were pooled and expressed as a percentage of control GH secretion (mean $\pm$ SEM). (Reprinted with permission from ref. 44, © The Endocrine Society, 1986.)

tions (42–45). Furthermore, insulin itself was shown to inhibit in vitro secretion of GH (22,53), to suppress the levels of pituitary GH mRNA sequences (54,55), and to inhibit the transcription of the GH gene (11).

The IGF-I effect on the pituitary is specific since other growth factors including epidermal growth factor (EGF) and fibroblast growth factor (FGF) do not alter GH gene expression (Fig. 6.3). IGF-I also does not alter the secretion of prolactin by the pituitary cells, further indicating that the inhibition of GH induced by IGF-I is indeed selective. Although GH secretion, mRNA levels and gene transcription have been shown to be inhibited by insulin in both normal pituitary cells and pituitary tumor cells, the dose response of IGF-I inhibition of GH gene expression is clearly more potent than that elicited by purified semisynthetic human insulin. Insulin receptor antiserum was able to neutralize the inhibitory effects of insulin on GH secretion, while the same antiserum did not alter IGF-I induced GH suppression (55). The concentration of IGF-I shown to inhibit GH expression correlates well with the above described displacement curves for IGF-I binding to its receptor sites on pituitary cells. Fifty

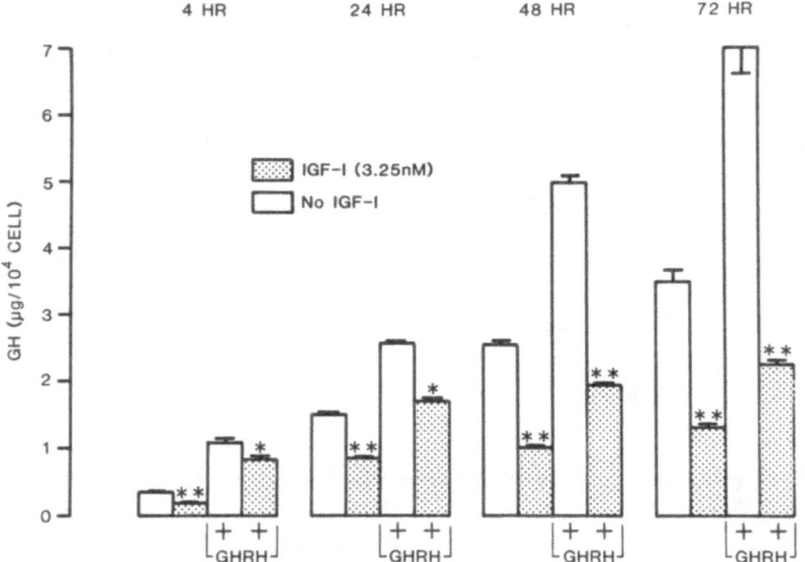

FIGURE 6.4. Effect of IGF-I on GHRH-stimulated GH secretion. Pituitary cells were grown in serum-free defined medium containing IGF-I (3.25 nM) with or without GHRH (1 nM) as indicated. Each bar represents the mean ± SEM of triplicate wells.*$p < 0.05$, **$p < 0.001$ (vs. respective IGF-I-free controls). (Reprinted with permission from ref. 44, © The Endocrine Society, 1986.)

percent displacement of ^{125}I-IGF-I was achieved by 4–20 ng/ml IGF-I, corresponding to 0.5–2.5 nM.

Effects of IGF-I on Stimulated GH

GHRH has been shown to stimulate GH secretion, and to increase mRNA levels and GH gene transcription (56,57). IGF-I inhibits the increase of GH mRNA levels, as well as GH gene transcription induced by GHRH in both rat and human pituitary cells (59,60) (Fig. 6.4). IGF-I also inhibits T_3-induced rise in GH mRNA levels and GH secretion in pituitary glands derived from hypothyroid rats (Figs. 6.5 and 6.6). Whereas IGF-I suppressed T_3-induction of GH gene expression in primary hypothyroid rat pituitary cell cultures (45), insulin suppressed T_3-induced GH secretion in the cloned GH3 pituitary cell line.

The intracellular site of the inhibitory action of IGF-I on GH gene expression may be either at transcriptional or posttranscriptional levels. IGF-I may alter mRNA stability or suppress GH translation. Furthermore, a distinct suppressive effect of IGF-I on acute GH release has also been demonstrated (58). Transcriptional run-off assays have provided additional insight into the effect of IGF-I on GH gene expression.

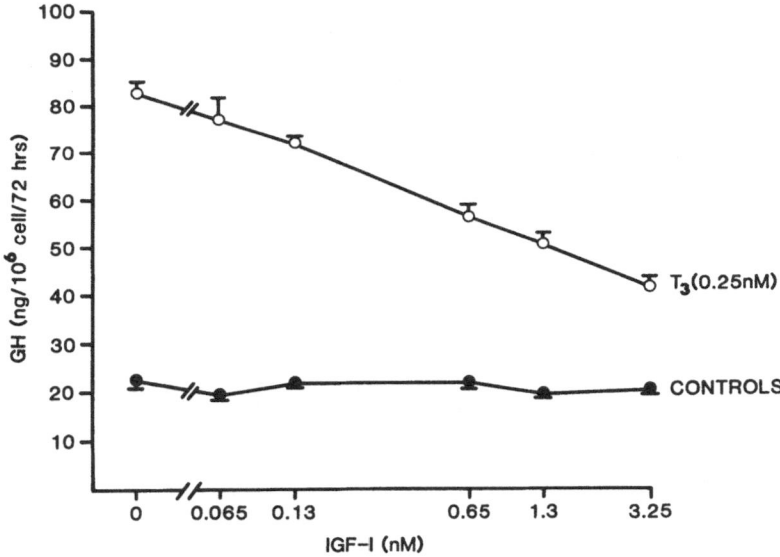

FIGURE 6.5. Effects of increasing doses of IGF-I on T_3-stimulated GH secretion. Pituitary cells derived from thyroidectomized rats were preincubated for 24 hours in medium depleted of thyroid hormone. Incubations were then continued for 72 h in the presence of the indicated doses of IGF-I with or without added T_3 (0.25 nM). Control wells received no added T_3. Each point represents the mean ± SEM of triplicate wells from four separate experiments. (Reprinted with permission from ref. 45, © The Endocrine Society, 1986.)

Transcriptional Effect of IGF-I on GH Gene Expression

IGF-I was added to primary cultures of rat anterior pituitary cells. Nuclei were then isolated and incubated in an in vitro transcriptional run-off assay with ^{32}P-labelled GTP (59). Newly synthesized radioactive GH mRNA was measured by hybridizations with immobilized excess rat GH cDNA. This nuclear run-off assay is a measure of in vitro elongation of nascent RNA chains and thus reflects transcriptional activity.

IGF-I treatment suppressed the transcriptional activity of the GH gene in a dose-responsive manner (Fig. 6.7). Maximal inhibition to 14% of control levels was achieved in the presence of 6.5 nM IGF-I, while 3.25 nM IGF-I maximally suppressed new GH mRNA synthesis by 70% at 4 hours. After 24 hours, GH gene transcription was still suppressed by about 50% when compared with control untreated cells. Prolactin (PRL) gene transcriptional activity was not altered by IGF-I (59).

The stimulation of new GH mRNA synthesis induced by GHRH was also blocked by IGF-I. Other growth factors, including EGF and FGF,

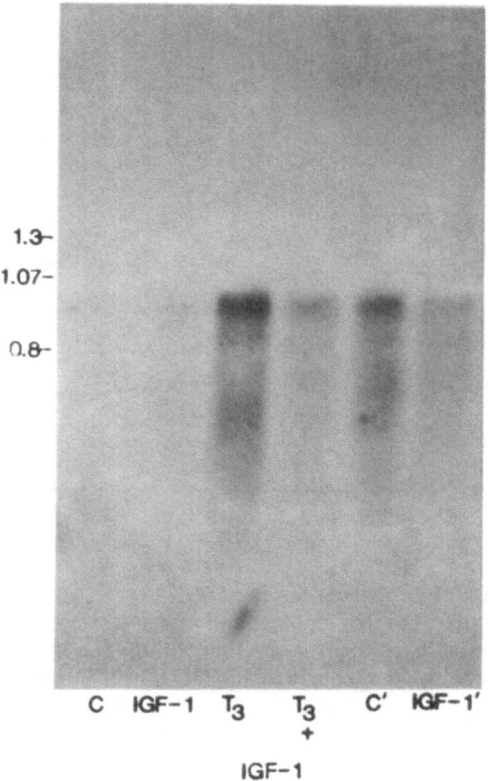

FIGURE 6.6. Agarose gel (1%) electrophoresis of pituitary RNA extracts. RNA 5 μg) was electrophoresed, transferred to nitrocellulose paper, hybridized with ^{32}P-labeled rGH cDNA (SA, ~10^8 cpm/μg DNA), and autoradiographed. Lanes (left to right) 1 and 2, control hypothyroid pituitary cells without added T_3 with (IGF-I) or without (C) IGF-I; lanes 3 and 4, hypothyroid pituitary cells incubated with 0.25 nM T_3 with or without IGF-I (3.25 nM); lanes 5 and 6, euthyroid pituitary cells with (IGF-I') or without (C') IGF-I (3.25 nM). (Reprinted with permission from ref. 45, © The Endocrine Society, 1986.)

did not alter GH transcription. Relatively high concentrations of insulin did suppress the GH gene transcriptional activity. This may indicate that insulin is acting via the IGF-I receptor, or alternatively, insulin may specifically inhibit GH gene transcription as it does in GH_3 cells (11).

Although SRIF inhibits the release of GH, this hypothalamic hormone has not been shown to regulate GH gene expression. The direct and selective inhibition of synthesis of nascent GH mRNA by IGF-I indicates the presence of a classic negative feedback loop in the regulation of pituitary GH gene transcription. Thus, like GHRH, T_3, and hydrocortisone, IGF-I participates in the balance of GH gene regulation by inhibiting GH gene transcription. This observation may explain the high

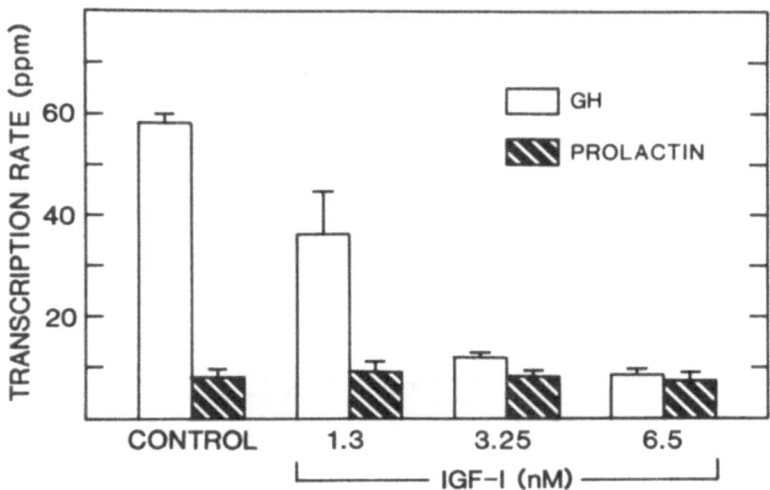

FIGURE 6.7. Dose-response in IGF-I effect on GH and prolactin gene transcription. Primary cultures of rat anterior pituitary cells were incubated in serum-free defined medium with or without added IGF-I. Nuclei were then isolated and incubated in an in vitro transcriptional runoff assay with ^{32}P-labeled GTP. New GH mRNA and prolactin mRNA were measured by hybridization against immobilized GH cDNA and prolactin cDNA, respectively. Bars represent mean and range of duplicate dots from a representative experiment. [Reproduced from *The Journal of Clinical Investigation* (1987); 79:449–482, by copyright of the American Society for Clinical Investigators.]

GH levels seen in starvation and protein-calorie malnutrition where IGF-I levels are suppressed.

HUMAN PITUITARY CELLS

Ceda et al. (21) showed that a partially purified somatomedin-C preparation inhibited the release of GH by human pituitary cell cultures. The suppression of human GH secretion and GH mRNA levels in human somatotropinoma cells was also studied using a purified recombinant human IGF-I analog, Thr-59-IGF-I (60). Somatotropinoma cells were incubated in serum-free medium in the presence of IGF-I, which inhibited GH secretion. The 50% stimulation of GH secretion induced by GHRH (1 n*M*) was also blocked by simultaneous exposure of the cells to IGF-I. In each tumor culture, the effects of IGF-I were prevented by addition of alpha-IR3, a specific antibody to the IGF-I receptor (60).

Effects of IGF-I on Human GH mRNA Levels

Relative levels of GH mRNA sequences were measured in adenoma cells by dot blot hybridization with radiolabeled hGH cDNA. Figure 6.8 shows the effect of varying IGF-I doses on GH mRNA levels. Maximal

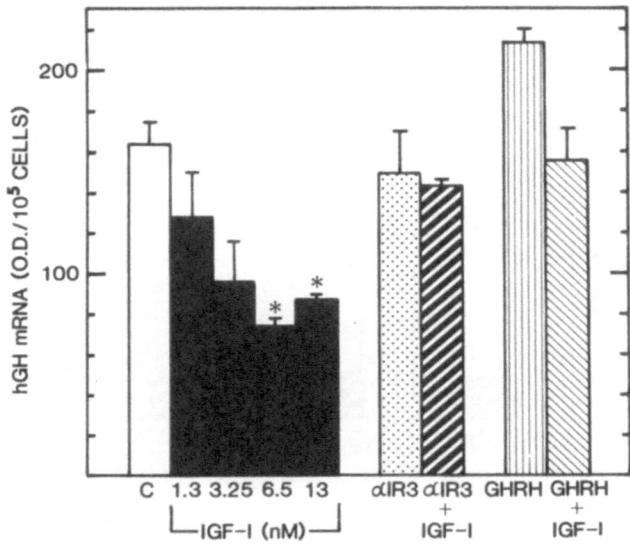

FIGURE 6.8. Effects of IGF-I, α-IR3, and GHRH on relative levels of GH mRNA sequences in cultured human somatotropinoma cells. After indicated treatment for 48 hours, extracted RNA was immobilized, hybridized with ^{32}P-cDNA for hGH (SA 10^8 cpm/μg DNA), and autoradiographed. Each bar represents mean ± SEM of triplicate wells. (Reprinted with permission from ref. 60, © The Endocrine Society, 1986.)

inhibition of GH mRNA (40% of basal value) was seen after treatment with 6.5 nM IGF-I. Alpha-IR3, either alone or with added IGF-I, did not alter the levels of GH mRNA. GHRH stimulated GH mRNA levels by about 35%, and this increase was abolished by simultaneous exposure of the cells to IGF-I. These data demonstrate the regulation of hGH mRNA in phenotypic human somatotroph cells (60).

Pituitary IGF-I

IGF-I gene expression within the pituitary itself has been convincingly shown by RIA, as well as by hybridization of specific mRNA with IGF-I cDNA probes (61,62). As circulating IGF-I is tightly and specifically bound to carrier proteins with an affinity constant comparable to the affinity of the pituitary IGF receptors, it is possible that locally produced pituitary IGF-I may be responsible for IGF-I bioactivity in the pituitary. Since GH and T_3 are known to induce IGF-I gene expression in the pituitary (62), the mechanism of the autoregulation of GH secretion may involve an interactive regulation of the expression of both the GH and IGF-I genes in neighboring pituitary cells (paracrine effect) or at a single cell level (autocrine effect). Thus, pituitary GH secretion may be under tonic inhibition by locally produced pituitary IGF-I.

Conclusion

The evidence summarized in this chapter indicates that IGF-I exerts a potent and selective feedback inhibition on rat and human GH gene expression. This feedback loop appears to be analogous to the negative feedback of thyroid hormone (63) and adrenal (64) steroids on their respective pituitary trophic hormone expression.

This information may explain the discordancy between serum IGF-I and GH levels during starvation, protein-calorie malnutrition, cachexia, and Laron dwarfism (65,66). The high GH levels in these clinical states may result from a protective lowering of peripheral IGF-I production leading to unrestrained GH secretion. Clearly, the central hypothalamic control of GH secretion is modulated by peripheral signal input of IGF-I, and both influences determine the final net effect on GH secretion. Recently, a direct effect of IGF-I has been demonstrated on the transfected human GH gene (67), suggesting the presence of cis-active regulatory elements responsive to the IGF-I signal on the somatotroph. The physiologic significance of the IGF-I feedback regulation on GH gene expression awaits clarification by further in vivo studies in man, which will be facilitated by the availability of large amounts of pure IGF-I.

References

1. Miller WL, Eberhardt NL (1983) Structure and evolution of the growth hormone gene family. Endocrinol Rev 4:97–130.
2. Martial JA, Hallewell RA, Baxter JD et al (1979) Human growth hormone: Complementary DNA cloning and expression in bacteria. Science 205:602–7.
3. Owerbach D, Rutter WJ, Martial JA et al (1980) Genes for growth hormone, chorionic somatomammotropin, and growth hormone-like gene on chromosome 17 in humans. Science 209:289–92.
4. Thorner MO, Peryman RL, Cronin MJ, et al (1982) Somatotroph hyperplasia: Successful treatment of acromegaly by removal of a pancreatic islet tumor secreting a growth hormone-releasing factor. J Clin Invest 70:965–77.
5. Guillemin RR, Brazeau P, Bohlen P et al (1982) Growth hormone-releasing factor from a human pancreatic tumor that caused acromegaly. Science 218:585–7.
6. Ling N, Zeytin F, Bohlen P et al (1985) Growth hormone releasing factors. Annu Rev Biochem 54:403–423.
7. Frohman LA, Jansson JO: (1986) Growth hormone-releasing hormone. Endocr Rev 7:223–253.
8. Brazeau P, Vale W, Burgus R et al (1973) Hypothalamic polypeptide that inhibits the secretion of immunoreactive pituitary growth hormone. Science 179:77–79.

9. Yaffe BM, Samuels HH (1984) Hormonal regulation of the growth hormone gene: Relationship of the rate of transcription to the level of nuclear thyroid hormone-receptor complexes. J Biol Chem 259:6284–91.

10. Evans RM, Birnberg NC, Rosenfeld MG (1982) Glucocorticoid and thyroid hormones transcriptionally regulate growth hormone gene expression. Proc Natl Acad Sci USA 79:7659–63.

11. Yamashita S, Melmed S: (1986) Insulin regulation of rat GH gene transcription. J Clin Invest 78:1008–1014.

12. Wehrenberg WB, Baird A, Ling N (1983) Potent interaction between glucocorticoids and growth hormone-releasing factor in vivo. Science 221:556–558.

13. Salmon WD, Daughaday W (1957) A hormonally controlled serum factor which stimulates sulfate incorporation by cartilage in vitro. J Lab Clin Med 49:825–836.

14. Rinderknecht E, Humbel RE (1978) The amino acid sequence of human insulin-like growth factor I and its structural homology with proinsulin. J Biol Chem 253:2769–2775.

15. Klapper DG, Svoboda ME, Van Wyk JJ (1983) Sequence analysis of somatomedin-C: Confirmation of identity with insulin-like growth factor I. Endocrinology 112:2215–2221.

16. D'Ercole AJ, Stiles AD, Underwood LE (1984) Tissue concentrations of somatomedin C: Further evidence for multiple sites of synthesis and paracrine or autocrine mechanisms of action. Proc Natl Acad Sci USA 81:935–939.

17. Fagin JA, Pixley S, Slanina S et al (1987) Insulin-like growth factor I gene expression in GH_3 rat pituitary cells mRNA content, immunocytochemistry and secretion. Endocrinology 120:2037–2043.

18. Rosenfeld RG, Ceda G, Wilson DM et al (1984) Characterization of high affinity receptors for insulin-like growth factors I and II on rat anterior pituitary cells. Endocrinology 114:1571–1575.

19. Goodyer CG, Stephano LD, Lai WH et al (1984) Characterization of insulin-like growth factor receptors in rat anterior pituitary hypothalamus and brain. Endocrinology 114:1187–1195.

20. Rosenfeld RG, Ceda G, Cutler CW et al (1985) Insulin and insulin-like growth factor receptors on cloned rat pituitary tumor cells. Endocrinology 117:2008–2014.

21. Ceda GP, Hoffman AR, Silverberg GD et al (1985) Regulation of GH release from cultured human pituitary adenomas by somatomedins and insulin. J Clin Endocrinol Metab 60:1204–1210.

22. Melmed S (1984) Insulin suppresses growth hormone secretion by rat pituitary cells. J Clin Invest 73:1425–1433.

23. Krulich L, McCann SM (1966) Influence of growth hormone on content of GH in the pituitaries of normal rats. Proc Soc Exp Biol Med 121:1114–1117.

24. Muller E, Pecile A (1966) Influence of exogenous growth hormone on endogenous growth hormone release. Proc Soc Exp Biol Med 122:1289–1291.

25. Macleod RM, DeWitt GW, Smith MC (1966) Hormonal properties of transplanted pituitary tumors and their relation to the pituitary gland. Endocrinology 79:1149–1156.

26. Peake GT, Mariz IK, Daughaday WH (1968) Radioimmunoassay of growth hormone in rats bearing somatotropin producing tumors. Endocrinology 83:714–720.

27. Sawano S, Arimura A, Bowers CY et al (1967) Effect of CNA depressants, dexamethasone and growth hormone on the response of growth hormone-releasing factor. Endocrinology 81:1410–1412.
28. Katz SH, Molitch ME, McCann SM (1969) Effects of hypothalamic implants of GH on anterior pituitary weight and GH concentration. Endocrinology 85:725–730.
29. Tannenbaum GS (1980) Evidence for autoregulation of GH secretion via the central nervous system. Endocrinology 107:2117–2220.
30. Voogt JL, Clemens JA, Negro-Vilar A et al (1971) Pituitary GH and hypothalamic GRF after median eminence implantation of ovine or human GH. Endocrinology 88:1363–1367.
31. Sakuma M, Knobil E (1970) Inhibition of endogenous growth hormone secretion by exogenous growth hormone infusion in the rhesus monkey. Endocrinology 86:890–894.
32. Abrams RL, Grumbach MM, Kaplan SL (1971) The effect of human growth hormone on the plasma growth hormone, cortisol, glucose and free fatty acid response to insulin: Evidence for growth hormone autoregulation in man. J Clin Invest 50:940–950.
33. Hagen TC, Lawrence AM, Kirsteins L (1972) Autoregulation of GH secretion in normal subjects. Metabolism 21:603–610.
34. Mendelson WB, Jacobs LS, Gillin JC (1983) Negative feedback suppression of sleep-related growth hormone secretion. J Clin Endocrinol Metab 56:486–488.
35. Sheppard MC, Kronheim S, Pimstone BL (1978) Stimulation by growth hormone of somatostatin release from the rat hypothalamus in vitro. Clin Endocrinol (Oxf) 9:583–586.
36. Hoffman DL, Baker BL (1977) Effects of treatment with growth hormone on somatostatin in the median eminence of hypophysectomized rats. Proc Soc Exp Biol Med 156:265–271.
37. Patel YC (1979) Growth hormone stimulates hypothalamic somatostatin. Life Sci 24:1589–1593.
38. Molitch ME, Hlivyak LE (1980) Growth hormone short-loop feedback: Anatomic specificity of growth hormone stimulation of hypothalamic somatostatin concentration. Horm Metabl Res 12:559–560.
39. Berelowitz M, Firestone SI, Frohman LA (1981) Effects of growth hormone excess and deficiency on hypothalamic somatostatin content and release and on tissue somatostatin distribution. Endocrinology 109:714–719.
40. Chihara K, Minamitani N, Kaji H et al (1981) Intraventricularly injected growth hormone stimulates somatostatin release into rat hypophysial portal blood. Endocrinology 109:2279–2286.
41. Berelowitz M, Szabo M, Frohman LA et al (1981) Somatomedin-C mediates growth hormone negative feedback by effects on both the hypothalamus and the pituitary. Science 212:1279–1281.
42. Brazeau P, Guillemin R, Ling N et al (1982) Inhibition by somatomedins of growth hormone secretion stimulated by hypothalamic growth hormone releasing factor (somatocrinin, GRF) on the synthetic peptide hpGRF. C R Acad Sci [D] (Paris) 295:651–654.
43. Goodyer CC, De Stephano L, Guyda HJ et al (1984) Effects of insulin-like growth factor on adult male rat pituitary function in tissue culture. Endocrinology 115:1568–1574.

44. Yamashita S, Melmed S (1986) Insulin-like growth factor I action on rat anterior pituitary cells: Suppression of growth hormone secretion and messenger ribonucleic acid levels. Endocrinology 118:176–182.
45. Melmed S, Yamashita S (1986) Insulin-like growth factor I action on hypothyroid rat anterior pituitary cells: Suppression of 3,5,3'-triiodothyronine-induced growth hormone secretion and messenger ribonucleic acid levels. Endocrinology 118:1483–1490.
46. Goodyer CG, Marcovitz S, Hardy J et al (1986) Effect of insulin-like growth factors on human foetal, adult normal and tumour pituitary function in tissue culture. Acta Endocrinol 112:49.
47. Yamashita S, Slanina S, Kado H et al (1986) Autoregulation of pituitary growth hormone messenger RNA levels in rats bearing transplantable mammosomatotrophic pituitary tumors. Endocrinology 118:915–8.
48. Ono M, Miki N, Miyoshi H (1985) Decreased content and release of hypothalamic GH-releasing factor in rats bearing GH producing GH_3 tumors. 67th Annual Meeting of the Endocrine Society, Baltimore, MD p 126 (Abstract).
49. Abe H, Molitch ME, Van Wyk JJ et al (1983) Human growth hormone and somatomedin-C suppresses the spontaneous release of growth hormone in unanesthetized rats. Endocrinology 113:1319–1324.
50. Tannenbaum GS, Guyda HJ, Posner BI (1983) Insulin-like growth factors: A role in growth hormone negative feedback and body weight regulation via brain. Science 220:77–79.
51. Peters MA, Lau EP, Snitman DL et al (1985) Expression of a biologically active analogue of somatomedin-C/insulin-like growth factor I. Gene 35:83–87.
52. Schalch D, Reisman D, Emler C et al (1984) Insulin-like growth factor I/somatomedin-C: Comparison of natural, solid phase synthetic and recombinant DNA analog peptides in two radio-ligand assays. Endocrinology 115:2490–2492.
53. Pertulla RV, Kohler PO, Frohman LA (1970) Effects of insulin on growth hormone induction by cortisol in rat pituitary tumor cells. Life Sci 9:805–813.
54. Melmed S, Neilson L, Slanina S (1985) Insulin suppresses rat growth hormone messenger ribonucleic acid in rat pituitary cells. Diabetes 34:409–412.
55. Yamashita S, Melmed S, (1986) Effects of insulin on rat anterior pituitary cells: Inhibition of GH secretion and mRNA levels. Diabetes 35:440–447.
56. Gick GG, Zeytin FN, Brazeau P et al (1984) Growth hormone-releasing factor regulates growth hormone mRNA in primary cultures of rat pituitary cells. Proc Natl Acad Sci USA 81:1553–1557.
57. Barinaga M, Yamamoto G, Rivier C et al (1983) Trancriptional regulation of growth hormone gene expression by growth hormone releasing factor. Nature 306:84–86.
58. Ceda GP, Davis RG, Rosenfeld RG et al (1987) Growth hormone releasing hormone-GH-somatomedin axis: Evidence for rapid inhibition of GHRH-elicited GH release by insulin-like growth factors I and II. Endocrinology (in press).
59. Yamashita S, Melmed S (1987) Insulin-like growth factor I regulation of growth hormone gene transcription in primary rat pituitary cells. J Clin Invest 79:449–482.

60. Yamashita S, Weiss M, Melmed S (1986) Insulin-like growth factor I regulates growth hormone secretion and messenger RNA levels in human pituitary tumor cells. J Clin Endocrinol Metab 63:730–735.
61. Binoux M, Hossenlop P, Lussarre C et al (1981) Production of insulin-like growth factors and their carrier by rat pituitary gland and brain explants in culture. FEBS Letts 124:178–184.
62. Fagin J, Melmed S (1987) Pituitary insulin-like growth factor I gene expression: In vitro and in vivo regulation by triiodothyronine and growth hormone. (In review).
63. Shupnick MA, Chin WW, Habener JF et al (1985) Transcriptional regulation of the thyrotropin subunit genes by thyroid hormone. J Biol Chem 260:2900–2905.
64. Eberwine JH, Roberts JL (1982) Glucocorticoid regulation of proopiomelanocortin gene transcription in the rat pituitary. J Biol Chem 259:2166–2170.
65. Clemmons DR, Van Wyk JJ (1984) Factors controlling blood concentration of somatomedin C. J Clin Endocrinol Metab 13:113–143.
66. Phillips LS, Vassilopoulou-Sellin R (1980) Somatomedins. N Engl J Med 302:371–380.
67. Yamashita S, Ong J, Melmed S (1987) Regulation of human GH gene expression by IGF-I in transfected cells. J Biol Chem 262:13254–13257.

Brain Adenosine and Purinergic Modulation of Central Nervous System Excitability

J.W. PHILLIS

Introduction

Evidence that adenosine can modify a variety of physiologic processes first appeared when Drury and Szent-Gyorgi (1) reported that the administration of adenosine to mammals caused a decrease in arterial blood pressure, dilatation of the coronary arteries, relaxation of the small intestine, reduction in locomotor activity, and induced sleep.

The possibility of purinergic neurotransmission in the central nervous system (CNS) was first considered in 1954, when Holton and Holton (2) suggested that adenosine triphosphate (ATP) could be the transmitter released at the central terminals of sensory afferent fibers. In the same year Feldberg and Sherwood (3) showed that administration of adenosine and ATP into the lateral cerebral ventricle could elicit muscular weakness, ataxia, and sleepiness in cats.

Intensive investigations into the role of adenosine in the control of cardiac, skeletal muscle, and eventually cerebral, blood flow have developed since the early sixties (4), and in 1972, Burnstock (5) published his purinergic nerve hypothesis, which postulated that ATP-releasing nerves constitute a third division of the autonomic nervous system (the "nonadrenergic, noncholinergic" division). Interest in the nucleoside, adenosine, was stimulated by the discovery that it could reduce the output of transmitter acetylcholine (ACh) from rat phrenic nerve endings. Shortly thereafter, adenosine and its nucleotides were shown to be remarkably active at depressing the spontaneous and evoked firing of neurons at many levels of the neural axis in both in vivo and vitro preparations (6–11).

The finding that adenosine and its nucleotides stimulated the formation of cyclic adenosine 3', 5'-monophosphate (cyclic AMP) in brain slices (12) was instrumental in developing interest in the neurochemical actions of these purines. The xanthine central stimulants, caffeine and theophylline, antagonize this effect of adenosine (12) and it was hypothesized that adenosine or ATP might regulate neuronal activity, and that the mild stimulatory actions of these methylxanthines are caused by antagonism of the effects of these purines.

More recently highly specific, high-affinity, binding sites for adenosine have been identified in the CNS and it is postulated that adenosine exerts many of its effects via an interaction with these extracellular receptors. In addition, adenosine and ATP release from a variety of in vivo and in vitro preparations have been demonstrated (13,14). In several instances, it has been possible to show that the release was calcium dependent.

To date, however, it has not been possible to identify conclusively adenosine or an adenine nucleotide as the transmitter at any central synapses, although there is evidence that ATP acts as an excitatory co-transmitter, together with norepinephrine, at certain peripheral autonomic synapses including those in the vas deferens (15). As adenosine has powerful depressant actions on the presynaptic release of norepinephrine from the vas deferens, it seems reasonable to suggest that ATP, released as an excitatory co-transmitter, provides a source of adenosine, which then controls or modulates the release of further excitatory transmitter from the presynaptic terminals. Extracellular formation from synaptically released ATP may provide one source of adenosine in the extracellular spaces of the CNS. Intracellular formation of adenosine as a result of ATP hydrolysis and its transport to the exterior of the cell by a facilitated diffusion mechanism may provide another route for adenosine to enter the extracellular space.

The primary intent of this chapter is to review some of the developments in the field of purinergic neurotransmission or neuromodulation in the CNS. For a further understanding of the potential roles of adenosine and its nucleotides, the reader can consult a number of detailed recent reviews and symposia proceedings (16–21).

Release of Adenosine and ATP

There is ample evidence to show that neuronal activity results in the release of adenosine, ATP, and other purines from central nervous tissues. In most of these studies, the released purines were generated from ^{3}H-labeled adenine or adenosine.

Adenosine release from brain slices was characterized by McIlwain and his colleagues. Electrical stimulation of cerebral slices or depolarization of synaptosomes with high potassium result in a large efflux of purines (22,23). The release of purine was found to be calcium-dependent. The bulk of the released material (50%) was identified as adenosine, whereas only about 8% was in the form of nucleotide. Similar results were obtained by Fredholm and Vernet (24) using electric stimulation or depolarization by high potassium or veratridine of hypothalamic synaptosomes. Release was, however, calcium-independent in this series of experiments.

As ATP is rapidly hydrolyzed extracellularly by ecto-ATPases and ecto 5'-nucleotidases (25,26) to form 5'-AMP, adenosine, inosine, and hypoxanthine, it can be debated whether the adenosine released from brain preparations by electric stimulation or elevated potassium actually reflects a release of ATP, which is subsequently degraded to adenosine. Fredholm and Hedqvist (16) suggest that purines are released primarily as nucleosides from isolated brain preparations. One of the observations upon which they base this conclusion is that, in hypothalamic synaptosomes, nucleotides accounted for only 6% of the total radioactive purine pool released by depolarization (24), but about 75% of the radioactivity released by a hypo-osmotic shock. A similar result was reported by Bender et al. (27). Adenosine constituted approximately 86 to 99% of the labeled purines released from rat brain synaptosomes during potassium depolarization, even though nucleotides accounted for 57 to 78% of the labeled material in the synaptosomes. Furthermore, in the presence of α, β-methylene ADP, an inhibitor of 5'-nucleotidase, stimulation-evoked release of adenine derivatives shows no change in the proportion of material released as adenosine (28). Similar observations were made by Pons et al. (29) using α, β-methylene ADP and guanosine 5'-monophosphate (GMP) to block the hydrolysis of ATP released during exposure of brain slices to veratridine. Veratridine-evoked release of labeled adenosine into perfusates of the cat caudate nucleus in vivo was unaffected by inhibition of ecto 5'-nucleotidase, suggesting that the released [³H]adenosine was not derived from labeled nucleotide (30).

To overcome some of the difficulties associated with the use of radiolabeled adenosine and adenine nucleotides in such studies (an assumption has to be made that the label has access to, and is incorporated into, all of the various metabolic and synaptic pools of adenosine and the adenine nucleotides and, furthermore, quantitation of the data is precluded because the specific activities of the released purines are not known), MacDonald and White (31) studied the release of endogenous adenosine from rat brain synaptosomes using an HPLC technique. In the presence of α, β-methylene ADP and GMP, the basal release of adenosine was reduced by 74%, while the release elicited by veratridine or potassium was reduced by 46% and 33%, respectively. Most of the basal accumulation of extrasynaptosomal adenosine, therefore, appears to be derived from nucleotide, probably ATP, but about half of the veratridine-evoked accumulation and most of that elicited by potassium may arise from adenosine released in its own right, rather than from a released nucleotide.

Despite the fact that adenosine constitutes a significant portion of the purine overflow from isolated brain preparations, ATP is clearly released in demonstrable amounts as well. Potter and White (32) used a luciferin–

luciferase system to measure the release of ATP from synaptosomal beds. Potassium-evoked release was calcium-dependent and unaffected by tetrodotoxin; veratridine-evoked release was blocked by tetrodotoxin and augmented in calcium-free media. A comparison of synaptosomes from different regions of the rat brain revealed that depolarization-induced release is not uniform throughout the brain. Potassium-evoked release was greatest in synaptosomes from the corpus striatum and cerebral cortex, and lowest in cerebellar preparations.

When purine release has been studied in the in situ brain, a somewhat different ratio of nucleotide/nucleoside release has been observed. Sulakhe and Phillis (33) studied the release of adenosine and its derivatives from the cat cerebral cortex after prelabeling with [^{3}H]adenosine. Adenosine, inosine, and hypoxanthine accounted for 55% of the material released from the unstimulated cortex, with adenine nucleotides contributing a further 33%. Direct electric stimulation of the cortex evoked a fivefold increase in the release of labeled adenosine and its derivatives with the proportion of the label incorporated into released nucleotide remaining constant. Studies with the in situ rat cerebral cortex indicate an even greater tendency for the label to be released in the nucleotide fraction. The bulk of the labeled material (50 to 70%) released spontaneously from the rat cerebral cortex after preincubation with [^{3}H]adenosine was associated with adenine nucleotides (34). Inosine accounted for most of the remaining labeled material (15 to 40%), with adenosine and hypoxanthine forming about 10% of total activity. A release of endogenous ATP from the in vivo rat cerebral cortex, which was increased by electric stimulation, has been demonstrated (35), using the luciferin–luciferase assay technique.

Comparable results were obtained in another study of the efflux of labeled purines from the in situ rat cerebral and cerebellar cortices (36). The pattern of spontaneous release from cerebral cortex was adenine 5%, adenosine 18%, inosine 25%, hypoxanthine 18% and adenine nucleotides 34% of total radioactivity. In the presence of an inhibitor of adenosine deaminase [erythro-9-(2-hydroxy-3-nonyl)adenine, EHNA] the proportion of material released as adenosine rose to about 33%, with a decline in the levels of inosine and hypoxanthine. Application of L-glutamate, L-aspartate, potassium, or veratridine to the cortex markedly stimulated the release of [^{3}H]adenosine derivatives from the cerebral cortex.

Adenosine release from the caudate nucleus has been studied using either a push-pull perfusion technique or brain dialysis. In the presence of [^{3}H]adenine, spontaneously released [^{3}H]adenosine was detectable in the superfusates of the caudate nucleus. Perfusion with elevated potassium, glutamate, or veratridine caused a marked and reversible increase in [^{3}H]adenosine release (30).

Recently a brain dialysis technique has been developed that allows

214   J.W. Phillis

direct sampling of cerebral interstitial fluid adenosine levels. Adenosine is released into dialysis fibers implanted in the rat caudate nucleus (37,38). Estimated basal interstitial fluid adenosine levels were 1 to 2 μM and could be increased fourfold by local potassium infusion. The results of experiments with the brain dialysis technique are particularly valuable in that they allow measurement of the actual concentrations of adenosine in extracellular fluid.

In concluding this section, it is worth emphasizing that the proportions of adenosine and the adenine nucleotides released by different CNS preparations will reflect both the metabolic status of the preparation and the nature of the releasing stimulus. Nucleoside and nucleotides may in fact be released from different pools; adenosine by a facilitated diffusion transport process which is only partially Ca^{2+}-dependent (27,39); the nucleotides by a Ca^{2+}-dependent exocytotic mechanism. A large proportion of the adenosine released from central tissues during potassium or veratridine depolarization probably arises from the utilization of ATP in a metabolic pool by membrane cation pumps and translocation of the adenosine to the extracellular space by the membrane transporter. ATP released during depolarization or synaptic activity may originate from a synaptic vesicle pool. The decreased release of nucleotides from in vitro slice or synaptosomal preparations may represent a consequence of their impaired metabolic status and depleted ATP stores (40).

In the absence of any direct evidence for the existence of adenosinergic or ATP-ergic neurons in the CNS, there has been considerable interest in the possibility of ATP release as a co-transmitter. ATP is known to be packaged in, and released from, cholinergic and noradrenergic nerve endings of the peripheral nervous system. As mentioned in the introduction, there is now convincing evidence that ATP acts as an excitatory co-transmitter at sympathetic nerve terminals (15,41). Silinsky (42) has demonstrated a calcium-dependent release of ATP from phrenic nerve preparations. The ATP is then hydrolyzed to adenosine by 5'-nucleotidase in order to produce its inhibitory effect on skeletal neuromuscular transmission (43).

ATP and adenosine may also be released from postsynaptic sites in some tissues. Israel et al. (44) have detected a postsynaptic release of endogenous ATP during individual discharges of the *Torpedo* electric organ. The release of labeled adenosine by frog paravertebral ganglia is also postsynaptic (45). Preganglionic stimulation enhanced ³H-purine release in a frequency dependent manner, but the release was blocked by curare plus atropine. A similar graded release of adenosine was induced by exposure of the ganglion to carbachol, and antidromic stimulation also caused a frequency-dependent release of ³H-purine, indicating that the release of adenosine was a result of depolarization of the postsynaptic neuron.

Inactivation of Released Purines

Once adenosine has been released into the extracellular space, or formed by the extracellular hydrolysis of released adenine nucleotides, it can be removed via uptake followed by resynthesis into adenine nucleotides or by conversion to less active metabolites such as inosine and hypoxanthine; or incorporation into S-adenosylhomocysteine.

High affinity uptake mechanisms for adenosine have been studied in several laboratories and have been the subject of a recent review (46). The K_m for adenosine uptake into rat cerebral cortical synaptosomes is approximately $1 \mu M$ (27). Since this approximates the normal extracellular concentrations of adenosine, it would appear that uptake plays a significant role in controlling the levels of adenosine in the interstitial space. Nucleotides are not substrates for the uptake system (47) and must first be converted to adenosine by ecto-ATPase and 5'-ectonucleotidases. Following its uptake, adenosine is converted to nucleotides by adenosine kinase. The proportion of nucleotides increases rapidly over time, with most of the adenosine ultimately appearing as nucleotide (48). Uptake into synaptosomes is inhibited by low temperature, reduced pH, and a variety of agents known to inhibit nucleoside transport in other tissues, including dipyridamole, hexobendine, dilazep, and nitrobenzylthioinosine. The transport process may be reversible, serving also to move adenosine out of the cell (49).

Adenosine accumulation has been described in several different tissue preparations including synaptosomes, brain slices, primary cultures of neurons and astrocytes, and glial cell lines (see 50 for references). Similarities of the adenosine uptake systems in astrocytes and neurons in primary cultures have been reported (50). Both cell types showed high-affinity uptake with K_m values of 6.5 μM in astrocytes and 6.1 μM in neurons. The intensity of uptake was about 60% higher in astrocytes than in neurons (V_{max} values of 0.16 and 0.105 nmol $\cdot$ min^{-1} $\cdot$ mg^{-1} protein, respectively).

The regional distribution of adenosine transport sites in rat brain has been studied using autoradiographic and receptor binding techniques and [3H] nitrobenzylthioinosine ([3H]NBI) or [3H]dipyridamole as ligand probes (51,52). The highest levels of [3H]NBI sites were found in the thalamus, followed by midbrain, superior colliculus, olfactory cortex, and hypothalamus. Cerebellum exhibited the lowest level of [3H]NBI binding.

The major enzyme that degrades adenosine is adenosine deaminase, which converts adenosine to inosine. It appears to be primarily a cytoplasmic enzyme with a K_m for rat cerebral cortical tissues of 54 to 57 μM (53). This relatively high K_m value suggests that adenosine deaminase may become important in situations in which intracellular adenosine levels are elevated, e.g., depolarization or hypoxia. Adenosine kinase is subject to potent substrate inhibition by relatively low adenosine concentrations (i.e., 10 μM) (54) and under these conditions deamination to inosine may become a significant factor in the disposal of adenosine.

Although adenosine deaminase is mainly localized in the cell cytosol, its presence as an ectoenzyme cannot be ruled out. The existence of an ectoadenosine deaminase would explain why the addition of adenosine deaminase inhibitors is often found to potentiate the effects of adenosine on central neurons and indeed, the potent behavioral effects of deaminase inhibitors (17,55–57).

A second deaminating enzyme involved in the metabolism of adenine nucleotides in brain is adenylate deaminase (58), which deaminates adenosine 5'-monophosphate (AMP) to inosine 5'-monophosphate (IMP). Inosine could then be generated by the degradation of 5'-IMP by 5'-nucleotidase. Adenylate deaminase has been identified as an ectoenzyme in skeletal muscle (59), suggesting that it may also exist as an ectoenzyme in the brain.

The properties of adenosine deaminase in the CNS of the rat have been the subject of intense scrutiny. The distribution of adenosine deaminase activity was measured by both immunohistochemical and neurochemical techniques. The K_m and V_{max} values for enzyme activity in whole brain homogenates were 47 μM and 107 nmol/mg protein/30 min (60). Among the 62 regions of the CNS examined, the highest activity was found in the posterior hypothalamic magnocellular nuclei and the lowest in the hippocampus. The uneven distribution of the adenosine deaminase activity, measured neurochemically, corresponded well with the immunohistochemical localization of the enzyme in discrete neuronal systems (61). However, the potential significance of the presence of adenosine deaminase in posterior hypothalamic magnocellular nuclei neurons as a predictor of their purinergic nature has been lessened by the discovery that many of these neurons also contain the transmitter synthesizing enzymes histidine decarboxylase (histamine), and glutamate decarboxylase (GABA); express monoamine oxidase activity; and display an ability to accumulate and decarboxylate 5-hydroxytryptophan (62). The validity of adenosine deaminase as a marker for purinergic transmission, therefore remains to be established.

Another potential mechanism for the removal of adenosine is its conversion into S-adenosylhomocysteine. In heart muscle, the enzyme S-adenosylhomocysteine synthetase is a cytoplasmic enzyme with a low K_m of 0.4 μM (63). McIlwain and Poll (64) have provided evidence that incorporation into S-adenosylhomocysteine is an important disposal route for adenosine formed in neocortical tissues depolarized by electric stimulation, glutamate, potassium, or hypoxia. Conversely, the production of adenosine from S-adenosylhomocysteine by the enzyme S-adenosylhomocysteine hydrolase may account for much of the adenosine production in normoxic tissues (63).

Binding of Ligands to Purinergic Receptors

Adenosine and Its Analogs

The search for suitable ligands to identify and characterize purinergic receptors by binding techniques was initiated during the 1970s and has achieved success with the identification of a number of labeled probes for a high-affinity receptor (A_1 site) and, with less certainty, for a lower affinity A_2 site.

Such efforts began with the use of [³H]adenosine (65–67). Binding sites with dissociation constants in the micromolar range and equivocal pharmacologic properties were described. High-affinity [³H]adenosine binding has been reported in NG108CC15 cells and brain tissue by Snell and Snell (68). In general, however, adenosine has not been favored as a ligand for either A_1 or A_2 receptors.

A source of complication with the use of labeled adenosine was the presence of endogenous adenosine, and possible degradation of the labeled probe by adenosine deaminase. An alternative approach, widely employed by other investigators, has been the use of radioligands which are resistant to deamination by adenosine deaminase, permitting the investigator to treat the membranes with adenosine deaminase to eliminate endogenous adenosine (69–71).

A number of tritiated ligands for adenosine A_1 receptors have now been extensively used, including N^6-[³H]cyclohexyladenosine ([³H]CHA); (R)-N^6-[³H]phenylisopropyladenosine (R-[³H]PIA), 2-[³H]chloroadenosine ([³H]CADO), and the adenosine antagonist 1,3-diethyl-8-phenylxanthine ([³H]DPX). The affinities of the binding sites are in the low nanomolar range and the relative potencies of adenosine analogs in displacing binding are consistent with interactions primarily at an A_1 receptor.

Sulfhydryl reducing and alkylating agents inhibit [³H]CHA binding, indicating that the binding site is a sulfhydryl-dependent protein (72). This property has been used to identify A_2 adenosine receptors in the rat striatum by inactivating A_1 receptors with N-ethylmaleimide (NEM) and using the potent A_2 agonist 5'-N-ethylcarboxamide [³H]-adenosine ([³H]NECA) as a ligand (73). In an alternative strategy, A_2 receptor binding can be studied in tissues exposed to 100 nM unlabeled CHA to occupy the A_1 receptors (74). In hippocampal membranes, which are apparently devoid of A_2 receptors, NECA binds with a K_D of 1.2 nM (73). NECA binds to rat striatal membranes at both A_1 and A_2 receptors. By using N-ethylmaleimide to inactivate the A_1 but not A_2 receptors, A_2 receptor binding can be studied selectively and a K_D of 17 nM has been observed for A_2 receptors in striatal membranes. An analysis of NECA binding in hippocampal (A_1 receptors) membranes and NEM-treated striatal membranes (A_2 receptors) shows that NECA is more active at A_2

sites than are (R)-PIA and CHA, whereas at A_1 sites, (R)-PIA and CHA are considerably more active than the carboxamide.

A_1 receptors labeled with [³H]CHA and [³H]DPX are modulated by guanine nucleotides and divalent cations (72,75). At micromolar concentrations, GTP decreases binding of the agonist [³H]CHA but not the antagonist [³H]DPX. However, GTP decreases the potency of agonists in competing for [³H]DPX binding. Removal of endogenous divalent cations with chelating agents inhibits [³H]CHA but not [³H]DPX binding, suggesting that endogenous divalent cations regulate agonist affinity at adenosine receptors. The action of guanine nucleotides and divalent cations on receptor binding are thought to reflect interactions with the GTP binding of "N" proteins that couple the receptor to membrane effector systems.

2′,5-Dideoxyadenosine (DDA) is an adenosine analog which is specific for an inhibitory intracellular adenosine receptor associated with the enzyme adenylate cyclase (the P site). DDA binds to rat cortical and striatal membranes with a dissociation constant of about 90 nM (76). Binding was insensitive to agonists and antagonists for A sites, including CADO, (R)-PIA, and theophylline, but was displaced by adenosine and 5′-methylthioadenosine. The nature of the DDA binding site remains unclear; the investigators raised the possibility that it may correspond to the binding site for adenosine involved in the uptake of adenosine.

Last, but not least, mention should be made of a novel class of micromolar affinity adenosine receptors characterized by [³H]CADO binding (77). Specific [³H]CADO binding was reversible and saturable with an apparent K_D of 9.1 μM and a B_{max} of 61 pmoles/mg protein. The membranes were not treated with adenosine deaminase. Specific [³H]CADO binding under these conditions was insensitive to (R)-PIA and CHA and to treatment with NEM. Specific binding was antagonized by theophylline and caffeine. The micromolar [³H]CADO binding site was observed in all brain regions with the highest levels of specific binding in the hippocampus and striatum. Although the relationship of this binding site to the pharmacologic actions of adenosine remains to be established, it is disturbing that (R)-PIA and CHA, which are potent depressants of neuronal firing and synaptically evoked activity (see Section VII. C), were inactive in displaying [³H]CADO binding.

Adenosine Triphosphate

Based on pharmacologic studies, Burnstock (78) proposed the existence of two types of purinergic receptor on the cell surface: a P_1 receptor, responsive to adenosine and blocked by the methylxanthines, and a P_2 receptor, which would respond to ATP, but not adenosine. An attempt has been made to use [³H]adenyl-5′-imidodiphosphate (AppNHp) as a ligand for a P_2 site in rat brain membranes (79). Although a K_D in the nanomolar range was observed, the binding of AppNHp was irreversible.

Localization of Adenosine Receptors

Autoradiographic techniques have permitted a microscopic localization of adenosine receptors in rat CNS (80–81). Results obtained from these studies indicate A_1-receptors to be heterogeneously distributed in the rat central nervous system. The cerebellum, thalamic nuclei, medial geniculate, hippocampus, superior colliculus, cerebral cortex, and caudate nucleus are areas relatively rich in [^{3}H]CHA receptors. Very low [^{3}H]CHA binding regions include the hypothalamus and fiber tracts. Binding in the cerebellar cortex is primarily localized in the molecular layer. In the cerebral cortex, adenosine receptors occur with considerable density in layers I, IV, VI, with lower levels in layers II, III, and V. Receptor density was high in the hippocampal molecular and polymorphic layers, and low in the pyramidal layer.

The distribution and ontogenesis of adenosine binding sites in the cat visual cortex have been studied autoradiographically using [^{3}H]CHA and [^{3}H]NECA as ligands (82). Binding sites for both ligands are concentrated in layers I to III and upper layer V in the visual cortex of adult cats. These laminar patterns change during postnatal development, showing the densest binding in the deep cortical layers in kittens younger than 30 days of age and a fairly homogeneous distribution in older kittens before achieving the adult distribution.

To localize further the adenosine receptors in the molecular layer of the cerebellum, Goodman et al. (83) made use of mutant mice that lack specific nerve types. "Weaver" mice, which lack 80% of the granule cells, also lack 70 to 80% of adenosine receptors, indicating that the adenosine receptors in the molecular layer are probably associated with parallel fibers of granule cells. This conclusion is supported by experiments with "Reeler" mice in which the neural migration of granule cells does not occur, so that granule cells and their axons remain in the external granular layer. In these mice the adenosine receptors are also restricted to the external granular layer. By contrast, the pattern and density of adenosine receptors in the cerebellar cortex of "nervous" mice which have a 90% decline of Purkinje cells was the same as those in litter mate controls. Accordingly, it is apparent that A_1 adenosine receptors in the cerebellum are localized to granule cells, especially their axons and terminals in the molecular layer; this conclusion is consistent with the neurophysiologic finding that adenosine blocks the synaptic effects of parallel, but not of climbing fibers on Purkinje cells of the cerebellar cortex (84).

In Vivo Regulation of Adenosine Receptor Density

The effects of chronic administration of caffeine or theophylline (adenosine antagonists) on adenosine receptors have been studied by several groups. Caffeine consumption of 50 to 100 mg/kg/day produced a significant elevation of mouse whole brain adenosine receptors labelled with [^{3}H]CHA and [^{3}H]DPX and a transient increase in benzodiazepine receptors (85). Similar increases in adenosine receptors have been observed in rat cortical membranes after chronic caffeine or theophylline administration (86–88). The largest increases in B_{max} were observed in rats administered 75 mg/kg of caffeine daily for 12 days (87). (R)-[^{3}H]PIA binding was up-regulated by 123% and even 30 days following the termination of caffeine administration, (R)-[^{3}H]PIA binding was still elevated (87%). The up-regulation of adenosine receptors during caffeine administration can therefore be of considerable duration. Two binding sites for [^{3}H]CHA were identified in rat cerebral cortical membranes with K_D values of 1.3 nM (88). Chronic caffeine treatment (100 mg/kg for 30 days) increased the B_{max} of the high-affinity receptor by 64% but did not change the B_{max} of the lower affinity site. The lower affinity [^{3}H]CHA binding site disappears in aged rats (89), which may explain the reported loss of the inhibitory effects of adenosine on acetylcholine release from cortical slices of aged rats (90). A down-regulation of hippocampal adenosine receptors following transient ischemia occurs in the CA$_1$ region of the hippocampus (91,92), which may be associated with injury to pyramidal neurons.

Adenosine Actions at the Cellular Level

Regulation of Adenylate Cyclase

Adenosine is an effective regulator of cAMP formation in brain tissues acting at membrane localized receptors (Table 7.1). Adenosine has been shown to affect adenylate cyclase via two extracellular receptors, one a high affinity inhibitory site (A$_1$, Ri), the other a lower affinity stimulatory site (A$_2$, Ra), and a low-affinity intracellular regulatory site (P site). Both classes of extracellular receptor occur in brain and both are antagonized by caffeine and theophylline.

TABLE 7.1 Actions of adenosine at the cellular level.

Inhibition (A$_1$) or stimulation (A$_2$) of adenylate cyclase
Depression of transmitter release
Inhibition of synaptic transmission
Reduction of Ca^{2+} channel conductance
Increase in K$^+$ channel conductance

The properties of the A_2 receptor have been explored in brain slices, homogenates, and cultured neurons. While the A_2 receptor sensitive cAMP-generating system occurs in slices from all brain regions [EC_{50} (concentration for 50% of maximal stimulation of cAMP formation by agent) for adenosine 10 to 20 μM], an A_2-adenosine receptor-mediated activation of adenylate cyclase has been detected only in membranes of striatum and other limbic regions (93). The reason that the ubiquitous A_2-receptor sensitive cyclic AMP system of brain slices is not generally detectable in brain membranes is not apparent, but it may be that striatal membranes contain a unique cAMP system. It has been suggested (94) that the slice and membrane responses represent activation of two different classes of adenosine receptors and that the relatively high affinity (EC_{50} 0.5 μM) A_2 system in striatal membranes makes little contribution to the A_2 receptor-mediated accumulation of cyclic AMP in striatal slices.

The existence of an A_1 inhibitory site linked to adenylate cyclase is a relatively recent discovery (95,96). The A_1 system can be detected in brain membranes (97), but it has proven difficult to demonstrate in brain slices. Fredholm et al. (98) have reported adenosine receptor-mediated inhibition of the cyclic AMP accumulation in hippocampal slices elicited by low concentrations of forskolin in the presence of a phosphodiesterase inhibitor, rolipram.

The two types of A-receptors differ markedly in their pharmacologic characteristics. The A_1 receptor is more sensitive to (R)-PIA than to NECA, which is more active than (S)-PIA. In contrast, the A_2 receptor shows greater sensitivity to NECA, and the stereospecificity of the PIA isomers is not as pronounced (Table 7.2). While modulation of adenylate cyclase by both receptors requires GTP, GppNHp, a nondegradable analog of GTP, can mimic the action of GTP in promoting the stimulation of adenylate cyclase but does not promote the inhibition of cyclase by the A_1 receptor.

The P site is located intracellularly on the catalytic subunit of adenylate cyclase. It can be activated by adenosine analogs such as 2',5'-dideoxyadenosine and xylofuranosyladenosine but not by PIA or CADO. Nor is it antagonized by the methylxanthines. There are indications that the P site can play a role in the pharmacologic actions of adenosine (99).

TABLE 7.2 Pharmacological criteria for differentiating P_1 (A_1 and A_2 adenosine) and P_2 (ATP) receptors.

	Agonist potencies	Antagonists
P_1	ADO > AMP > ADP > ATP	
A_1	R-PIA, CHA > CADO > S-PIA, NECA	Methylxanthines
A_2	NECA > CADO > R-PIA, CHA > S-PIA	Methylxanthines
P_2	ATP > ADP > AMP > ADO	2,2'-Pyridilisatogen
		Arylazido Aminopropionyl ATP

Effects of Adenosine on Transmitter Release

Adenosine and the adenine nucleotides inhibit transmitter release from a variety of peripheral nerve terminals. The actions of adenosine on sympathetic, parasympathetic, and skeletal muscle nerve terminals have been reviewed in detail elsewhere (16,100) and the present discussion is largely restricted to the action of adenosine on central nerve terminals.

There is ample evidence that adenosine, in concentrations similar to those found in the extracellular fluid, depresses the release of neurotransmitters from brain tissue. This was first demonstrated in studies on acetylcholine release from intact and isolated cerebral cortical tissue (101,102). Theophylline and caffeine antagonize these actions of adenosine and enhance in situ acetylcholine release from the cerebral cortex (101,103). Adenosine depressed the K^+-induced release of labeled norepinephrine from cerebral cortical slices (104), and (R)-PIA depressed its release from hippocampal slices (105). Release of dopamine, serotonin, and GABA from striatal preparations, and of GABA from cerebral cortical slices was depressed by adenosine (106,107). The release of glutamate, but not of GABA, from slices of rat dentate gyrus was inhibited by the adenosine agonist CADO (108).

In some instances, adenosine analogs appear to be able to have both inhibitory and facilitatory effects on transmitter release. The A_1 agonist CHA has been reported to inhibit acetylcholine release from rat cortical slices, whereas the A_2 agonist NECA-stimulated release (109). Both effects were blocked by aminophylline. Calcium-dependent K^+-evoked release of dopamine from guinea pig striatal vesicular preparations was inhibited by CADO and facilitated by (R)-PIA and adenosine 5'-cyclopropylcarboxamide (CPCA) (110). This latter observation indicates that the sites of action of adenosine analogs on transmitter release do not necessarily appear to have the structural requirements expected of A_1 (high-affinity) or A_2 (low-affinity) adenosine receptors.

Adenosine depresses the spontaneous and evoked acetylcholine release at skeletal neuromuscular junctions (110). Various mechanisms have been considered to explain the inhibitory action of adenosine on transmitter release at the motor nerve terminal. Evidence has accumulated that adenosine interferes with the entry of calcium into the nerve terminal and depresses excitation-secretion coupling. Experiments on brain synaptosomes have demonstrated a reduced influx of $^{45}Ca^{2+}$ into K^+-depolarized brain synaptosomes in the presence of adenosine derivatives (111,112). Similar results were observed in cholinergic synaptosomes of *Torpedo* electric organ (113). The effects of adenosine on acetylcholine release and calcium uptake were also examined in a synaptosomal fraction prepared from guinea pig ileum myenteric plexus (114). Adenosine reduced $^{45}Ca^{2+}$ uptake in both potassium and electrically stimulated synaptosomes. The effect of adenosine was much more pronounced in the electrically stimulated preparations, with a complete abolition of $^{45}CA^{2+}$ uptake. The adenosine evoked inhibition of [^{3}H]ACh release was also more pronounced in the electrically stimulated synaptosomes than in potassium-depolarized ones. Shinozuka et al. (114) suggest that the depolarization and calcium influx in the presence of high potassium could be so intense as to partially mask the inhibitory action of adenosine.

The proposal that adenosine depresses transmitter output by suppressing the influx of Ca^{2+} into depolarized nerve terminals is further supported by electrophysiologic evidence. Adenosine depresses Ca^{2+}-dependent potentials in sympathetic ganglia (115), dorsal root ganglion cells (116,117), patch-clamped neuroblastoma cells (118), and hippocampal pyramidal cells (119), although Halliwell and Schofield (120) have suggested that the reduction in Ca^{2+}-dependent action potentials in hippocampal neurons may have been secondary to an increase in K^+-conductance.

The role of cyclic AMP in the purinergic effects on membrane Ca^{2+} fluxes remains controversial. The initial proposal was that adenosine, by raising cyclic AMP levels in the nerve terminal, would reduce transmitter output (121). This concept is difficult to reconcile with more recent evidence that suggests that if cyclic AMP has an effect, it is to facilitate transmitter release at nerve terminals (122).

An inhibitory effect of adenosine receptor activation on adenylate cyclase, mediated by the regulatory GTP-binding N_i protein, has been considered as the mechanism of purinergic depression of transmitter release. Pertussis toxin, which inactivates the N_i protein, antagonized the inhibitory effect of adenosine receptor activation on glutamate release from cultured neurons (123). N-Ethylmaleimide had similar effects on (R)-PIA-evoked inhibition of norepinephrine release from hippocampal slices (124). It is still unclear whether the N-protein involved in the prejunctional effects of adenosine is the N_i, or perhaps the ubiquitous GTP-binding N_o protein, that is also a substrate for pertussis toxin or NEM. Alternatively the N_i-protein may be coupled to effectors other than adenylate cyclase.

Electrophysiologic Action of Adenosine

Initial studies on the electrophysiologic actions of locally applied adenosine demonstrated that this purine had a marked depressant action on the spontaneous firing of neurons in virtually all brain regions tested (6,17). The effects of adenosine on cell firing were enhanced by adenosine uptake inhibitors or by inhibitors of adenosine deaminase, and antagonized by the methylxanthine adenosine antagonists. Caffeine and theophylline enhanced neuronal activity, presumably by antagonism of endogenous adenosine. Depression of synaptically evoked potentials by adenosine has been observed in slice preparation from several brain regions (see 17,21 for references). In both in vivo and in vitro brain preparations, decreases in spontaneous firing and of synaptically evoked excitatory potential amplitude were accompanied by membrane hyperpolarizations with no significant change in membrane resistance, or in the ability of the membrane to generate action potentials (17,21,125).

In addition to effects on excitatory synaptic transmission, there have been reports that adenosine can antagonize transmission in inhibitory circuits (21,125). Whether this action represents depression of the release of inhibitory transmitter or inhibition of the synaptic excitation of the inhibitory interneurons remains unclear.

Excitatory actions of iontophoretically applied ATP on central neurons have also been described (126–129). The physiologic significance of these excitations has been debated, as they can be mimicked by other purine and pyrimidine polyphosphates, and may reflect a Ca^{2+} chelating ability.

The possibility that ATP acts as a synaptic transmitter at the central terminals of primary fibers (2) has been examined by applying it iontophoretically to cat spinal dorsal horn neurons. ATP selectively excited a subset of non-nociceptive neurons (130,131). The finding that ATP excited only a particular group of neurons is consistent with the observation that pressure-applied ATP caused a rapid and marked depolarization of a subset of cultured dorsal horn neurons (132). The remaining neurons in the culture were unaffected by ATP, even at high concentrations ($10^{-4}M$). The disodium, magnesium, and calcium salts of ATP were equally effective in depolarizing dorsal horn neurons. Furthermore, the chelators EDTA and inorganic pyrophosphate were without effect, implying that the action of ATP was unlikely to have been mediated by chelation of divalent cations. Marked depolarizations were observed only with ATP, adenosine tetraphosphate, and the slowly hydrolyzable ATP analogs AppNHp and β,γ-methylene ATP. 8-Phenyltheophylline did not antagonize the effects of ATP, providing further evidence that P_2 rather than P_1 (adenosine) receptors were involved.

The issue of pre- versus postsynaptic actions of adenosine, and the extent to which they contribute to the depression of synaptic transmission remains unresolved at this time. Although there is considerable evidence that adenosine acts presynaptically to decrease transmitter release, the possibility that the decrease in excitatory postsynaptic potential (EPSP) amplitude by adenosine might be primarily a postsynaptic event cannot be ignored.

The finding of adenosine-induced membrane hyperpolarizations, seen in conjunction with small reductions of membrane resistance, provided indications that adenosine could increase membrane potassium conductances (133–136). These effects persisted even when transmitter release in the slices was blocked by perfusion with low Ca^{2+}-high Mg^{2+} solutions (136). The postsynaptic effects of adenosine are, however, observed at higher concentrations of adenosine than those required to depress EPSP amplitude (133). Yet another action of adenosine, namely prolongation of a calcium-dependent long duration afterhyperpolarization has been described (137), which could account for a reduction in cell firing after exposure to adenosine. The mechanism responsible for the change in the time course of the afterhyperpolarization was both calcium- and voltage-insensitive and may reflect a slower rate of return to baseline of intracellular calcium concentrations following neuronal activity.

The difficulties in separating pre- versus postsynaptic sites of action of adenosine are compounded by the fact that the small diameter of presynaptic nerve terminals makes it very difficult to undertake intracellular recordings. Hence, electrophysiologists have resorted to studying the actions of adenosine on dorsal root ganglion cells (116,117) as surrogates for their nerve terminals. As already mentioned, adenosine inhibits Ca^{2+}-dependent currents in these neurons. A more direct measurement of the differential effects of adenosine on pre- and postsynaptic calcium fluxes in hippocampal slices has been obtained by measuring the changes in extracellular Ca^{2+}-concentrations ($[Ca^{2+}]_o$) with ion-sensitive microelectrodes (138). After synaptic transmission had been blocked by reducing extracellular Ca^{2+}, orthodromic stimulation of hippocampal region CA1 afferents elicited a marked decrease in extracellular Ca^{2+}, attributable to Ca^{2+} entry into activated axon terminals. This decrease was significantly depressed by adenosine. 4-Aminopyridine (4-AP), which as been proposed as an adenosine antagonist, enhanced the decreases in extracellular Ca^{2+} during orthodromic stimulation. Adenosine also depressed these 4-AP enhanced Ca^{2+} signals. In the presence of 4-AP, the orthodromically elicited decrease in $[Ca^{2+}]_o$ was considerably increased when endogenously released adenosine was removed by the addition of adenosine deaminase to the perfusion fluid. Antidromic stimulation of hippocampal pyramidal cells also evoked reductions in $[Ca^{2+}]_o$, which were reduced by adenosine, but to a smaller degree than the presynaptic fluxes. These experiments offer rather convincing evidence that the depressant effects of adenosine on synaptic transmission are exerted through a reduction in Ca^{2+} entry into presynaptic nerve terminals, and that this represents the primary site of action in depressing synaptic transmission.

Other indications of the lack of postsynaptic effects of adenosine at concentrations which depress synaptic transmission can be observed with agents that directly excite the postsynaptic neuron. Thus, cholinergic synaptic excitation of spinal Renshaw cells can be depressed by adenosine, even though responses to locally applied ACh are unaffected (139). Studies with the excitant amino acid, L-glutamate, have yielded similar conclusions. Responses to locally applied L-glutamate are very resistant to adenosine. This is apparent in Fig. 7.1 where adenosine, in amounts sufficient to depress spontaneous firing, apparently enhanced the response to glutamate. The enhancement was more likely a result of the

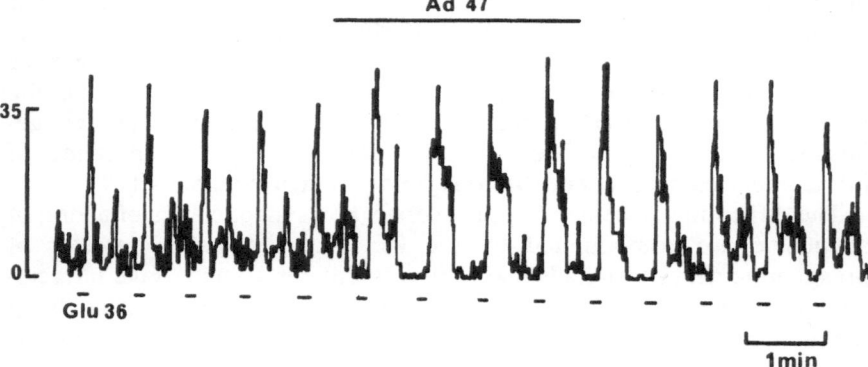

FIGURE 7.1. Firing frequency record of a spontaneously firing rat cerebral cortical neuron. This is a rate meter recording with the number of action potentials on the ordinate scale. Horizontal bars above and below the recording indicate periods of drug application. L-Glutamate (Glu, 36 nA) applied repetitively caused periodic bursts of firing. During the application of adenosine (Ad, 47 nA) spontaneous firing was greatly reduced, whilst the glutamate-induced excitations were enhanced.

reduction in spontaneous firing rather than of any potentiative effect of adenosine on L-glutamate excitation.

It has been suggested (140) that adenosine might modulate synaptic transmission via two distinct mechanisms, a presynaptic A_1-receptor-mediated depression of transmitter release and a postsynaptic A_2-receptor-mediated hyperpolarization. We have attempted to investigate this suggestion by comparing the actions of adenosine, which has a greater potency at A_1 receptors and NECA, which is approximately equipotent at A_1 and A_2 receptors, on glutamate-evoked firing of cerebral cortical neurons. As is evident in Fig. 7.2, in amounts that were sufficient to reduce the spontaneous background firing, neither NECA nor adenosine had an effect on glutamate-evoked firing. These results reinforce the conclusion that adenosine's effects are primarily exerted at the presynaptic terminal and do little to support the suggestion of an A_2-mediated postsynaptic hyperpolarization of cerebral cortical neurons.

An attempt was made to investigate the involvement of GTP-binding proteins in the inhibitory effects of adenosine on neuronal firing. Unfortunately, pertussis toxin, while it is much more selective than NEM, has the drawback that if it is active at all with intact cells, it requires a long time to inactivate the N-proteins (123). NEM acts rapidly (within 5 min at 100 μM), to block N_i proteins (124). Locally applied NEM potentiated, rather than antagonized, the effects of iontophoretically applied adenosine (Fig. 7.3). This outcome was undoubtedly a result of NEM's ability to competitively inhibit adenosine uptake (K_1, 180μM, 47).

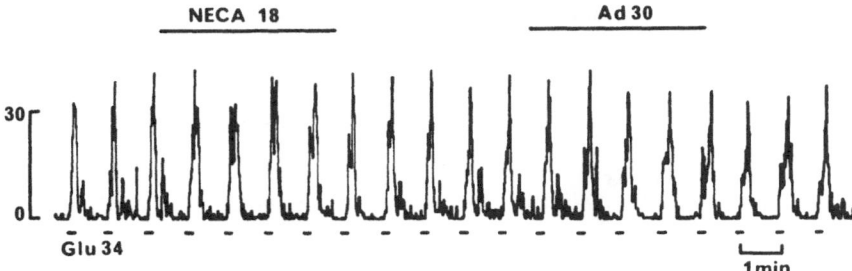

FIGURE 7.2. 5'-N-ethylcarboxamidoadenosine (NECA, 18 nA) and adenosine (Ad, 30 nA) both depressed the spontaneous firing of this rat cerebral cortical neuron without affecting the excitatory responses to L-glutamate (Glu, 34 nA).

Behavioral Actions of Adenosine

When administered by parenteral injection, adenosine and its analogs have been shown to have a variety of behavioral actions including inhibition of spontaneous motor activity, motor incoordination, analgesia, hypnotic activity, changes in respiration, anticonvulsant activity, and alterations in food intake (Table 7.3). Adenosine also influences numerous peripheral systems. Its cardiovascular actions result in profound falls in blood pressure; it causes bronchial constriction, hypothermia, affects renal blood flow and renin release, and depresses acetylcholine and norepinephrine release at autonomic nervous system terminals.

A problem arising in many of the behavioral studies to date has been that, following peripheral administration, it has been difficult to determine whether the behavioral responses are of central origin or secondary to the

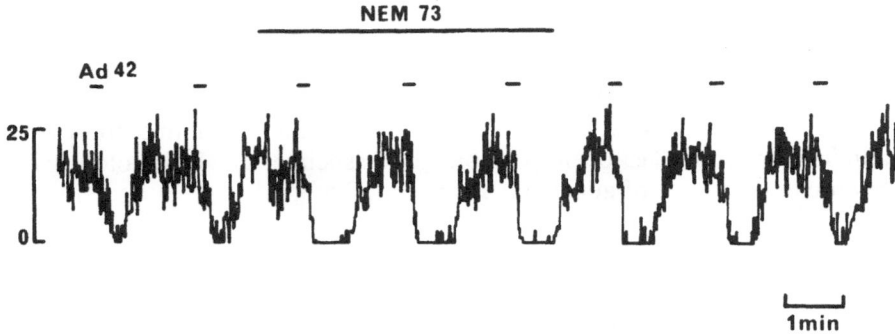

FIGURE 7.3. N-Ethylmaleimide (NEM, 73 nA) potentiated the depressant action of adenosine (Ad, 42 nA) on the spontaneous firing of the rat cerebral cortical neuron.

TABLE 7.3 Behavior actions of adenosine.

Depression of locomotor activity
Decreased schedule-controlled responding
Antinociceptive
Sedative/hypnotic
Anticonvulsant
Suppression of food intake
Hypotensive
Hypothermic
Stimulation of respiration

decreases in blood pressure and/or body temperature. The cardiovascular actions of adenosine analogs were already emphasized in 1929 by Drury and Szent-Györgi (1) and the hypothermic effect was first recognized by Bennet and Drury (141). It is now known that peripherally administered adenosine can also elicit an urge to breathe deeply and feelings of suffocation in man (142) which, in animals, could affect behavioral responses.

Recent studies by Dunwiddie and his colleagues have shown that (R)-PIA and NECA probably do not cross the intact blood–brain barrier in the hippocampus in pharmacologically active amounts (143). These results suggest that many of the reported behavioral actions that have been ascribed to peripherally administered adenosine analogs may have been mediated indirectly via actions at peripheral sites, or in regions of the brain where the blood–brain barrier is more permeable, such as the area postrema. The dangers inherent in overlooking the peripheral effects of adenosine in any assessment of its purported central actions are exemplified in a recent report that maintenance of body temperature of mice given intraperitoneal CADO or NECA can abolish the anticonvulsant actions of these compounds (144). The hypotensive and hypothermic actions of adenosine are likely to be caused primarily by peripheral actions, with inhibition of central compensatory mechanisms as an additional factor (145,146).

With these factors in mind, the subsequent account will emphasize the results obtained with direct or intracerebroventricular (icvt) administration of adenosine and its analogs into the brain. With this route of administration it has been possible to dissociate the locomotor and hypotensive effects of adenosine (147).

Effects on Locomotor Activity

Drury and Szent-Györgi (1) were the first to describe locomotor depressant and hypnotic actions, accompanied by respiratory stimulation following peripheral (subcutaneous) administration of adenosine in unanesthetized rabbits. These investigators clearly foresaw the dilemma that still confronts us today by declaring that "Whether the low blood pressure, the heart block, and the loss of intestinal movement are sufficient to bring about the observed condition or whether an action on the nervous system must be considered we are unable to state."

A decrease in spontaneous locomotor activity following icvt injections of adenosine and ATP was noted in early experiments on cats (3) and rats (148). In an extensive series of experiments on icvt administration of adenosine analogs in mice, NECA was found to be the most potent depressant of locomotor activity with a number of N^6-substituted analogs also displaying considerable activity. Depression of locomotor activity by NECA and (R) -PIA was antagonized by caffeine (57). Evidence for a neuromodulatory role of endogenously released adenosine in the brain was sought in experiments with inhibitors of adenosine transport and adenosine deaminase. Three potent transport inhibitors, dipyridamole, dilazep, and papaverine inhibited locomotor activity when injected icvt. The effects of dilazep and papaverine, but not of dipyridamole, were antagonized by caffeine. EHNA was used as an inhibitor of adenosine deaminase. This compound also depressed locomotor activity and its actions were antagonized by caffeine.

A paradoxic effect was observed when caffeine administration preceeded an icvt injection of adenosine, in that low doses of adenosine now stimulated locomotor activity (57). In contrast, when mice were given NECA after caffeine, there were no dose combinations which produced stimulation. The observations with adenosine find a parallel in an earlier study reporting that, after pretreatment with caffeine, peripherally administered (R-)PIA in doses which previously depressed activity, now caused a pronounced locomotor stimulation (149). Low doses (0.005 to 0.07 mg/kg) of (R-)PIA can stimulate locomotor activity, even in the absence of caffeine (150).

Somewhat unexpected findings were also observed with isobutyl-methylxanthine (IBMX). This substance depressed locomotor activity in mice when administered icvt (57), even though it is an adenosine antagonist. Indeed IBMX was able to antagonize the locomotor depressant actions of adenosine. Its failure to stimulate locomotor activity may be related to its phosphodiesterase (PDE) inhibiting properties, as a variety of PDE inhibitors can cause behavorial depression in mice (151). Support for this explanation comes from experiments with forskolin, which stimulates the enzyme adenylate cyclase, and enhances cyclic AMP levels, and depresses locomotor activity when injected icvt (152) or intraperitoneally (151).

Another type of motor activity affected by adenosine is the rotational behavior associated with modulation of striatal dopaminergic function. Local applications of NECA into the striatum induced ipsilateral rotation of rats following systemic apomorphine injection (153). (R)-PIA was less effective than NECA, whereas DDA antagonized the response to NECA, as did theophylline. Rotation behavior induced by apomorphine in rats with unilateral lesions of the nigro-striatal pathway was inhibited by (R)-PIA, by the PDE inhibitor rolipram, and by EHNA (154). By contrast theophylline and 8-phenyltheophylline caused a potentiation of rotation behavior, presumably by antagonizing endogenous adenosine.

Peripherally administered analogs of adenosine decrease schedule-controlled behavior of rats and squirrel monkeys (155–157). When the behavioral effects of a series of metabolically stable analogs of adenosine were studied in squirrel monkeys responding under a fixed interval schedule of stimulus-shock termination, all of the drugs depressed the responding rate. NECA and CADO were as potent as, or more effective than, (R)-PIA, CHA, and cyclopentyladenosine, suggesting an action mediated by A_2 receptors.

Tolerance develops to the decreasing effects of administered (R)-PIA on response rate of animals responding on a fixed-ratio schedule (158). Tolerance to (R)-PIA conferred tolerance to CHA, but no cross-tolerance was observed for the rate-suppressing effects of 4, 5, 6, 7-tetrahydro-isoxazolo [5, 4-c] pyridin-3-ol (THIP; a GABA-mimetic), diazepam, pentobarbital, ketamine, clonidine, amphetamine, and caffeine. The lack of cross-tolerance to THIP, diazepam, ketamine, and pentobarbital is interesting in that these agents are effective sedative-hypnotics; producing similar gross behavioral signs to (R)-PIA. While the absence of cross-tolerance suggests that these agents are not acting via central purinergic mechanisms, a final interpretation of this data will have to be withheld until it is convincingly established that the behavioral effects of peripherally administered adenosine analogs are mediated via central (and not peripheral) receptors. If tolerance to the actions of centrally administered adenosine (and analogs) can be demonstrated; cross-tolerance studies would appear to offer exciting possibilities for the evaluation of an involvement of central adenosine receptors in the action of other sedative hypnotic agents.

The development of tolerance to caffeine has been evaluated in rodent locomotor paradigms. After a 2-week exposure of rats to caffeine, via their drinking water, the stimulatory actions of caffeine on mesencephalic reticular neurons was abolished and locomotor stimulation was attenuated (159). This was accompanied by an increase in the number of CHA binding sites without any change in receptor affinity. Similar results were observed in mice after chronic administration of caffeine. The dose response curves of (R)-PIA for both analgesia and locomotor depression were shifted to the left and the B_{max} values for (R)-PIA and diethylphenylxanthine were increased without any alterations in affinity (160). By contrast, in mice given (R)-PIA chronically for 14 days, the dose–response curves for both analgesia and locomotor depression were shifted to the right. The dose–response curve for the locomotor effects of caffeine was shifted to the left, and caffeine exhibited greater antagonist activity against the analgesic action of *(R)*-PIA. Interestingly, following *(R)*-PIA administration, there was no change in the brain K_D or B_{max} values of either labelled *(R)*-PIA or DPX. This last result strongly hints at the possibility that *(R)*-PIA may have failed to gain access to central neurons in sufficient concentrations to affect their receptors. Caffeine enters the brain with relative facility (17) and would thus affect both central and peripheral adenosine receptors.

Antinociception

Adenosine analogs can induce analgesia. Systemic injections of *(R)*-PIA cause analgesia in mice (148,161,162) and rats (163). Yarbrough and McGuffin-Clineschmidt (164) have characterized analgesic actions of adenosine analogs after injection into the cisterna magna. Papaverine and EHNA potentiated the analgesic action of adenosine, whereas caffeine and theophylline antagonized the actions of the adenosine analogs. Adenosine had a similar analgesic action when injected into the lateral cerebral ventricle of rats tested with the tail flick paradigm (165). Theophylline and DDA antagonized these antinociceptive actions of adenosine.

The pain threshold lowering effect of aminophylline and the antinociceptive effect of dibutyryl cyclic AMP administered intrathecally were described by Jurna (166). Intrathecally administered adenosine analogs have an analgesic action in mice and rats (167–169), which are antagonized by the methylxanthines.

Spinal adenosine involvement in morphine-induced analgesia is suggested by studies using intrathecal injections of methylxanthines to antagonize analgesia induced by systemic or intrathecal injections of morphine (166, 170). Several investigations have indicated that one mechanism of action of opioids involves an opioid-induced release of adenosine (170,171), which would account for the numerous reports of methylxanthine-antagonism of opioid actions. Adenosine A_2 receptors, as measured by ^{3}H-DPX binding, are up-regulated in morphine-dependent animals (172), possibly as a result of a decreased release of endogenous adenosine during the development of tolerance.

Hypnotic Actions

Adenosine administered into the cerebral ventricles induces a sleep state in animals and birds (3,173,174). Parenteral administration of adenosine analogs to rats increased deep slow-wave sleep and, paradoxically, rapid eye movement sleep (175). These effects were antagonized by caffeine. Parenterally administered deoxycoformycin, an adenosine deaminase inhibitor, had similar effects on deep slow-sleep and rapid eye movement sleep (176) suggesting that endogenous adenosine is involved in sleep regulation.

Intracerebral adenosine reduced wakefulness, and increased total sleep time and deep slow-wave sleep time in rats (177). Adenosine also significantly reduced the time to onset of rapid eye movement sleep. Intracerebroventricular injections of forskolin or DDA failed to alter the amount or quality of sleep, suggesting that generalized stimulation or inhibition of adenylate cyclase is insufficient to modulate sleep (178).

Anticonvulsant Activity

Anticonvulsant effects of adenosine and adenosine analogs in intact animals have been reported by a number of investigators. Intraperitoneally administered adenosine has a protective action against audiogenic seizures, which is still evident after arterial blood pressure has recovered (179). In an extensive study, Dunwiddie and Worth (180) compared the activity of several adenosine analogs and observed anticonvulsant actions in mice and rats against a variety of convulsants.

Adenosine antagonizes the electrically evoked afterdischarges induced by penicillin in hippocampal slices (181), and reduces the frequency of generation of burst potentials in bicuculline-treated slices (182). Evidence for a role of endogenous adenosine as a modulator of slice excitability has been obtained in penicillin-treated hippocampal slices. Both adenosine deaminase and theophylline increased the rate of interictal spiking in these slices (183).

In the kindling model of epilepsy, adenosine has anticonvulsant effects, whether administered peripherally or centrally. Adenosine, adenosine analogs, and the uptake inhibitor papaverine reduce the severity and duration of amygdala-kindled seizures (184-186), and *(R)*-PIA prolongs the phase of postictal depression following a seizure (187). The adenosine antagonists prolong seizures and reduce seizure threshold.

Brain levels of adenosine rise rapidly during seizures induced by bicuculline and electroshock (188,189). These increases in adenosine are a result of ATP catabolism and appear to be important for the termination of seizure discharges. Burley and Ferrendelli (190) report that the uptake blocker dipyridamole prolongs the seizure refractory period following maximal electroshock in rats.

Adenosine and Food Intake

Adenosine, and to a lesser extent inosine, produced a significant suppression of food intake following subcutaneous administration (191). Intracerebroventicular administration of adenosine, adenine, but not inosine, suppressed food and water intake in rats (192).

Methylxanthine adenosine antagonists can enhance taste in humans

(193). Taste potentiation is thought to be due to local modulation of endogenously released adenosine at the level of the taste receptors themselves or the nerve endings of the chorda tympani associated with taste cells of the anterior tongue. Human subjects taking aminophylline or theophylline can have an increased sensitivity to both taste and odors, that is reversed when the drugs are discontinued (193). Increased taste and smell sensitivity with the methylxanthines may be one of the factors that leads the elderly to consume greater amounts of tea and coffee than younger individuals. Caffeine excreted in the saliva would enhance taste sensation, and could thus influence appetite. Caffeine has been reported to increase food consumption by rats (194), but whether this reflects central or peripheral sites of action is unknown.

Neuroendocrine Effects of Adenosine

Caffeine Studies

The involvement of adenosine receptors in the secretion and actions of endocrine hormones has only recently become a target for systematic investigation in animals and humans. Most of the studies to date have utilized caffeine, which probably exerts most of its pharmacologic effects through antagonism of adenosine receptors (17), as an indirect probe of adenosine-modulated events. High doses of caffeine in rats (30 mg/kg), but not lower doses, provoke a stress-like response, decreasing serum levels of GH, and thyroid-stimulating hormone (TSH), and increasing those of cortisol and serum β-endorphin-like activity (195,196). Caffeine (500 mg) elevated plasma levels of β-endorphin-like immunoreactivity in humans but had no significant effect on plasma levels of cortisol, TSH, GH, or prolactin in one study (196).

Central Effects of Progesterone, Adrenal Steroids, and Adenosine

The depressant effects of progesterone on brain excitability have long been recognized. In pharmacologic doses, progesterone exerts an anesthetic effect on rats (197) and at lower doses it has demonstrable anticonvulsant actions (198) and raises the threshold for electroshock seizure (199). A depressant effect of progesterone, administered in physiologic amounts, on the generation of spikes from a penicillin focus in the cat cerebral cortex has been reported (200). There is now evidence that adenosine may be involved in such central actions of progesterone.

The relationship between steroid hormones and brain function has generally been considered from the point of view of feedback regulation of brain and anterior pituitary activity by the secretory products of endocrine glands. It is now accepted that steroid hormones exert this form of feedback control upon pituitary secretion of their respective trophic hormones by binding to intracellular receptors and altering gene transcription in target cells within specific neuroendocrine structures. Intravenously or locally administered steroids can exert very rapid changes in neuronal excitability, which are likely to be exerted at plasma membrane-bound receptors rather than by genomic mechanisms, and there are intriguing indications that glucocorticoids can affect responses to putative transmitter receptors (210).

Progesterone induced rapid changes in the firing patterns of cerebral cortical neurons and depressed synaptic activity evoked by sensory stimuli (202). A potential involvement of adenosine in this depressant action of progesterone on cortical neurons was indicated by the observation that various steroids, including progesterone, were potent inhibitors of adenosine uptake into rat brain cortical synaptosomes (203). In related experiments, an ability of locally applied progesterone, and of other progestational agents (pregnenolone, cyproterone, and norethindrone), to potentiate the depressant effects of adenosine on cerebral cortical neurons has been established (203,204). The depressant effects of these steroids on cerebral cortical neurons were antagonized by caffeine, confirming that activation of an adenosine receptor was involved.

The luteal phase of the menstrual cycle, when progesterone levels are high, is associated with reduced central nervous system activity (205). It has also been alleged that falling progesterone levels are involved in the etiology of premenstrual tension (206), and the instability of mood, anxiety, depression, and insomia at term and following pregnancy. Changes in the incidence of seizure frequency in epileptics during the menstrual cycle have been ascribed to progesterone. Seizures were fewer when progesterone levels were high and increased as progesterone levels fell (catamenial epilepsy; 207). The potentiating effects of progesterone on purinergic neuromodulation described above suggest that premenstrual tension and catamenial epilepsy may occur as a result of the withdrawal of adenosine-mediated depression of synaptic transmission, consequent upon falling progesterone levels. This observation suggests the possibility of controlling mood fluctuations during and following the luteal phase of the menstrual cycle by the appropriate administration of an adenosine antagonist followed by an uptake inhibitor during periods of high and low progesterone formation respectively.

Adrenal cortical steroids can also inhibit adenosine transport. Cortisone was estimated to have an IC_{50} of 40 μM for adenosine uptake by rat brain cortical synaptosomes (A.S. Bender and J.W. Phillis, unpublished observations). Potentiation of endogenously released adenosine could therefore be involved in the depressant actions of adrenal steroids on central neurons (208,209), and the anticonvulsant properties of ACTH (210), which triggers steroid hormone release by the adrenal gland.

Adrenal steroids are released in greater than usual amounts in Cushing's disease and depressive illness (211), and these changes may result in disturbances of purinergic neuromodulation and thus trigger the emotional disturbances associated with these diseases.

Adenosine and Prolactin Secretion

The regulation of prolactin (PRL) secretion has been shown to involve neuronal control over the secretion of hypothalamic hypophyseal hormones and neurotransmitters which then act on the pituitary to inhibit or stimulate hormone release. A number of transmitters, but mainly dopamine and serotonin, have been implicated in the regulation of PRL secretion. In a study of the effects of adenosine analogs on prolactin release, neither NECA nor (R)-PIA affected release from dispersed rat anterior pituitary cells, nor did they alter the dopamine-induced decrease in release from cultured pituitary cells. In vivo administration of NECA and (R)-PIA caused increases in rat serum PRL, an effect which was antagonized by aminophylline. The dopaminergic agents L-DOPA and bromocriptine also antagonized the NECA and (R)-PIA increase in serum prolactin (212). It was postulated that adenosine may increase prolactin secretion, in part, by inhibiting central dopamine release. The failure of the analogs to alter in vitro PRL secretion argues against an action at the pituitary level.

Adenosine and Vascular Responses in Experimental Diabetes

There is considerable evidence that patients with diabetes mellitus are prone to congestive heart failure and cerebrovascular inadequacies. Studies of the coronary circulation indicate that coronary flow may be reduced (213). The coronary bed of diabetic lambs is much less responsive to adenosine than that of nondiabetics, and insulin replacement rapidly reverses the adenosine subsensitivity in these animals (214). Adenosine has been proposed as a major regulator of coronary artery blood flow (4). The lack of effect of adenosine in insulinopenic animals and the reversal of this phenomenon by insulin administration indicates a central role of this hormone in the regulation of coronary vascular sensitivity to adenosine, and may account for the abnormalities in coronary blood flow in diabetics. Modulation of adenosine receptor number or affinity would be one possible mechanism by which insulin acts in this system. Alternatively, insulin may regulate adenosine transport or its metabolizing enzymes.

As adenosine appears to be critically involved in vasodilatory cerebrovascular responses to hypoxia and hypercapnia (215,216), it is possible that the abnormal cerebrovascular reactivity seen in diabetes mellitus (217) is a result of a similar failure of the cerebral vessels to dilate with adenosine.

Loss of sensitivity to adenosine is not, apparently, an inevitable consequence in diabetic animals. Streptozotocin-induced diabetes in rats did not impair the response of adipocytes to *(R)*-PIA-induced increases in glucose utilization (218). Unfortunately adenosine, itself, was not used in this study, and it is possible that its actions might have been attenuated if insulin exerts its effects on the transport or metabolism of adenosine rather than on the receptors. *(R)*-PIA is a stable analog of adenosine and not subject to transport or rapid metabolism.

Conclusions

The preceding sections have presented a summary of the literature on the central actions of adenosine and those agents which potentiate or antagonize its actions. It has become apparent that adenosine has actions at every level of the neural axis, leaving little doubt that endogenously released adenosine is an important and ubiquitous, regulator of neuronal excitability. Whether adenosine is a neurotransmitter or neuromodulator remains unclear, and indeed it may ultimately transpire that it serves both roles. Within the past few years, it has become apparent that certain cells have higher levels of immunohistochemically demonstrable adenosine deaminase or of autoradiographically localizable adenosine transport capacity. Even though metabolic pools of adenosine occur in all cells, "adenosinergic" neurons might be identifiable by such characteristics. Neurons that concentrate adenosine itself have recently been identified in the CNS (219), and in these neurons it may function as a neurotransmitter. Adenosine release from other neurons may accompany enhanced levels of neuronal activity, especially where this is associated with a localized or generalized state of hypoxia. Thus one important function of adenosine may be the maintenance of a balance between neuronal activity and metabolic demand.

Another source of adenosine is by hydrolysis of ATP in the extracellular space. Evidence has been presented that ATP may be an excitatory neurotransmitter in the CNS and its release would also furnish a source of adenosine.

Unresolved problems which are alluded to in this chapter concern the identification of adenosine receptor sub-types (A_1, A_2, etc) involved in the actions of adenosine, the role of adenylate cyclase in the generation of responses to adenosine and the extent to which peripherally administered adenosine and adenosine analogs exert their actions at central, as opposed to peripheral, receptors. With many new adenosine receptor agonists and antagonists now becoming available, it can be anticipated that more selective ligands for the various adenosine receptors will soon be developed. Using such compounds, it should become possible to resolve some of these issues.

References

1. Drury AN, Szent-Györgyi A (1929) The physiological activity of adenine compounds with especial reference to their action upon the mammalian heart. J Physiol (Lond) 68:213–237.
2. Holton FA, Holton P (1954) The capillary dilator substances in dry powders of spinal roots: a possible role of adenosine triphosphate in chemical transmission from nerve endings. J Physiol (Lond) 126:124–140.
3. Feldberg W, Sherwood SL (1954) Injections of drugs into the lateral ventricle of the cat. J Physiol (Lond) 123:148–167.
4. Baer HP, Drummond GI (1979) Physiological and regulatory functions of adenosine and adenine nucleotides. Raven Press, New York.
5. Burnstock G (1978) A basis for distinguishing two types of purinergic receptor. In: Bolis L, Straub RW (eds) Cell Membrane Receptors for Drugs and Hormones: A Multidiciplinary Approach. Raven Press, New York, pp 107–118.
6. Phillis JW, Kostopoulos GK, Limacher JJ (1974) Depression of corticospinal cells by various purines and pyrimidines. Can J Physiol Pharmacol 52:1226–1229.
7. Phillis JW, Kostopoulos GK (1975) Adenosine as a putative transmitter in the cerebral cortex: Studies with potentiators and antagonists. Life Sci 17:1085–1094.
8. Kostopoulos GK, Phillis JW (1977) Purinergic depression of neurons in different areas of the rat brain. Exp Neurol 55:719–724.
9. Kuroda Y, Saito M, Kobayashi K (1976) Concomitant changes in cyclic AMP level and postsynaptic potentials of olfactory cortex slices induced by adenosine derivatives. Brain Res 109:196–201.
10. Schofield CN (1978) Depression of evoked potentials in brain slices by adenosine compounds. Br J Pharmacol 63:239–244.
11. Dunwiddie TV, Hoffer BJ (1980) Adenine nucleotides and synaptic transmission in the *in vitro* rat hippocampus. Br J Pharmacol 69:59–68.
12. Sattin A, Rall TW (1970) The effect of adenosine and adenine nucleotides on the cyclic adenosine 3′, 5′-phosphate content of guinea pig cerebral cortex slices. Mol Pharmacol 6:13–23.
13. White TD (1985) Release of ATP from central and peripheral nerve terminals. In: Stone TW (ed) Purines: Pharmacology and Physiological Roles. MacMillan Press, London, pp 95–105.
14. Su C (1983) Purinergic neurotransmission and neuromodulation. Annu Rev Pharmacol Toxicol 23:397–411.
15. Sneddon P, Westfall DP (1984) Pharmacological evidence that adenosine triphosphate and noradrenaline are co-transmitters in the guinea-pig vas deferens. J Physiol (Lond) 347:561–580.
16. Fredholm BB, Hedqvist P (1980) Modulation of neurotransmission by purine nucleotides and nucleosides. Biochem Pharmacol 29:1635–1643.
17. Phillis JW, Wu PH (1981) The role of adenosine and its nucleotides in central synaptic transmission. Progr Neurobiol 16:187–239.

238   J.W. Phillis

18. Snyder SH (1985) Adenosine as a neuromodulator. Annu Rev Neurosci 8:103–124.
19. Stone TW (1985) Purines: Pharmacology and Physiological Roles. MacMillan Press, London.
20. Stefanovich V, Rudolphi K, Schubert P (1985) Adenosine: Receptors and Modulation of Cell Function. IRL Press, Oxford.
21. Dunwiddie TV (1985) The physiological role of adenosine in the central nervous system. Int Rev Neurobiol 27:63–139.
22. Pull I, McIlwain H (1973) Output of [^{14}C]adenine nucleotides and their derivatives from cerebral tissues. Tetrodotoxin-resistant and calcium requiring components. Biochem J 136:893–901.
23. Kuroda Y, McIlwain H (1974) Uptake and release of [^{14}C] adenine derivatives at beds of mammalian cortical synaptosomes in a superfusion system. J Neurochem 22:691–699.
24. Fredholm BB, Vernet L (1979) Release of ^{3}H-nucleosides from ^{3}H-adenine labelled hypothalamic synaptosomes. Acta Physiol Scand 106:97–107.
25. Nagy A, Shuster TA, Rosenberg MD (1983) Adenosine triphosphatase activity at the external surface of chicken brain synaptosomes. J Neurochem 40:226–234.
26. Nagata H, Mimori Y, Nakamura S et al (1984) Regional and subcellular distribution in mammalian brain of the enzymes producing adenosine. J Neurochem 42:1001–1007.
27. Bender AS, Wu PH, Phillis JW (1981) Some biochemical properties of the rapid adenosine uptake system in rat brain synaptosomes. J Neurochem 37:1282–1290.
28. Daval JL, Barberis C (1981) Release of radiolabelled adenosine derivatives from superfused synaptosome beds. Evidence for output of adenosine. Biochem Pharmacol 30:2559–2567.
29. Pons F, Bruns RF, Daly JW (1980) Depolarization-evoked accumulation of cyclic AMP in brain slices: The requisite intermediate adenosine is not derived from hydrolysis of released ATP. J Neurochem 34:1319–1323.
30. Barberis C, Guibert B, Daudet F et al (1984) In vivo release of adenosine from cat basal ganglia-studies with a push pull cannula. Neurochem Int 6:545–551.
31. MacDonald WF, White TD (1985) Nature of extrasynaptosomal accumulation of endogenous adenosine evoked by K^{+} and veratridine. J Neurochem 45:791–797.
32. Potter P, White TD (1980) Release of adenosine 5'-triphosphate from synaptosomes from different regions of rat brain. Neurosci 5:1351–1356.
33. Sulakhe PV, Phillis JW (1975) The release of ^{3}H-adenosine and its derivatives from cat sensorimotor cortex. Life Sci 17:551–555.
34. Phillis JW, Jiang ZG, Chelack BJ et al (1980) The effect of morphine on purine and acetylcholine release from rat cerebral cortex: Evidence for a purinergic component in morphine's action. Pharmacol Biochem Behav 13:421–427.
35. Wu PH, Phillis JW (1978) Distribution and release of adenosine triphosphate in rat brain. Neurochem Res 3:563–571.
36. Jhamandas K, Dumbrille A (1980) Regional release of [^{3}H]adenosine derivatives from rat brain in vivo: Effect of excitatory amino acids, opiate agonists, and benzodiazepines. Can J Physiol Pharmacol 58:1262–1278.

37. Zetterstrom T, Vernet L, Ungerstedt U et al (1982) Purine levels in the intact rat brain: Studied with an implanted perfused hollow fibre. Neurosci Lett 29:111–115.

38. Van Wylen DGL, Park TS, Rubio R et al (1986) Increases in cerebral interstitial fluid adenosine concentration during hypoxia, local potassium infusion, and ischemia. J Cerebr Blood Flow Metab 6:522–528.

39. Bender AS, Wu PH, Phillis JW (1980) The characterization of ^{3}H-adenosine uptake into rat cerebral cortical synaptosomes. J Neurochem 35:629–640.

40. Lipton P, Whittingham TS (1984) Energy metabolism and brain slice function. In: Dingledine R (ed) Brain Slices. Plenum Press, New York, pp 113–153.

41. Allcorn RJ, Cunnane TC, Kirkpatrick K (1986) Actions of $\alpha\beta$-methylene ATP and 6-hydroxydopamine on sympathetic neurotransmission in the vas deferens of the guinea-pig, rat, and mouse: support for co-transmission. Br J Pharmacol 89:647–659.

42. Silinsky EM (1975) On the association between transmitter secretion and the release of adenine nucleotides from mammalian motor nerve terminals. J Physiol (Lond) 247:145–162.

43. Ribeiro JA, Sebastião AM (1986) 5′-Nucleotidase and neuromuscular transmission. Br J Pharmacol 89:621P.

44. Israel M, Lesbats B, Meunier FM et al (1976) Postsynaptic release of adenosine triphosphate induced by single impulse transmitter action. Proc Roy Soc Lond [Biol] 193:461–468.

45. Bencheriff M, Rubio R, Berne RM (1985) Mechanisms and site of release of adenosine in nervous tissue. Physiologist 28:360.

46. Wu PH, Phillis JW (1984) Uptake by central nervous tissues as a mechanism for the regulation of extracellular adenosine concentrations. Neurochem Int 6:613–632.

47. Bender AS, Wu PH, Phillis JW (1981) Some biochemical properties of the rapid adenosine uptake system in rat brain synaptosomes. J Neurochem 37:1282–1290.

48. Barberis C, Minn A, Gayet J (1981) Adenosine transport into guinea-pig synaptosomes. J Neurochem 36:347–354.

49. Green RD (1980) Adenosine transport by a variant of C1300 murine neuro-blastoma cells deficient in adenosine kinase. Biochem Biophys Acta 598:366–374.

50. Bender AS, Hertz L (1986) Similarities of adenosine uptake systems in astrocytes and neurons in primary cultures. Neurochem Res 11:1507–1524.

51. Geiger JD, Nagy JI (1984) Heterogeneous distribution of adenosine transport sites labelled by [^{3}H]nitrobenzylthioinosine in rat brain: An autoradiographic and membrane binding study. Brain Res Bull 13:657–666.

52. Bisserbe JC, Deckert J, Marangos P (1986) Autoradiographic localization of adenosine uptake sites in guinea pig brain using [^{3}H]dipyridamole. Neurosci Lett 66:341–345.

53. Pull I, McIlwain H (1974) Rat cerebral-cortex adenosine deaminase activity and its subcellular distribution. Biochem J 144:37–41.

54. Fisher MN, Newsholme EA (1984) Properties of rat heart adenosine kinase. Biochem J 221:521–528.

55. Mendelson WB, Kuruvilla A, Watlington T et al (1983) Sedative and electroencephalographic actions of erythro-9-(2-hydroxy-3-nonyl)-adenine

(EHNA): Relationship to inhibition of brain adenosine deaminase. Psychopharmacology 79:126–129.

56. Radulovacki M, Virus RM, Djuricic-Nedelson M et al (1983) Hypnotic effects of deoxycoformycin in rats. Brain Res 271:392–395.

57. Phillis JW, Barraco RA, DeLong RE et al (1986) Behavioral characterization of centrally administered adenosine analogs. Pharmacol Biochem Behav 24:263–270.

58. Conway EJ, Cooke R (1939) Deaminases of adenosine and adenylic acid in blood and tissues. Biochem J 33:479–492.

59. Mannery JF, Dryden EE (1979) Ecto-enzymes concerned with nucleotide metabolism. In: Baer HP, Drummond GI (eds) Physiological and Regulatory Function of Adenosine and Adenine Nucleotides. Raven Press, New York, pp 323–339.

60. Geiger JD, Nagy JI (1986) Distribution of adenosine deaminase activity in rat brain and spinal cord. J Neurosci 6:2707–2714.

61. Nagy JI, Labella LA, Buss M et al (1984) Immunohistochemistry of adenosine deaminase: Implications for adenosine neurotransmission. Science 224:166–168.

62. Staines WA, Yamamoto T, Daddona PE et al (1986) Neuronal colocalization of adenosine deaminase, monoamine oxidase, galanin, and 5-hydroxytryptophan uptake in the tuberomammillary nucleus of the rat. Brain Res Bull 17:351–365.

63. Schrader J (1983) Metabolism of adenosine and sites of production in the heart. In: Berne RM, Rall TW, Rubio R (eds): Regulatory Function of Adenosine. Martinus Nijhoff, Boston. pp 133–156.

64. McIlwain H, Poll JD (1986) Adenosine in cerebral homeostatic role: Appraisal through actions of homocysteine, colchicine, and dipyridamole. J Neurobiol 17:39–49.

65. Schwabe U, Kiffe H, Puchstein C et al (1979) Specific binding of ^{3}H-adenosine to rat brain membranes. Naunyn-Schmiedeberg's Arch Pharmacol 310:59–67.

66. Newman ME, Patel J, McIlwain H (1981) The binding of [^{3}H] adenosine to synaptosomal and other preparations from mammalian brain. Biochem J 194:611–620.

67. Traversa U, Puppini P, de Angelis L et al (1984) Regional distribution of high affinity binding of ^{3}H-adenosine in rat brain. Pharm Res Commun 16:589–603.

68. Snell PH, Snell CR (1983) [^{3}H]Adenosine binding sites on 108CC15 neuroblastoma × glioma hybrid cell line and rat brain membranes. Neurochem Int 5:245–252.

69. Bruns RF, Daly JW, Snyder SH (1980) Adenosine receptors in brain membranes: Binding of N^6-cyclohexy[^{3}H]adenosine and 1,3-diethyl-8-[^{3}H]phenylxanthine. Proc Natl Acad Sci USA 77:5547–5551.

70. Williams M, Risley EA (1980) Biochemical characterization of putative purinergic receptors by using 2-chloro[^{3}H] adenosine: a stable analog of adenosine. Proc Natl Acad Sci USA 77:6892–6896.

71. Wu PH, Phillis JW (1982) Adenosine receptors in rat brain membranes: Characterization of high affinity binding of [^{3}H]-2-chloroadenosine. Int J Biochem 14:399–404.

72. Patel J, Marangos PJ, Stivers J et al (1982) Characterization of adenosine

receptors in brain using N^6-cyclohexyl[^{3}H] adenosine. Brain Res 237:203–214.

73. Yeung SMH, Green RD (1984) [^{3}H]5'-N-ethylcarboxamide adenosine binds to both Ra and Ri adenosine receptors in rat striatum. Naunyn-Schmiedebergs Arch Pharmacol 325:218–225.

74. Kalaria RN, Harik SI (1986) Adenosine receptors of cerebral microvessels and choroid plexus. J Cereb Blood Flow Metab 6:463–470.

75. Goodman RR, Cooper MJ, Gavish M et al (1982) Guanine nucleotide and cation regulation of binding of [^{3}H]cyclohexyladenosine and [^{3}H] diethylphenylxanthine to adenosine A_1 receptors in brain membranes. Mol Pharmacol 21:329–335.

76. Nimit Y, Law J, Daly JW (1982) Binding of 2',5'-dideoxyadenosine to brain membranes. Comparison to P-site inhibition of adenylate cyclase. Biochem Pharmacol 31:3279–3287.

77. Chin JH, DeLorenzo RJ (1986) A new class of adenosine receptors in brain. Characterization by 2-chloro[^{3}H]adenosine binding. Biochem Pharmacol 35:847–856.

78. Burnstock G, Buckley NJ (1985) The classification of receptors for adenosine and adenine nucleotides. In: Paton DM (ed) Methods in Pharmacology, Volume 6, Methods used in adenosine research. Plenum Press, New York. pp 193–212.

79. Williams M, Risley EA (1980) Binding of ^{3}H-adenyl-5'-imidodiphosphate (AppNHp) to rat brain synaptic membranes. Fed Proc 39:1009.

80. Goodman RR, Snyder SH (1982) Autoradiographic localization of adenosine receptors in rat brain using [^{3}H]cyclohexyladenosine. J Neurosci 2:1230–1241.

81. Geiger JD, Labella FS, Nagy JI (1984) Characterization and localization of adenosine receptors in rat spinal cord. J Neurosci 4:2303–2310.

82. Shaw C, Hall SE, Cynader M (1986) Characterization, distribution, and ontogenesis of adenosine binding sites in cat visual cortex. J Neurosci 6:3218–3228.

83. Goodman RR, Kuhar MJ, Hester L et al (1983) Adenosine receptors: Autoradiographic evidence for their location on axon terminals of excitatory neurons. Science 220:967–968.

84. Kocsis JD, Eng DL, Bhisitkul RB (1984) Adenosine selectively blocks parallel-fiber-mediated synaptic potentials in rat cerebellar cortex. Proc Natl Acad Sci USA 81:6531–6534.

85. Boulenger J-P, Patel J, Post RM et al (1983) Chronic caffeine consumption increases the number of brain adenosine receptors. Life Sci 32:1135–1142.

86. Fredholm BB (1982) Adenosine actions and adenosine receptors after 1 week treatment with caffeine. Acta Physiol Scand 115:283–286.

87. Wu PH, Coffin VL (1984) Up-regulation of brain [^{3}H] diazepam binding sites in chronic caffeine-treated rats. Brain Res 294:186–189.

88. Corradetti R, Pedata F, Pepeu G et al (1986) Chronic caffeine treatment reduces caffeine but not adenosine effects on cortical acetylcholine release. Br J Pharmacol 88:671–676.

89. Corradetti R, Kiedrowski L, Nordstrom O et al (1984) Disappearance of low affinity adenosine binding sites in aging rat cerebral cortex and hippocampus. Neurosci Lett 49:143–146.

90. Pedata F, Slavikova J, Kotas A et al (1983) Acetylcholine release from rat

cortical slices during postnatal development and aging. Neurobiol Aging 4:31–35.

91. Onodera H, Kogure K (1985) Autoradiographic visualization of adenosine A_1 receptors in the gerbil hippocampus: Changes in receptor density after transient ischemia. Brain Res 345:406–408.

92. Lee KS, Tetzlaff W, Kreutzberg GW (1986) Rapid down regulation of hippocampal adenosine receptors following brief anoxia. Brain Res 380: 155–158.

93. Premont J, Perez M, Blanc G et al (1979) Adenosine-sensitive adenylate cyclase in rat brain homogenates: Kinetic characteristics, specificity, topographical, subcellular, and cellular distribution. Mol Pharmacol 16:790–804.

94. Daly JW, Butts-Lamb P, Padgett W (1983) Subclasses of adenosine receptors in the central nervous system: Interaction with caffeine and related methylxanthines. Cell Mol Neurobiol 3:69–80.

95. Londos CD, Cooper MF, Wolff J (1980) Subclasses of external adenosine receptors. Proc Natl Acad Sci USA 77:2551–2554.

96. Van Calker D, Muller M, Hamprecht B (1979) Adenosine regulates via two different types of receptors, the accumulation of cyclic AMP in cultured brain cells. J Neurochem 33:999–1005.

97. Ebersolt C, Premont J, Prochiantz A, et al (1983) Inhibition of brain adenylate cyclase by A_1 adenosine receptors: pharmacological characteristics and locations. Brain Res 267:123–129.

98. Fredholm BB, Jonzon B, Lindstrom K (1986) Effect of adenosine receptor agonists and other compounds on cyclic AMP accumulation in forskolin-treated hippocampal slices. Naunyn-Schmiedebergs Arch Pharmacol 332:173–178.

99. Collis MG, Brown CM (1983) Adenosine relaxes the aorta by interacting with an A_2-receptor and an intracellular site. Eur J Pharmacol 96:61–69.

100. Ribeiro JA, Sebastião AM (1986) Adenosine receptors and calcium: Basis for proposing a third (A_3) adenosine receptor. Progr Neurobiol 26:179–209.

101. Jhamandas K, Sawynok J (1976) Methylxanthine antagonism of opiate and purine effects on the release of acetylcholine. In: Kosterlitz HW (ed) Opiates and Endogenous Opioid Peptides. Elsevier, North Holland Biomedical Press, Amsterdam, pp 161–168.

102. Pedata F, Antonelli T, Lambertini L et al (1983) Effect of adenosine, adenosine triphosphate, adenosine deaminase, dipyridamole, and aminophylline on acetylcholine release from electrically-stimulated brain slices. Neuropharmacology 22:609–614.

103. Phillis JW, Siemens RK, Wu PH (1980) Effects of diazepam on adenosine and acetylcholine release from rat cerebral cortex: Further evidence for a purinergic mechanism in action of diazepam. Br J Pharmacol 70:341–348.

104. Harms HH, Wardeh G, Mulder AH (1978) Adenosine modulates depolarization-induced release of ^{3}H-noradrenaline from slices of rat brain neocortex. Eur J Pharmacol 49:305–308.

105. Jonzon B, Fredholm BB (1984) Adenosine receptor mediated inhibition of noradrenaline release from slices of rat hippocampus. Life Sci 35:1971–1979.

106. Harms HH, Wardeh G, Mulder AH (1979) Effects of adenosine on depolarization-induced release of various radiolabelled neurotransmitters from slices of rat corpus striatum. Neuropharmacology 18:577–580.

107. Hollins C, Stone TW (1980) Adenosine inhibition of γ-aminobutyric acid release from slices of rat cerebral cortex. Br J Pharmacol 69:107–112.

108. Dolphin AC, Archer ER (1983) An adenosine agonist inhibits and a cyclic AMP analogue enhances the release of glutamate but not GABA from slices of rat dentate gyrus. Neurosci Lett 43:49–54.

109. Spignoli G, Pedata F, Pepeu G (1984) A_1 and A_2 adenosine receptors modulate acetylcholine release from brain slices. Europ J Pharmacol 97: 341–342.

110. Ebstein RP, Daly JW (1982) Release of norepinephrine and dopamine from brain vesicular preparations. Effects of adenosine analogues. Cell Mol Neurobiol 2:193–204.

111. Ribeiro JA, Sa-Almeida AM, Namorado JM (1979) Adenosine and adenosine triphosphate decrease ^{45}Ca uptake by synaptosomes stimulated by potassium. Biochem Pharmacol 28:1297–1300.

112. Wu PH, Phillis JW, Thierry DL (1982) Adenosine receptor agonists inhibit K^+-evoked Ca^{2+} uptake by rat brain cortical synaptosomes. J Neurochem 39:700–708.

113. Quintana J (1986) Adenosine and related nucleotides alter calcium uptake in depolarized synaptosomes of *Torpedo* electric organ. J Neural Transm 64:271–284.

114. Shinozuka K, Maeda T, Hayashi E (1985) Effects of adenosine on ^{45}Ca uptake and [^3H] acetylcholine release in synaptosomal preparation from guinea-pig ileum myenteric plexus. Eur J Pharmacol 113:417–424.

115. Henon BK, McAfee DA (1983) The ionic basis of adenosine receptor actions on post-ganglionic neurones in the rat. J Physiol (Lond) 336:607–620.

116. Dolphin AC, Forda SR, Scott RH (1986) Calcium-dependent currents in cultured rat dorsal root ganglion neurones are inhibited by an adenosine analogue. J Physiol 373:47–61.

117. MacDonald RL, Skerritt JH, Werz MA (1986) Adenosine agonists reduce voltage-dependent calcium conductance of mouse sensory neurones in cell culture. J Physiol 370:75–90.

118. Kuroda Y (1985) Modulation of calcium channels through different adenosine receptors: ADO-1 and ADO-2. In: Stefanovich V, Rudolphi K, Schubert P (eds) Adenosine: Receptors and Modulation of Cell Function. IRL Press, Oxford. pp. 233–240.

119. Proctor WR, Dunwiddie TV (1983) Adenosine inhibits calcium spikes in hippocampal pyramidal neurons in vitro. Neurosci Lett 35:197–201.

120. Halliwell JV, Scholfield CN (1984) Somatically recorded Ca-currents in guinea-pig hippocampal and olfactory cortex neurones are resistant to adenosine action. Neurosci Lett 50:13–18.

121. Ginsborg BL, Hirst GDS (1972) The effect of adenosine on the release of the transmitter from the phrenic nerve of the rat. J Physiol (Lond) 224: 629–645.

122. Drummond GI (1983) Cyclic nucleotides in the nervous system. Adv Cyclic Nucleotide Res 15:373–494.

123. Dolphin AC, Prestwich SA (1985) Pertussis toxin reverses adenosine inhibition of neuronal glutamate release. Nature 316:148–150.

124. Fredholm BB, Lindgren E (1986) Possible involvement of the N_i-protein in the prejunctional inhibitory effect of a stable adenosine analogue (R-PIA) on

noradrenaline release in the rat hippocampus. Acta Physiol Scand 126: 307–309.

125. Okada Y, Ozawa S (1980) Inhibitory action of adenosine on synaptic transmission in the hippocampus of the guinea pig in vitro. Eur J Pharmacol 68:483–492.

126. Galindo A, Krnjevíc K, Schwartz S (1967) Micro-iontophoretic studies on neurones in the cuneate nucleus. J Physiol (Lond) 192:359–377.

127. Hoffer BJ, Siggins GR, Oliver AP et al (1971) Cyclic AMP mediation of norepinephrine inhibition in rat cerebellar cortex: A unique class of synaptic responses. Ann NY Acad Sci 185:531–549.

128. Phillis JW, Edstrom JP, Kostopoulos GK et al (1979)Effects of adenosine and adenine nucleotides on synaptic transmission in the cerebral cortex. Can J Physiol Pharmacol 57:1289–1312.

129. Salt TE, Hill RG (1983) Excitation of single sensory neurones in the rat caudal trigeminal nucleus by iontophoretically applied adenosine 5'-triphosphate. Neurosci Lett 35:53–57.

130. Fyffe REW, Perl ER (1984) Is ATP a central synaptic mediator for certain primary afferent fibers from mammalian skin? Proc Natl Acad Sci USA 81:6890–6893.

131. Salter MW, Henry JL (1985) Effects of adenosine 5'-monophosphate and adenosine 5'-triphosphate on functionally identified units in the cat spinal dorsal horn. Evidence for a differential effect of adenosine 5'-triphosphate on nociceptive vs non-nociceptive units. Neuroscience 15:815–825.

132. Jahr CE, Jessell TM (1983) ATP excites a subpopulation of rat dorsal horn neurones. Nature 304:730–733.

133. Siggins GR, Schubert P (1981) Adenosine depression of hippocampal neurons in vitro: An intracellular study of dose-dependent actions on synaptic and membrane potentials. Neurosci Lett 23:55–60.

134. Segal M (1982) An intracellular analysis of a postsynaptic action of adenosine in the rat hippocampus. Eur J Pharmacol 79:193–199.

135. Trussell LO, Jackson MB (1985) Adenosine-activated potassium conductance in cultured striatal neurons. Proc Natl Acad Sci USA 82:4857–4861.

136. Shefner SA, Chiu TH (1986) Adenosine inhibits locus coeruleus neurons: An intracellular study in a rat brain slice preparation. Brain Res 366:364–368.

137. Greene RW, Haas HL (1985) Adenosine actions on CA1 pyramidal neurones in rat hippocampal slices. J Physiol (Lond) 366:119–127.

138. Schubert P, Heinemann U, Kolb R (1986) Differential effect of adenosine on pre- and postsynaptic calcium fluxes. Brain Res 376:382–386.

139. Lekíc D (1977) Presynaptic depression of synaptic response of Renshaw cells by adenosine 5'-monophosphate. Can J Physiol Pharmacol 55:1391–1393.

140. Reddington M, Lee KS, Schubert P (1982) An A_1-adenosine receptor, characterized by [^{3}H] cyclohexyladenosine binding, mediates the depression on evoked potentials in a rat hippocampal slice preparation. Neurosci Lett 28:275–279.

141. Bennet DW, Drury AN (1931) Further observations relating to the physiological activity of adenine compounds. J Physiol (Lond) 72:288–320.

142. Watt AH, Routledge PA (1985) Adenosine stimulates respiration in man. Br J Clin Pharmacol 20:503–506.

143. Brodie MS, Lee K, Fredholm BB et al (1987). Central versus peripheral

mediation of responses to adenosine receptor agonists: evidence against a central mode of action. Brain Res. *415*, 323–330.

144. Bowker HM, Chapman AG (1986) Adenosine analogues. The temperature-dependence of the anticonvulsant effect and inhibition of ^{3}H-D-aspartate release. Biochem Pharmacol 35:2949–2953.

145. Jonzon B, Bergquist A, Li Y-O et al (1986) Effects of adenosine and two stable adenosine analogues on blood pressure, heart rate, and colonic temperature in the rat. Acta Physiol Scand 126:491–498.

146. Evoniuk GE, von Borstel RW, Wurtman RJ (1986) 8-p-Sulphophenyltheophylline reverses adenosine-induced hypotension by acting at a peripheral site. Soc Neurosci Abstr 12:524.

147. Barraco RA, Aggarwal AK, Phillis JW et al (1984) Dissociation of the locomotor and hypotensive effects of adenosine analogs in the rat. Neurosci Lett 48:139–144.

148. Buday PV, Carr CJ, Miya TS (1961) A pharmacologic study of some nucleosides and nucleotides. J Pharm Pharmacol 13:290–299.

149. Snyder SH, Katims JJ, Annau Z et al (1981) Adenosine receptors and behavioral actions of methylxanthines. Proc Natl Acad Sci USA 78:3260–3264.

150. Katims JJ, Annau Z, Snyder SH (1983) Interactions in the behavioral effects of methylxanthines and adenosine derivatives. J Pharmacol Exp Ther 227:167–173.

151. Wachtel H, Loschmann PA (1986) Effects of forskolin and cyclic nucleotides in animal models predictive of antidepressant activity: Interactions with rolipram. Psychopharmacol 90:430–435.

152. Barraco RA, Phillis JW, Altman HJ (1985) Depressant effect of forskolin on spontaneous locomotor activity in mice. Gen Pharmacol 16:521–524.

153. Green RD, Proudfit HK, Yeung SMH (1982) Modulation of striatal dopaminergic function by local injection of 5'-N-ethylcarboxamide adenosine. Science 218:58–61.

154. Fredholm BB, Herrera-Marschitz M, Jonzon B et al (1983) On the mechanism by which methylxanthines enhance apomorphine-induced rotation behavior in the rat. Pharmacol Biochem Behav 19:535–541.

155. Coffin VL, Carney JM (1986) Effects of selected analogs of adenosine on schedule-controlled behavior in rats. Neuropharmacology 25:1141–1147.

156. Spealman RD, Coffin VL (1986) Behavioral effects of adenosine analogs in squirrel monkeys: Relation to adenosine A_2 receptors. Psychopharmacol 90:419–421.

157. Coffin VL, Spealman RD (1985) Modulation of the behavioral effects of chlordiazepoxide by methylxanthines and analogs of adenosine in squirrel monkey. J Pharmacol Exp Ther 235:724–728.

158. Spencer DG, Caldwell P, Emmett-Oglesby MW (1984) Tolerance to N^6-(L-phenylisopropyl)adenosine. Contribution of behavioral mechanisms and cross-tolerance profile. Neuropharmacology 23:671–676.

159. Chou DT, Khan S, Forde J et al (1985) Caffeine tolerance: Behavioral, electrophysiological, and neurochemical evidence. Life Sci 36:2347–2358.

160. Ahlijanian MK, Takemori AE (1986) Cross-tolerance studies between caffeine and (−)-N^6-(phenylisopropyl)-adenosine (PIA) in mice. Life Sci 38:577–588.

161. Vapaatalo H, Onken D, Neuvonen PJ et al (1975) Stereospecificity in some central and circulating effects of phenylisopropyl-adenosine (PIA). Arzneim Forsch (Drug Res) 25:407–410.

162. Ahlijanian MK, Takemori AE (1985) Effects of (−)-N^6-(R-phenylisopropyl)adenosine (PIA) and caffeine on nociception and morphine-induced analgesia, tolerance, and dependence. Eur J Pharmacol 112:171–179.

163. Holmgren M, Hedner T, Nordberg G et al (1983) Anti-nociceptive effects in the rat of an adenosine analogue, N^6-phenylisopropyladenosine. J Pharm Pharmacol 35:679–680.

164. Yarbrough GG, McGuffin-Clineschmidt JC (1981) In vivo behavioral assessment of central nervous system purinergic receptor. Eur J Pharmacol 76:137–144.

165. Haulică I, Nemtu C, Petrescu GH et al (1984) The influence of adenosine upon thermoalgesic sensitivity. Physiologie Bucarest 21:167–172.

166. Jurna I (1984) Cyclic nucleotides and aminophylline produce different effects on nociceptive motor and sensory responses in the rat spinal cord. Naunyn Schmiedeberg's Arch Pharmacol 327:23–30.

167. Post C (1984) Antinociceptive effects in mice after intrathecal injection of 5'-N-ethylcarboxamide adenosine. Neurosci Lett 51:325–330.

168. Holmgren M, Hedner J, Mellstrand T et al (1986) Characterization of the antinociceptive effects of some adenosine analogs in the rat. Naunyn Schmiedeberg's Arch Pharmacol 334:290–293.

169. Sawynok J, Sweeney MI, White TD (1986) Classification of adenosine receptors mediating antinociception in the rat spinal cord. Br J Pharmacol 88:923–930.

170. Delander GE, Hopkins CJ (1986) Spinal adenosine modulates descending antinociceptive pathways stimulated by morphine. J Pharmacol Exp Ther 239:88–93.

171. Wu PH, Phillis JW, Yuen H (1982) Morphine enhances the release of ^{3}H-purines from rat brain cerebral cortical prisms. Pharmacol Biochem Behav 17:749–755.

172. Ahlijanian MK, Takemori AE (1986) Changes in adenosine receptor sensitivity in morphine-tolerant and age-dependent mice. J Pharmacol Exp Ther 236:615–620.

173. Marley E, Nistico G (1972) Effects of catecholamines and adenosine derivatives given into the brain of fowls. Br J Pharmacol Chemother 46:619–636.

174. Haulică I, Ababei L, Brănisteanu D et al (1973) Preliminary data on the possible hypnogenic role of adenosine. J Neurochem 21:1019–1020.

175. Radulovacki M, Virus RM, Djuricic-Nedelson M et al (1984) Adenosine analogs and sleep in rats. J Pharmacol Exp Ther 228:268–274.

176. Virus RM, Djuricic-Nedelson M, Radulovacki M et al (1983) The effects of adenosine and 2'-deoxycoformycin on sleep and wakefulness in rats. Neuropharmacology 22:1401–1404.

177. Radulovacki M, Virus RM, Rapoza D et al (1985) A comparison of the dose response effects of pyrimidine ribonucleosides and adenosine on sleep in rats. Psychopharmacology 87:136–140.

178. Glaum SR, Yanik GM, Porter NM et al (1985) Effects of intracerebroventricular administration of forskolin and 2',5'-dideoxyadenosine on sleep in rats. Fed Proc 11:749.

179. Maitre M, Ciesielski L, Lehmann A et al (1974) Protective effect of adenosine and nicotinamide against audiogenic seizure. Biochem Pharmacol 23: 2807–2816.
180. Dunwiddie TV, Worth T (1982) Sedative and anticonvulsive effects of adenosine analogs in mouse and rat. J Pharmacol Exp Ther 220:70–76.
181. Lee KS, Schubert P, Heinemann U (1984) The anticonvulsive action of adenosine: A postsynaptic, dendritic action by a possible endogenous anticonvulsant. Brain Res 321:160–164.
182. Ault B, Wang CM (1986) Adenosine inhibits epileptiform activity arising in hippocampal area CA3. Br J Pharmacol 87:695–703.
183. Dunwiddie TV (1980) Endogenously released adenosine regulates excitability in the in vitro hippocampus. Epilepsia 21:541–548.
184. Albertson TE, Stark LJ, Joy RM et al (1983) Aminophylline and kindled seizures. Exp Neurol 81:703–713.
185. Barraco RA, Swanson TH, Phillis JW et al (1984) Anticonvulsant effects of adenosine analogs on amygdaloid-kindled seizures in rats. Neurosci Lett 46:317–322.
186. Dragunow M, Goddard GV, Laverty R (1985) Is adenosine an endogenous anticonvulsant? Epilepsia 26:480–487.
187. Rosen JB, Berman RF (1985) Prolonged postictal depression in amygdala-kindled rats by the adenosine analog, L-phenylisopropyladenosine. Exp Neurol 90:549–557.
188. Schultz V, Lowenstein JM (1978) The purine nucleotide cycle: Studies of ammonia production and interconversions of adenine and hypoxanthine nucleotides and nucleosides by rat brain in situ. J Biol Chem 253:1938–1943.
189. Winn HR, Welsh JE, Rubio R et al (1980) Changes in brain adenosine during bicuculline-induced seizures in rats. Effects of hypoxia and altered systemic blood pressure. Circ Res 47:568–577.
190. Burley ES, Ferrendelli JA (1984) Regulatory effects of neurotransmitters in electroshock and pentylenetetrazol seizures. Fed Proc 43:2521–2524.
191. Capogrossi MC, Francendese A, Digirolamo M (1979) Suppression of food intake by adenosine and inosine. Am J Clin Nutr 32:1762–1768.
192. Levine AS, Morley JE (1983) Effect of intraventricular adenosine on food intake in rats. Pharmacol Biochem Behav 19:23–26.
193. Schiffman SS, Gill JM, Diaz C (1985) Methylxanthines enhance taste: Evidence for modulation of taste by adenosine receptor. Pharmacol Biochem Behav 22:195–203.
194. Wager-Srdar S, Levine AS, Morley JE (1984) Food intake: Opioid/purine interactions. Pharmacol Biochem Behav 21:33–38.
195. Spindel ER, Wurtmann RJ, McCall DB et al (1984) Neuroendocrine effects of caffeine in normal subjects. Clin Pharmacol Ther 36:402–407.
196. Arnold MA, Carr DB, Togasaki DM et al (1982) Caffeine stimulates β-endorphin release in blood but not in cerebrospinal fluid. Life Sci 32:1017–1024.
197. Selye H (1942) Studies concerning the correlation between anesthetic potency, hormonal activity, and chemical structure among steroid compounds. Anesth Analges 21:41–47.
198. Costa PJ, Bonnycastle DD (1952) The effect of DCA, compound E, testosterone, progesterone, and ACTH in modifying "Agene-induced" convulsions in dogs. Arch Int Pharmacodyn Ther 91:330–338.

199. Woolley DE, Timiras PS (1962) Estrous and circadian periodicity and electroshock convulsions in rats. Am J Physiol 202:379–382.
200. Landgren S, Bäckström T, Kalistratov G (1978) The effect of progesterone on the spontaneous interictal spike evoked by the application of penicillin to the cat's cerebral cortex. J Neurol Sci 36:119–133.
201. Majewska MD, Bisserbe JC, Eskay RL (1985) Glucocorticoids are modulators of $GABA_A$ receptors in brain. Brain Res 339:178–182.
202. Komisaruk BR, McDonald PG, Whitmoyer DI et al (1967) Effects of progesterone and sensory stimulation on EEF and neuronal activity in the rat. Exp Neurol 19:494–507.
203. Phillis JW, Bender AS, Marszalec W (1985) Estradiol and progesterone potentiate adenosine's depressant action on rat cerebral cortical neurons. Gen Pharmacol 16:609–612.
204. Phillis JW (1986) Potentiation of the depression by adenosine of rat cerebral cortical neurones by progestational agents. Br J Pharmacol 89:693–702.
205. Bäckström T, Sanders D, Leask R et al (1983) Mood, sexuality, hormones, and the menstrual cycle II. Hormone levels and their relationship to the premenstrual syndrome. Psychosom Med 45:503–507.
206. Bäckström T (1983) Premenstrual tension syndrome. In: Bardin CW, Milgrom E, Mauvaiis-Jarvis P (eds) Progesterone and Progestins. Raven Press, New York. pp 203–217.
207. Mattson RH, Cramer JA (1985) Epilepsy, sex hormones, and antiepileptic drugs. Epilepsia 26 (Suppl 1):S40-S51.
208. Mor G, Saphier D, Feldman S (1986) Inhibition by corticosterone of paraventricular nucleus multiple-unit activity responses to sensory stimuli in freely moving rats. Exp Neurol 94:391–399.
209. Dubrovsky B, Illes J, Birmingham MK (1986) Effect of 18-hydroxydeoxy-corticosterone on central nervous system excitability. Experientia 42:1027–1028.
210. Woodbury DM (1954) Effect of adrenocortical steroids and adrenocorticotrophic hormone on electroshock seizure threshold. J Pharmacol Exp Ther 105:27–36.
211. Gold PW, Loriaux DL, Roy A et al (1986) Responses to corticotropin-releasing hormone in the hypercortisolism of depression and Cushing's disease. N Engl J Med 314:1329–1335.
212. Stewart SF, Pugsley TA (1985) Increase of rat serum prolactin by adenosine analogs and their blockade by the methylxanthine aminophylline. Naunyn-Schmiedeberg's Arch Pharmacol 331:140–145.
213. Lee JC, Downing SE (1979) Coronary dynamics and myocardial metabolism in the diabetic lamb. Am J Physiol 237:H118–H124.
214. Downing SE (1985) Restoration of coronary dilator action of adenosine in experimental diabetes. Am J Physiol 249:H102–H107.
215. Phillis JW, Preston G, DeLong RE (1984) Effects of anoxia on cerebral blood flow in the rat brain: Evidence for a role of adenosine in autoregulation. J Cereb Blood Flow Metab 4:586–592.
216. Phillis JW, DeLong RE (1987) An involvement of adenosine in cerebral blood flow regulation during hypercapnia. Gen Pharmacol 18:133–139.
217. Dandona P, James IM, Newbury PA et al (1978) Cerebral blood flow in diabetes mellitus: Evidence of abnormal cerebrovascular reactivity. Br Med J 2:325–326.

218. Won EHA, Smith JA, Jarett L (1985) Adenosine effects of glucose oxidation of adipocytes isolated from streptozotocin-diabetic rats. Biochem J 232: 301–304.
219. Braas KM, Newby AC, Wilson VS et al (1986) Adenosine-containing neurons in the brain localized by immunocytochemistry. J Neurosci 6:1952–1961.

Index